ELECTRICAL THEORY AND CONTROL SYSTEMS IN HEATING AND AIR-CONDITIONING TECHNOLOGY

Electrical Theory and Control Systems in Heating and Air-Conditioning Technology

Robert F. Dorner

Montclair State College

Martin L. Greenwald

Montclair State College

Delmar Publishers Inc.™

I(T)P™

NOTICE TO THE READER

Publisher does not warrant or guarantee any of the products described herein or perform any independent analysis in connection with any of the product information contained herein. Publisher does not assume, and expressly disclaims, any obligation to obtain and include information other than that provided to it by the manufacturer.

The reader is expressly warned to consider and adopt all safety precautions that might be indicated by the activities described herein and to avoid all potential hazards. By following the instructions contained herein, the reader willingly assumes all risks in connection with such instructions.

The publisher makes no representations or warranties of any kind, including but not limited to, the warranties of fitness for particular purpose or merchantability, nor are any such representations implied with respect to the material set forth herein, and the publisher takes no responsibility with respect to such material. The publisher shall not be liable for any special, consequential or exemplary damages resulting, in whole or in part, from the readers' use of, or reliance upon, this material.

DELMAR STAFF

Senior Administrative Editor: Vernon Anthony
Project Editor: Eleanor Isenhart
Production Coordinator: Dianne Jensis
Art/Design Coordinator: Heather Brown

For information, address Delmar Publishers Inc.
3 Columbia Circle, Box 15-015
Albany, New York 12212-5015

Printed in the United States of America.
Published simultaneously in Canada
by Nelson Canada,
a division of The Thomson Corporation

1 2 3 4 5 6 7 8 9 10 xxx 00 99 98 97 96 95 94

Library of Congress Cataloging-in-Publication Data

Dorner, Robert F.
     Electrical theory and control systems in heating and air
-conditioning technology / Robert F. Dorner, Martin L. Greenwald.
          p.   cm.
     Includes index.
     ISBN 0-8273-5749-4
     1. Heating—Control. 2. Air conditioning—Control. 3. Electric
engineering. I. Greenwald, Martin L., 1943-   . II. Title.
TH7466.5.D67 1994
697.9'32—dc20   93-3396
                                                       CIP

Contents

Preface

Electrical Theory and Control Systems in Heating and Air-Conditioning Technology is intended to provide the concepts and practices that will enable the technical, vocational, or community college student to become an HVAC professional rather than a laborer or a mere "parts changer." The authors have carefully selected the basic science theory that explains and supports climate control systems. Through examples, comparisons, and many illustrations, a "common-sense" understanding is provided. The examples are drawn directly from actual HVAC systems, while the comparisons are made with experiences that are common to current culture. The illustrations in this text are sharply focused to develop a concept; in later chapters, real-life HVAC systems, circuits, and appliances are shown. Irrelevant theory has been excluded, but, where necessary, some difficult concepts are presented. The few formulas and mathematics included are developed in a step-by-step fashion. All formula parts are explained, then put together like a machine. Throughout this text, the primary emphasis is on true understanding through the use of common sense.

The authors believe that the technical advances in the climate control field, as well as environmental concerns, have resulted in more complex HVAC systems. These more efficient and cleaner systems depend on electronics, sensors, actuators, and controls that are "smart." This text provides the analytical "how it works" explanation of these components and systems that is essential to understanding their installation and repair. The "smart" HVAC technician is in great demand.

Chapter 1 contains a general overview of common HVAC systems. It provides the student with a framework and the foundation upon which the remainder of the text is built. Section 1 (Chapters 2-10), the first story, deals with theory and control, while Section 2, the second story, presents actual HVAC systems.

Section 1 begins with the principles of energy conversion and heat transfer. This provides a practical understanding of the fundamentals of thermodynamics (heat movement), including the first and second laws, heat conversion, heat transfer, measuring systems, and calculations for system sizing. In Chapter 3, the focus shifts to the nature of electricity and answers the question, "What is electricity and why do we use it?" The dynamics of electric, mechanical, and heat energy are tied together in Chapter 4. The focus is on the ratio or lever idea and its connection with all types of energy systems. Chapter 5 discusses sources of electricity and the effects of resistive loads and magnetism, and offers a brief introduction to the inductor and capacitor. Electric power distribution and AC effects are explained in Chapter 6. This requires careful study, as does Chapter 7, which presents simple AC loads, as well as more complex motors. Chapter 8 develops the idea of series and parallel circuits, and then shows how common meters are used to measure circuit values. Electronic components and circuits are described in Chapter 9, which concludes with the fundamental ideas of digital microprocessors. The information provided in previous chapters of Section 1 is brought together in Chapter 10, which presents sensor, actuator, and control theory.

Section 2 focuses on heating and cooling control and application technology. Chapter 11 discusses oil-fired combustion technology, including operation and efficiency testing, oil supply systems, and electro-mechanical and computerized controls. Chapter 12 covers gas-fired combustion, from the combustion characteristics of natural and synthetic gases through control devices. Hydronic and steam heating applications are examined in Chapter 13. This chapter also introduces solid-fuel and solar heating as part of conventional hydronic and steam systems. Forced warm-air heating systems

and associated airflow concepts are the focus of the last chapter, Chapter 15.

It is important to recognize that much of the material covered here is also applicable to various other technical fields. As a result, the student who studies this text thoroughly and also completes an appropriate corollary course will then be equipped with a sound foundation for participating in many other technically related areas.

Acknowledgments

The authors wish to express their thanks to Kathie and Reesa for their encouragement, assistance and understanding during the preparation of this text.

Dedication

To Roby, Meridith, and Wendy Dorner

and

Benjamin and David Greenwald

INTRODUCTION

CHAPTER 1

Introduction to Residential Heating, Air-Conditioning, and Heat-Pump Technology

THE CHALLENGE OF EFFICIENTLY PRODUCING AND MOVING HEAT AND ENERGY

This text has two objectives in its examination of heating and air conditioning. The first is to focus on specific basic concepts that relate to energy and power—what energy and power are and how they are generated and converted. The second aim is to give the reader an insight into and an understanding of the relationship between energy and power, as well as the resulting applications and associated control systems that are used in most heating and air-conditioning applications.

This material is divided into two sections. The first section begins with an examination of electrical theory—what electricity is and what it can do. Topics then branch out to include electrical sources and loads, power distribution systems, single- and multiphase alternating current loads, circuitry and instrumentation, and control theory and electronic fundamentals.

The second section of the book builds upon the theory and concepts presented in the first section, examining operating systems and application technology. Written from this perspective of heating and cooling system design and control, an

emphasis is placed on how things work, as well as on the technology involved in controlling, in the most efficient manner, all system functions.

Our discussion begins with a description of the basic characteristics of common heating and air-conditioning systems. Included in this discussion are both stand-alone and interfaced operating solar and solid-fuel heating systems.

HEATING SYSTEM CLASSIFICATIONS

The majority of residential and commercial applications rely on four basic types of heating systems: hot air, hot water, steam, and electric resistance. Included in this examination are solar- and solid-fuel based technologies.

Forced Warm-Air Heating

Forced warm-air heating systems use a centrally located furnace to provide heat through fuel combustion that takes place within a heat exchanger. Air is circulated by a motor-driven fan in a duct system that runs throughout the house. As it passes over the heat exchanger, the air picks up heat from the combustion process. The basic configuration of a forced warm-air, oil-fired heating system is shown in figure 1-1.

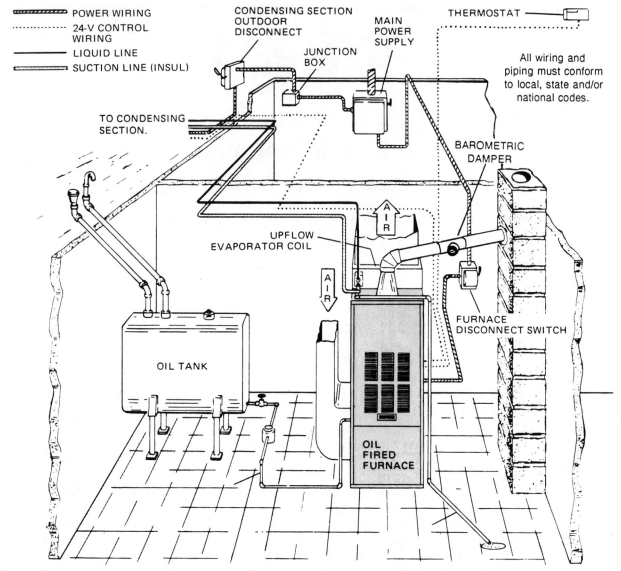

FIGURE 1-1 Warm-air heating system. The system shown here is also equipped for central air conditioning. (Courtesy of York International, Inc.)

Warm-air furnaces are large heating units, since large heat exchanger surfaces are required to ensure effective heat transfer. A residential warm-air furnace is illustrated in figure 1-2.

Warm-air systems can be zoned to provide heat for specific areas or rooms in the home. Also, since all of the air in the home is channeled through a duct distribution network,

warm-air heating systems allow for the installation of power-humidification, air-conditioning, and air-cleaning devices. These will be examined in greater detail in Chapter 12.

Hydronic (Hot-Water) Heating

A hot-water, or hydronic, heating system consists of a centrally located boiler that

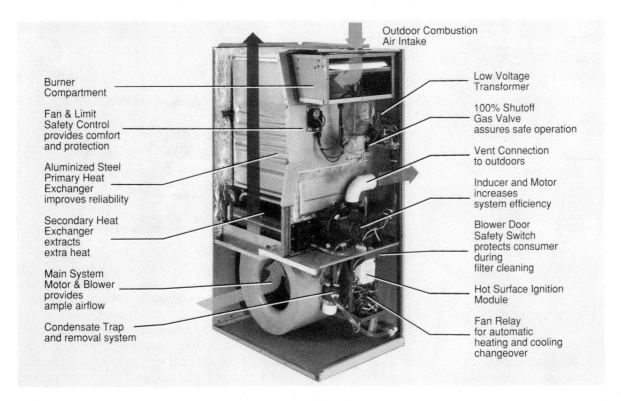

Outdoor Combustion
Air Intake

Burner
Compartment

Fan & Limit
Safety Control
provides comfort
and protection

Aluminized Steel
Primary Heat
Exchanger
improves reliability

Secondary Heat
Exchanger
extracts
extra heat

Main System
Motor & Blower
provides
ample airflow

Condensate Trap
and removal system

Low Voltage
Transformer

100% Shutoff
Gas Valve
assures safe operation

Vent Connection
to outdoors

Inducer and Motor
increases
system efficiency

Blower Door
Safety Switch
protects consumer
during
filter cleaning

Hot Surface Ignition
Module

Fan Relay
for automatic
heating and cooling
changeover

FIGURE 1-2 Component arrangement of residential warm-air furnace. The arrangement of specific components will vary based upon airflow patterns, Btu capacity, and furnace manufacturer. (Courtesy of York International, Inc.)

delivers hot water for heating purposes through a piping system. Terminal heating units are fabricated with a series of metal fins clamped around the hot water pipe. This assembly is enclosed within a metal cabinet and is available in different lengths. Since these baseboard units heat via conductive heat transfer from the finned pipe to the room air, they are sometimes referred to as baseboard convectors. These fins provide a large surface area for heat exchange to take place between the circulating room air and the hot water in the piping system. Figure 1-3A illustrates a the typical component arrangement of an oil-fired hydronic boiler. Figure 1-3B shows the baseboard convector assembly in a metal cabinet enclosure.

Single- and multiple-zone systems: Hydronic heating systems are installed either as a single-zone circuit (figure 1-4) in small structures, or in a multiple-zone configuration that is common to

larger homes and many commercial applications (figure 1-5). Designing heating systems in multiple-zone circuits enables the building owner to put the heat where it is needed and to turn it off when and where it is not needed. Options available to the installer within all of these categories, along with associated control systems, are examined in greater detail in Chapter 11.

Radiant panel and slab heating: In a radiant slab heating application, a piping system or electric resistance cable is buried in the concrete floor of the structure. Rather than using finned radiators for heat dispersal, this system uses the mass of the concrete floor to store and radiate heat to the interior rooms. In this way, the floor acts as a thermal mass or heat sink. Heat is transferred by conduction from either the piping circuit or resistance cables to the concrete and then radiated within the building. One distinctive feature of radiant slab heating is the even nature of

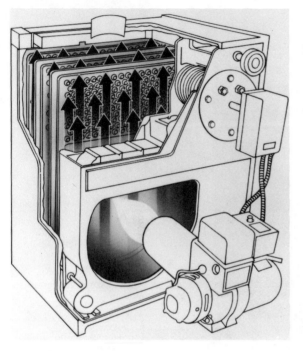

FIGURE 1-3A Hydronic oil-fired boiler illustrating typical component arrangement. (Courtesy of Slant/Fin Corp.)

FIGURE 1-3B Finned baseboard convector. The fins surrounding the water pipe aid in dispersing heat to the surrounding currents of air that enter and leave the baseboard enclosure cabinet. (Courtesy of Slant/Fin Corp.)

the heat distribution. The slab is kept at a constant temperature (approximately 85°F to 90°F). At this temperature, the slab is comfortable to

walk on, and also is able to radiate sufficient heat to keep interior temperatures comfortably balanced. These systems can be zoned in a manner similar to that of conventional baseboard radiator configurations.

In hydronic radiant panel heating, specially constructed and coated water tubing is woven between the floor joists and walls, or placed on

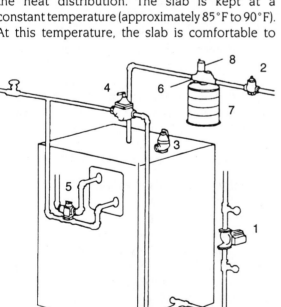

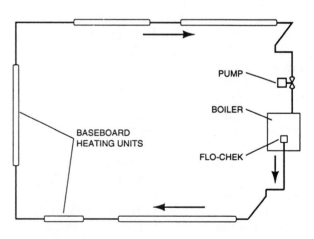

1	CIRCULATOR	5	TEMPERING VALVE
2	FLO-CHEK	6	AIR-SCOOP
3	RELIEF VALVE	7	EXPANSION TANK
4	REDUCING VALVE	8	AIR VENT

FIGURE 1-4 Single-zone heating configuration illustrating boiler component arrangement and typical single-circuit piping arrangement. (Courtesy of Taco, Inc.)

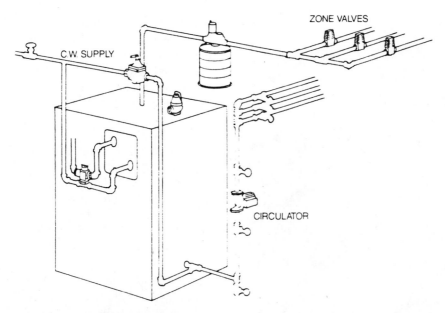

FIGURE 1-5 Multiple-zone heating configuration. This boiler is set up for three-zone heating with the zoned circuits controlled by electric zone valves. (Courtesy of Taco, Inc.)

a specially constructed subflooring and covered with either concrete, mortar, or gypsum, depending on the thickness of the floor and the manufacturer of the radiant system (fig. 1-6A). This type of radiant heating uses lower water temperatures than are used in conventional hydronic baseboard or radiator applications.

The circuit configuration for a typical low-voltage electric radiant slab system is illustrated in figure 1-6B.

Steam Heating Systems

Steam heating systems employ a hot water boiler to provide low-pressure steam as the heat source. Steam travels through a distribution piping system to a series of radiators, where it condenses back to water. The heat generated in this condensing process, through the latent heat of condensation, is transferred to the radiators in the heating system and from there throughout the house (for a more detailed explanation of the latent heat of condensation, see Chapter 2). The terminal radiators in these systems are made up of series of tubular columns, a type of construction that affords a large surface area from which to radiate heat.

One- and two-pipe systems: Two types of piping configurations are used in steam heating systems. In the one-pipe system, both live steam and condensate travel in the same pipe. These pipes are oversized in order to accommodate water and steam travel in two directions. A two-pipe system features a separate return line for the condensate to return to the boiler. Figure 1-7 illustrates a single-pipe boiler configuration for steam systems. Note that component "I" can be used for a two-pipe system in which a separate return line from the radiators is used.

Although steam heating systems were popular in the early 20th century, their installation in new residential construction has steadily declined in recent years as high-efficiency, low-profile baseboard hydronic systems have proliferated. Steam heating systems and their associated controls are examined in more detail in Chapter 11.

Baseboard and radiant panels: All-electric buildings feature two primary types of heating systems. The first type uses electric baseboard heaters, which are similar in appearance to their hydronic counterparts; see figure 1-8.

Electric baseboard heaters require separate 208/240-volt circuits for each room and line-

FIGURE 1-6A Hydronic radiant slab configuration. The piping, constructed from specially coated rubber- or plastic-based materials designed to resist corrosion, is either embedded within a slab or installed on a specially constructed sub-floor. Depending on the manufacturer of the in-floor radiant system, regular concreted, mortar beds, or gypsum-based materials are used to cover the pipe. (In-floor heating systems by Gyp-Crete Corp.)

voltage thermostats for controlling the power to the baseboard heaters (many appliances have both voltages identified in view of the use of three-phase utility power).

A second type of heating system in all-electric homes relies on heating elements built into wall and ceiling panels. Special heating panels replace standard Sheet-rock and wall-board room panels and are set in place during the construction of the building. The panels are wired together and their operation is controlled through a line-voltage thermostat for each

room, whose operating characteristics are similar to those of the thermostat used to control electric baseboard heaters. During operation, the panel is energized and radiates heat into the room through either the walls or ceiling, depending on the type of panel selected. A heating system of this type is illustrated in figure 1-9.

Given current economics, the cost of providing heat with electric resistance heating systems can be substantial in many parts of the country. As we will examine in later chapters, although the conversion from electricity to heat within the electric resistance unit is very efficient, other inefficiencies in the generating and delivery sys-

FIGURE 1-6B Electric resistance radiant heating systems rely on low-wattage electric cables embedded within a slab or sub-floor to radiate heat in a manner similar to that of a hydronic radiant system. Similar materials can be used to overlay the heating cables. (In-floor heating systems by Gyp-Crete Corp.)

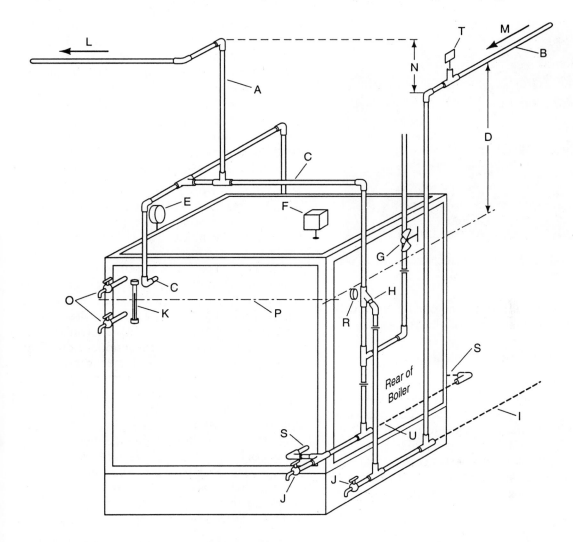

A Main
B Dry return
C Horizontal supply tappings
D End of dry return not less than 18 inches
   above water line of boiler
E Pressure gage
F Safety valve
G Globe valve in make-up water line
H Hartford loop connection. A "Y" fitting
   is preferred
I Wet return if used
J Drain cocks
K Gage glass

L Pitch main downward in direction of arrow not less
   than 1/4 inch in 10 feet
M Return pitches downward in direction of arrow not
   less than 1/4 inch in 10 feet
N End of dry return not less than 2 inches below top
   of vertical supply main
O Try cocks
P Boiler water line
R Low water cut-off
S Boiler return tappings
T Main vent
U Yoke return in both return tappings, if recommended
   by the manufacturer

FIGURE 1-7 One-pipe steam system. The use of steam in residential heating has declined as the popularity of other types of heating alternatives has increased.

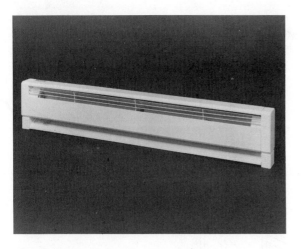

FIGURE 1-8 Electric baseboard heating unit. (Courtesy of Marley Electric Heating Co.)

tems must be paid for by the consumer, which increases the final cost of delivered energy.

Active Solar Space Heating

Many people believe that harnessing solar energy for heating, cooling, and a variety of other applications holds much potential. Although solar energy has been used for many years in residential and commercial heating and cooling applications, its market penetration has been limited by both high initial installation costs and relatively low system efficiencies. For

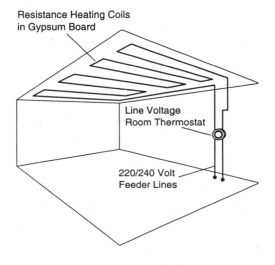

Resistance Heating Coils
in Gypsum Board

Line Voltage
Room Thermostat

220/240 Volt
Feeder Lines

FIGURE 1-9 Electric resistance heating panels.

purposes of this text, we will examine *active* as differentiated from *passive* solar heating technology.

Passive solar systems focus on the architectural design of the building to maximize incoming solar energy while limiting heat loss through the structure. Well-designed passive solar homes make a distinctive architectural statement and can significantly reduce heating and cooling costs as compared with homes that are more conventionally designed and constructed.

Principles of operation: Active solar heating incorporates the use of solar collectors installed on a roof or in some other suitable area that is close to the home. The collectors must be situated so as to receive unobstructed incoming solar energy for a major portion of the day. This incoming radiant solar energy is converted to heat in the solar collector and is transferred via a circulating fluid to a water storage tank. Heat from the storage tank is then available for use in a variety of conventional fossil-fuel heating systems, such as domestic hot-water, hydronic, or hot-air heating systems. The basic solar system configuration is shown in figure 1-10.

Note from the illustration that the solar collectors transfer heat via a piping circuit that runs through a heat exchange between the solar collectors and storage tank. This allows for the use of a transfer fluid, such as antifreeze, that enables the solar system to be operated throughout the year in climates that are subject to freezing temperatures.

System interfacing with conventional heating applications: The number and size of the storage tank(s) in any well-designed solar system are a function of the size and number of units in the solar collector array. Heat from the storage tanks is transferred by interconnecting the storage tank with a conventional heating system, either directly, as in hydronic applications, or by use of a heat exchanger in hot-air heating systems. Figure 1-11 illustrates a typical solar system interconnected to a hydronic heating system, and figure 1-12 a solar system interface with a forced hot-air heating system.

Although solar systems can be designed to handle 100% of a normal heating load in specific applications, this type of design can be

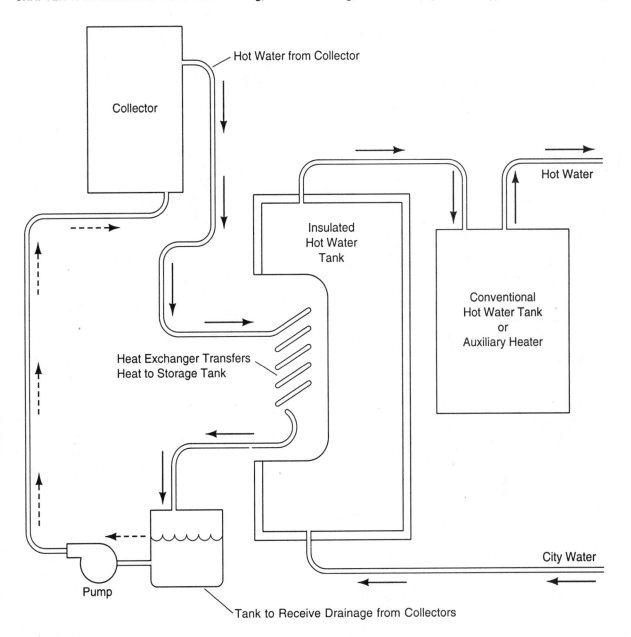

Hot Water from Collector

Collector

Hot Water

Insulated
Hot Water
Tank

Conventional
Hot Water Tank
or
Auxiliary Heater

Heat Exchanger Transfers
Heat to Storage Tank

City Water

Pump

Tank to Receive Drainage from Collectors

FIGURE 1-10 Operating logic of a solar heating system. The solar collectors heat a storage tank. Heat from the storage tank can then be transferred to other heating devices.

very expensive and usually requires excessive amounts of space for solar storage tanks. Ideally, solar heating systems are sized to provide between 25% and 60% of the heating load of a home, subject to the geographical location of the solar system and its cost of installation. Most domestic hot-water heating installations can carry between 50% and 70% of the heating load. Information regarding solar system applications and integration with conventional heating and cooling systems is found throughout this text.

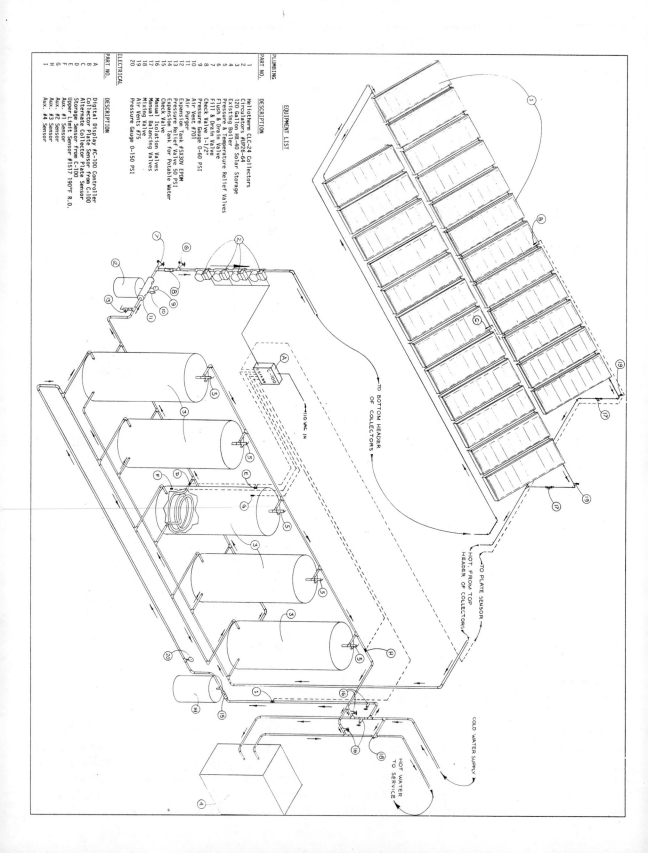

EQUIPMENT LIST

PLUMBING

PART NO.	DESCRIPTION
1	Heliotherm CLC-24 Collectors
2	Circulators #UP26-64
3	120 Gallon HX-40 Solar Storage
4	Existing Boiler
5	Pressure & Temperature Relief Valves
6	Flush & Drain Valve
7	Fill & Drain Valve
8	Check Valve 1-1/2"
9	Pressure Gauge 0-60 PSI
10	Air Vent #701
11	Air Purger
12	Expansion Tank #SX30V EPDM
13	Pressure Relief Valve 50 PSI
14	Expansion Tank for Potable Water
15	Check Valve
16	Manual Isolation Valves
17	Manual Balancing Valves
18	Mixing Valve
19	Air Vents #75
20	Pressure Gauge 0-150 PSI

ELECTRICAL

PART NO.	DESCRIPTION
A	Digital Display #C-100 Controller
B	Collector Plate Sensor from C-100
C	Alternate Collector Plate Sensor
D	Storage Sensor from C-100
E	Upper Limit Sensor #1517 190°F R.O.
F	Aux. #1 Sensor
G	Aux. #2 Sensor
H	Aux. #3 Sensor
I	Aux. #4 Sensor

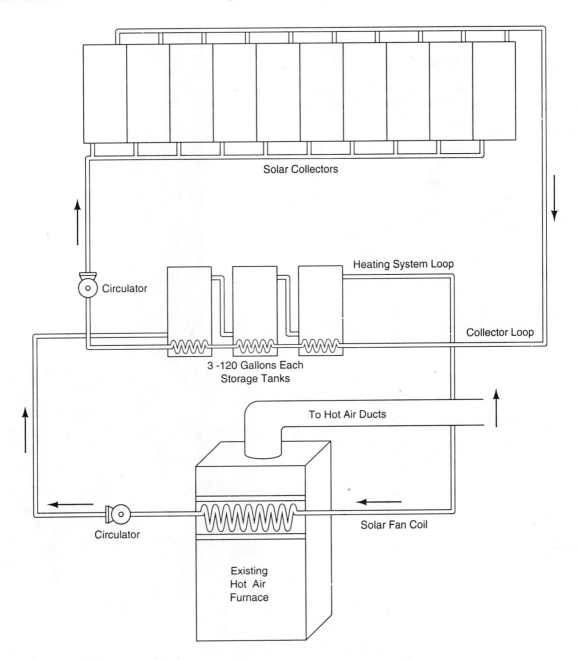

FIGURE 1-11 (Opposite page) Solar heating system intergrated with a conventional hot-water hydronic boiler and baseboard system. (Courtesy of Monitor Data Corp.)

FIGURE 1-12 (Above) Solar heating system Integated with a conventional hot-air heating system. Hot water from the solar storage tanks is circulated through an auxiliary fan coil installed in the existing furnace whenever there is a call for heat in the building. Air circulated by the furnace blower flows through the fan coil where it is heated prior to circulating throughout the building. A low-temperature cut-out aquastat stops the flow of solar-heated water through the fan coil if the storage tank temperature falls below a pre-defined temperature.

SOLID-FUEL HEATING SYSTEMS

Solid Fuels as an Alternative

The rising prices of conventional fossil fuels during the 1970s witnessed a revival of the use of wood, coal, and multifuel heating units. During the late 1970s, fuel oil prices rose to over $1.30 per gallon in many parts of the country, and the price of natural gas increased in a similar manner. The cost of electricity, given the shortage of conventional fossil fuels, resulted in heating bills for all-electric customers that often exceeded monthly mortgage costs. Given this pricing, many people turned to wood and coal heaters for a cost-effective alternative to conventional fossil fuels and utility-generated power. With the use of increasing numbers of solid-fuel units also came increased air pollution, and this has resulted in interesting changes in design and engineering advances that make today's modern solid-fuel devices quite different from their counterparts of just 25 years ago.

Types of Solid-Fuel Heaters

For most applications, solid-fuel heaters can be placed in one of three categories: radiant stoves, solid/multifuel boilers, and solid/multifuel furnaces. Each of these classifications is distinctive in terms of design features and end-use applications.

Radiant stoves: Radiant stoves are the most popular type of solid-fuel heaters in use in the United States. Installation of these units is critical with respect to adequate clearance to combustibles and proper venting (chimney) configurations. The stoves are made from either welded steel or cast iron, and combustion takes place in a brick-lined firebox that radiates heat in direct relationship to the amount and type of wood being burned and to the design and physical size of the stove. A typical radiant solid-fuel stove is illustrated in figure 1-13.

Solid- and multifuel boilers: Solid- and multifuel boilers are designed to burn either wood or coal, as well as fuel oil or gas. Although

FIGURE 1-13   Solid-fuel radiant stove. (Courtesy of Vermont Castings, Inc.)

individual manufacturers use different engineering configurations, most multifuel units feature two separate combustion chambers—one for the wood or coal and the other for the combustion of oil or gas. Both combustion chambers vent into a single flue for removal of the

FIGURE 1-14 Add-on solid-fuel boiler. These units have become popular with the resurgence of wood and coal heating. The unit illustrated is designed to burn both wood and coal as a primary fuel as well as fuel oil, natural gas, or propane as a secondary, back-up fuel. (Courtesy of Northland Corp.)

exhaust gases. Solid-fuel boilers, such as the unit in figure 1-14, are ideal as add-on units for existing heating systems. Such units use only one combustion chamber, which is designed to burn either wood or coal.

Several different piping methods are used to integrate the solid-fuel boiler with a conventional heating system. In these arrangements, the solid-fuel unit is incorporated into the heating system design as a backup, or add-on, heating unit. Operating in this way, as long as a fire is able to maintain a low-limit water temperature within the heating system, the fossil fuel boiler will not be energized. Should

either the fire go out in the add-on or the unit be unable to keep up with the heat loss in the home, the conventional burner will fire to maintain interior heating levels. The various piping configurations and control devices used to operate solid-fuel units, both as stand-alone and integrated heating units are discussed in greater detail throughout this text where appropriate.

Solid- and multifuel furnaces: These units are similar in design and operating characteristics to their hydronic solid- and multifuel counterparts, except that the medium of heat transfer is air rather than water. Most multifuel furnaces feature two separate combustion chambers—one for wood/coal and the other for either gas or oil—and vent from either a single or separate exhaust flues. These furnaces are larger than their hot-water counterparts, since large heat exchange surfaces are required for efficient forced convection heat transfer.

AIR-CONDITIONING AND HEAT-PUMP COOLING TECHNOLOGY

All air conditioners and heat pumps are based on the operating principle of the refrigeration cycle. In this cycle, heat is removed from one area, the inside of a house, for example, and deposited in a warmer area, such as the outside air on a warm summer day. In other words, in this process, heat is moved from an initial area to a second area that is warmer than the first. Consider the household refrigerator, which removes heat from the cold chamber of the refrigerator and deposits it in a warmer area, the kitchen. To accomplish this task, a compressor is used to circulate a refrigerant fluid between an evaporator, where heat is absorbed, and a condenser, where heat is removed. The intricacies of this cycle will be examined in detail in Chapter 15.

Two devices are commonly used to accomplish most air-conditioning processes: the air conditioner and the heat pump. A conventional air conditioner moves heat in one direction only, and is used strictly for room or home

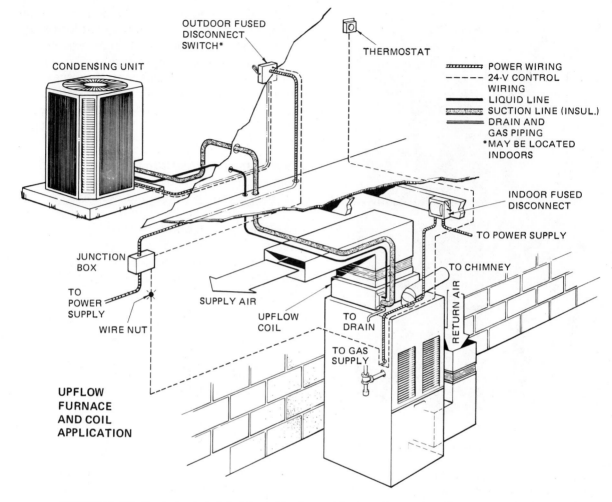

FIGURE 1-15 Central warm-air heating with air conditioning. (Courtesy of York International, Inc.)

cooling. A heat pump is a reversible air conditioner. It can be used for both heating and cooling. It accomplishes this by reversing the direction of flow of refrigerant within the system, depending on the necessity for it to provide either heat or cooling. Heat pumps deliver their most efficient performance in milder climates where winter temperatures do not ordinarily fall below 40°F., although they are also used in colder climates in place of electric baseboard heating, offering the air-conditioning capability that the latter cannot provide.

Air conditioners and heat pumps are in-stalled either as stand-alone units or in conjunc-tion with existing forced convection heating systems. Dual-purpose heating and cooling sys-tems share the same air ducts to distribute both hot and cold air throughout the home. A system of this type is illustrated in figure 1-15.

SUMMARY

This chapter has introduced the basic classi-fications of heating systems, including some of the terminology associated with them. In future chapters, an in-depth examination of these systems highlights the operating technologies of the various heat-transfer systems, as well as their many similarities in operating theory and control technology. For example, many of these systems can be combined; hot-air-heating and air-conditioning systems use the same ducts for air distribution and can share the same furnace and operating controls. The design of air ducts and hydronic piping systems is based on the same assumptions concerning heat-transfer characteristics; only the method of distribution varies with the difference in heat-transfer mediums. These common principles of operation and control technology will be identified throughout as we move on to a discussion of the terminology and concepts that form the basis for understanding all heating and cooling system applications.

PROBLEM-SOLVING ACTIVITIES

Review Questions

1. Highlight the differences between forced hot-air and hydronic heating systems.

2. Describe how both solid-fuel and solar-powered heating systems can be used as either stand-alone or integrated heating systems alongside conventional fossil-fuel systems.

SECTION ONE

Basic Electrical Theory
and Concepts

CHAPTER 2

Principles of Energy Conversion
and
Heat Transfer

A BRIEF HISTORY OF ENERGY

Somewhere between 12 and 20 billion years ago, the universe is believed to have been infinitely hot and to have had zero size. Referred to by physicists as the big bang singularity, this incredible collection of dense energy exploded, and the universe began to expand. During this time, the universe contained a collection of subatomic particles known as photons, electrons, nutrinos, and other primary forces (which will be described in greater detail in Chapter 3). Later these particles formed atoms of the lightest element, hydrogen. Large clouds of hydrogen gas collected and eventually began to compress under the influence of gravity. This compression decreased the size of the hydrogen clouds and caused an increase in both the temperature and pressure of the gases within the clouds. Eventually these extreme pressures initiated the process of atomic fusion, causing the hydrogen atoms to fuse and create small amounts of heavier elements while releasing a tremendous amount of energy in the form of heat. In this way, the first stars were born. Many of these early stars passed their lives by balancing the inward forces of gravity with the outward forces of nuclear fusion. Some stars, however, were not able to maintain the balance between these

two forces. This resulted in an explosion of the star, called a supernova. In this process, heavier elements and the remnants produced by these explosions would be sent back into the galaxy, providing some of the raw material for the next generation of stars. In this manner, our sun, which is a second- or third-generation star, together with the revolving planets, was born.

Energy is indestructible, although it may be converted to matter, and vice versa. All of the energy we have today can be traced back to the big bang singularity at the beginning of the universe. In general, the longer an energy source remains undisturbed, the more energy it contains. For example, nuclear fusion is the process of squeezing hydrogen to form heavier elements. Nuclear fusion is responsible for powering our sun, as well as the process used in the production of thermonuclear weapons. Its origin goes back to the formation and lifecycle of first-generation stars. We have yet to solve the considerable technical problems associated with controlling nuclear fusion reactions. Another nuclear reaction, fission, involves the splitting of heavy atoms such as uranium. Nuclear fission produces great amounts of energy, although less than is generated in fusion reactions.

Solid-fuel energy sources on the earth are all organic in origin. Petroleum, for example, was produced millions of years ago from the remains of marine plant and animal life that was abundant in the warm seas that covered most of the earth during that time. These organic remains were sealed by layers of sedimentary deposits. Eventually these remains decomposed into vast amounts of oil and gas. Coal deposits trace their origin to the residues of ancient forests and other organic material that was formed over 300 million years ago. Although coal, oil, and gas are excellent fuels, their energy density is less than that of the geologically older nuclear fuel sources. Fuel sources such as hard and soft woods release less energy than coal, and year-old vegetation produces even less. Radiant solar energy, perhaps the most benign, as well as the most recent (geologically), of all energy sources, is the most difficult to utilize. While plentiful on a planetary scale, it is diffuse and available only in low energy densities for most point-of-use applications.

Of all the energy sources, few are directly nonsolar in nature. One of these, geothermal, uses the earth's molten core are a source of energy for space heating and the generation of electric power. The other, tidal energy, relies on a significant difference that occurs in the heights of high and low tides in a few locations around the world for generating electricity. These tidal differences are based on changing gravitational fields between the earth and moon and are not directly solar related.

Given this background, our discussion now focuses on the specific terminology associated heat and energy as they relate to heating and air-conditioning technology.

TEMPERATURE

Everyone seems to know what temperature is. References to temperature, such as "a hot car" or "a cold person," imply a general understanding. However, a more detailed understanding of temperature is required by HVAC technicians.

To convert celsius to Fahrenheit, use the formula:

$T_F = 9/5 \; T_C + 32$

Where: T_F = Fahrenheit temperature
T_C = Celsius temperature

To convert Fahrenheit temperature to Celsius, use the formula:

$T_C = 5/9 \; (F - 32)$

Example 1: What is the Fahrenheit temperature equivalent for 150 degrees Celsius?

$T_C = 150$
$T_F = 9/5 \; (T_C + 32)$
$\quad 9/5 \; (150) + 32 = 270 + 32 = 302$ degrees Fahrenheit

Example 2: What is a temperature of 97 degrees Fahrenheit in Celsius?

$T_F = 97$
$T_C = 5/9 \; (F - 32)$
$T_C = 5/9 \; (97 - 32) = 5/9 \; (65) = 36$ degrees Celsius

FIGURE 2-1 Relationship between Fahrenheit and Celsius scales.

Temperature is indeed a measure of "hotness." For example, water boils at 212 degrees on the Fahrenheit scale and at 100 degrees on the Celsius scale. Although these measurements describe the temperature level of substances or areas, they do not describe the quantity of heat available. For example, the temperature of a drop of water sizzling on a stove may be the same as that of thousands of gallons of water boiling in the turbine of an electrical generating plant. Although both have the same temperature, the quantity of heat in the steam plant is far greater.

Fahrenheit was a thermometer maker who was so well respected that his scale became a common measuring standard. On the Fahrenheit scale, water freezes at 32 degrees and boils at 212 degrees, a difference between the freezing and boiling points of water of 180 degrees. The Celsius, or metric, scale is designed on a base of multiples of ten. Water freezes at 0 degrees and boils at 100 degrees. Although both scales use water as the calibrating standard, they differ on their respective freezing points (0 and 32) and boiling points (100 and 212). Conversion from one scale to the other is easily accomplished by using the conversion formulas in figure 2-1.

HEAT, BTU'S, AND CALORIES

Throughout the ages, humankind has developed an increasingly sophisticated technology that has been applied to solving a wide variety of problems. In spite of how clever these solutions and technologies might appear to be, the fact remains that our survival as a species on the planet is inexorably tied to a medium-sized star we call the sun. Solar energy, either stored in the form of fossil fuels and plants or made available on a daily basis as radiant energy, powers the planet and all life forms. Understanding heat measurement and transfer is, therefore, central to any discussion of energy and climate control technology. Two measurements are used to quantify heat energy: the calorie and the British thermal unit (Btu).

The calorie, also referred to as the gram calorie or small calorie, is the standard unit of heat energy in the metric system. The gram calorie or small calorie is often respresented by a lower case c and is the amount of heat required to raise one gram of water 1° Celsius. This calorie is commonly used in standard technical or scientific writing. The foods industry uses a special calorie called the large calorie to indicate the energy food provides. The large calorie is represented by a capital C and is equal to 1000 gram or small calories. This is the Calorie that people on diets count.

In countries such as the United States where the metric system is not widely used in most conventional activities, the British thermal unit (Btu) is used. One Btu is the amount of heat required to raise one pound of water 1° Fahrenheit. One large calorie equals 3.97 Btu.

MEASURES OF HEAT AND ENERGY

Knowledge of Btu's and calories allows for the determination of the heat content available in a variety of fuels and related energy sources. It also enables the climate control engineer to calculate the energy that is required for any specific application. For our purposes, Btu's will be used in all discussions of heating calculations. The formula used to determine the quantity of heat required to raise the temperature of a specific quantity of water is:

$Q = W(T^2 - T^1)$, where:

Q = quantity of heat required
W = weight of water
T^1 = initial temperature of the water
T^2 = final temperature of the water

Therefore, if we wish to determine how many Btu's are required to heat 80 gallons of water from 40°F to 140°F (the typical amount of energy required to completely cycle an 80-gallon hot-water heater), we apply the formula as follows (assuming a weight of 8 lbs. per gallon of water).

$Q = W(T^2 - T^1)$
$Q = 80 \times 8 (140 - 40)$
$Q = 640 (100)$
$Q = 64,000$ Btu's Required

To apply this information to predict actual fuel consumption for a particular application, assume that a residence has a propane-fired water heater. Using the information provided in figure 2-2, we can calculate the amount of propane needed to produce 80 gallons of hot water as follows:

Quantity of propane = 64,000 Btu/93,000 Btu
gallon propane = .68 gallon of propane required

Using these calculations, it is possible to accurately determine energy costs for almost all types of energy consumption and climate control devices in order to make accurate cost comparisons.

Fuel Type	Standard Unit	Btu Value
Anthracite coal	Pound	12,500
Electricity	Kilowatt-hour (1000 watts)	3,412
Fuel Oil	Gallon	140,000
Liquified petroleum gas (propane)	Gallon	93,500
	Cubic foot	2,500
Natural gas	100 cubic feet	1,000,000
	Cubic foot	1,000
Hardwoods (assorted)	Cord (4ft × 4ft × 8ft)	24,000,000
Softwoods (assorted)	Cord (4ft × 4ft × 8ft)	15,000,000

FIGURE 2-2 Btu values of selected common fuels.

FIRST AND SECOND LAWS OF THERMODYNAMICS

All processes that involve the generation or movement of heat are subject to two basic universal laws of physics known as the laws of thermodynamics. In general terms, these laws govern the limits concerning how energy can be converted and changed from one form to another, as well as the end-use efficiencies in theoretical, closed systems.

The First Law of Thermodynamics

The first law of thermodynamics is known as the law of conservation of energy. This law states that energy can be neither created nor destroyed, but only changed from one form to another. Within this closed system, the total amount of energy has remains constant.

To illustrate this point, consider the operation of the familiar electric baseboard room heater. The process begins at the utility plant where either fossil fuels or nuclear fuel generates the heat necessary to produce high-pressure steam for electrical production. The electricity is transmitted through the utility network and into the home, where it travels through a series of resistance elements located in the baseboard heater. In the baseboard heater, electricity is converted to radiant heat for purposes of raising the temperature of the dwelling. Throughout this entire process, there are many inefficiencies. For example, combustion efficiencies of conventional fossil-fuel-burning utility plants range between 30% and 40%. Therefore, between 60% and 70% of the potential energy within the fossil fuel is rejected as waste heat in the combustion process. Although nuclear fuel reactions are more thermally efficient than fossil-fuel combustion (oxidation) processes, the overall electrical generating efficiencies of both processes are roughly equivalent. This is so because nuclear powered utility plant efficiency calculations also must take into account the amount of energy used for the extraction, refining, and processing of the nuclear fuel, as well as the downtime (capacity factor) experienced by most nuclear power plants for refueling and other safety procedures.

When considering any of these systems as a whole, it is noted that the total amount of energy within the system has remained constant. The amount of heat generated by the baseboard heater is equivalent to the potential energy available within the fuel prior to its consumption at the utility plant, minus all of the heat and associated energy losses that occur at every stage of the production and conversion process. Energy is conserved.

The Second Law of Thermodynamics

The second law of thermodynamics deals with energy quality through the introduction of a term known as entropy. In its most basic form, the second law states that heat, when left to itself (without any outside forces or work being performed), will always flow from the hotter object to the colder object. In order to reverse this process, work must be performed. By virtue of the performance of work, waste heat will be rejected, ensuring that the useful energy available at the end of the process is always less than the potential energy that was available from the fuel source at the beginning of the process. This principle limits the efficiency of any production system to less than 100%. Entropy is a measure of both heat and the random distribution of molecular activity.

Entropy as a descriptive term is often used in a number of different situations, and is not limited to a description of physical or heat-generating systems. As a political descriptor, consider a highly ordered democratic society (a low-entropy system) as opposed to a society in an anarchistic state (a high-entropy/highly disordered state). In an environmental sense, one can use the example of a sand pit that was quarried and later abandoned. After it has been neglected a time, nature begins to reclaim the pit. Erosion fills in the sand pit areas (a highly ordered, low entropy-area), bringing, in the process, an assortment of trees, dirt, and other miscellaneous material (a disordered high-entropy arrangement) that will eventually fill in the once highly ordered quarry. A pond can be dug and turned into a beautiful, highly ordered recreation area (low entropy), but, if neglected, will eventually fill in with silt. Nature, left to its own devices, will turn the highly ordered pond into a random assortment of dirt, annual and perennial weeds, and brush, a high-entropy environment.

Entropy is most often used, however, to describe energy quality, or the randomness of a particular system. When dealing with energy quality, entropy refers to waste heat. For example, natural gas extracted from the wellhead is an ordered, high-quality source of energy. Once it has been burned, whether in a power plant or a gas-fired boiler, this highly ordered source of energy is almost totally rejected as waste heat, eventually to become randomly spread throughout the universe. It can be neither recaptured nor reconstituted back to its original highly ordered state. When we consider the prolific use of all types of nonrenewable sources of energy, the second law illustrates that the randomness of the universe will always either remain constant or increase, but can never decrease.

TEMPERATURE QUANTITY AND QUALITY

To help clarify concepts relating to heat discussed up to this point, consider a comparison of the heat values of a burning match and of a grassy field bathed in sunlight. A single burning match is much hotter than the solar radiation falling on the field. But although the match is higher in temperature than any place on the field, the quantity of heat energy or Btu's in the field is far greater than that available in the burning match. Heat in the match is more concentrated and of a higher quality than the more diffuse solar radiation. These concepts have wide applications in the heating and air-conditioning field. For example, if a room in a house is cold, is it a temperature or Btu problem? The room could be cold because of excessive drafts, insufficient baseboard heating unit, marginal water temperature in the heating system, too small a combustion head in the burner mechanism, and so on. Diagnosing problems relating to the distribution and quantity of heat requires a clear understanding of these concepts.

MEASUREMENT OF HEAT

Although the concept of heat might seem uncomplicated at first glance, it soon becomes apparent that heat production and manipulation during the climate control process require clear and accurate terminology. What follows is an explanation of the characteristics of heat and the terms that are used to desribe these characteristics.

Specific Heat

The specific heat of any substance is the number of Btu's required to raise one pound of that particular substance 1°F. Water is so abundant that it is used as the standard benchmark for specific heat calculations. With a specific heat of water at 1.00, 1 Btu is required to raise one pound of water 1°F. Specific heat can also be viewed as the ability of a substance to store heat energy. This ability has been referred to as "thermal inertia." Different substances have different specific heats; however, water has the greatest specific heat of any of the commonly used substances. If two one-pound samples of different materials are both heated to the same temperature and then dropped into equal amounts of water at the same temperature, measuring the temperature increases in each water sample will illustrate the different specific heats of each material.

Sensible Heat

Sensible heat is the amount of heat added to any substance that will raise its temperature without changing its state. For example, heating a pot of water from 70°F to 90°F results in an increase of 20°F of sensible heat in the water. Most temperature changes with which we ordinarily come into contact are sensible heat changes.

Latent Heat

Latent heat is the energy required to change the state of a substance without changing its temperature. For example, when water reaches 212°F, energy in the form of heat must be added to the water in order to change it from a liquid state to a vapor, steam. In order to do this, a substantial amount of energy is consumed in the process. Conversely, when steam vaporizes back to a liquid, energy is released. There are several different types of latent heat measurements.

1. *Latent heat of Vaporation*—the amount of energy required to vaporize a specific liquid. For example, latent heat of vaporization is the cooling process that is used by the human body—the production of sweat which evaporates, thereby lowering body temperature.

2. *Latent heat of condensation*—the amount of energy released when a change of state occurs from a vapor back to a liquid. This process is evident when one suffers a burn after being exposed to steam. The steam condenses from a vapor to a liquid, releasing its latent heat of condensation, which can cause severe burns.

3. *Latent heat of fusion*—the amount of heat energy released when a substance changes from a liquid to a solid state. This process is illustrated during a metalcasting process when iron or steel solidifies in a mold.

4. *Latent heat of melting*—the amount of energy absorbed when ice, for example, melts back into a liquid state. It should be noted that the heat of fusion and the heat of melting are essentially the same process occurring in different directions.

For reference, consult figure 2-3 for the Btu values that apply to various changes of state of water.

Substance	Amt	From Temp	To Temp	Temp Rise	Spec Heat	Amt of Heat	Type of Heat	Name of Change
Ice (solid)	1 lb	0°	32° =	32° at	1/2 BTU =	16 BTU's	Sensible	(Heating)
Ice (solid)	1 lb	32°	0° =	-32° at	1/2 BTU =	-16 BTU's	Sensible	(Cooling)
Ice to Water	1 lb	32°	32° =	0° at	144 BTU =	144 BTU's	Latent	(Melting)
Water to Ice	1 lb	32°	32° =	0° at	-144 BTU =	-144 BTU's	Latent	(Freezing)
Water (liquid)	1 lb	32°	212° =	180° at	1 BTU =	180 BTU's	Sensible	(Heating)
Water (liquid)	1 lb	212°	32° =	-180° at	1 BTU =	-180 BTU's	Sensible	(Cooling)
Water to Vapor	1 lb	212°	212° =	0° at	970 BTU =	970 BTU's	Latent	(Vaporization)
Vapor to Water	1 lb	212°	212° =	0° at	-970 BTU =	-970 BTU's	Latent	(Condensation)
Vapor (gas)	1 lb	212°	222° =	10° at	1/2 BTU =	5 BTU's	Sensible	(Heating)
Vapor (gas)	1 lb	222°	212° =	-10° at	1/2 BTU =	-5 BTU's	Sensible	(Cooling)

FIGURE 2-3 Btu values for various changes of state of water. (Courtesy of Williamson Corp.)

Super Heat

Super heat is sensible heat that, when added to a vapor above its boiling point, will raise the temperature of the vapor. No change of state is involved in superheating. Thus any heat that is added to steam that is initially at a temperature of 212°F will superheat the steam, raising its temperature. Pressure in the system determines the temperature of the superheated steam. Processes that use super heat often require special piping and handling provisions for the superheated vapor. High operating pressures are also usually encountered in superheated system applications.

MEASUREMENT OF WORK AND POWER

Heat is rarely generated in total isolation from other energy systems or mechanical devices. One basic purpose of generating and capturing heat is to use it to perform some kind of work. In its most basic sense, work involves the application of some type of force, F, which is utilized to move an object a certain distance. The formula for work is:

Work (*W*) = Force (*F*) × Distance (*D*)

In this formula, work is measured in units called *joules*, force acting on the body is given in *newtons*, and distance is measured in *meters*. The resulting work is, therefore, given in *newton-meters*. A force of one newton is equivalent to approximately 3 1/2 ounces. Therefore, if a person were to lift a weight of 3 1/2 ounces one meter high, one joule of work will have been performed.

Of equal importance is the concept of power. Power quantifies the amount of work done over a specific period and is given by the formula:

Power (*P*) = Work/Time

For example, compare the effort expended when a person slowly lifts a 100-pound carton into a truck as opposed to quickly lifting and loading the carton. In each instance, the amount of work performed is identical—the force applied to lift the box a specified height onto the truck. However, the power used in the second instance is greater, since the work was accomplished in less time. The standard unit of power is the watt (W). The watt is defined as doing work at the rate of one joule per second. One kilowatt equals 1000 watts. The English unit of power is the horsepower, which is equivalent to lifting a 550-pound weight one foot in one second. One kilowatt is equivalent to 1.34 horsepower. See figure 2-4 for energy and work equivalents.

Work = Force × Distance
$$W = F \times D$$

In the International System (SI), work is expressed in joules
Joules = Newton/meters

In the British System, work is expressed in foot-pounds
Work = Foot/pounds (ft-lb)

1 joule per second = 1 watt
746 watts = 1 horsepower = 550 ft-lb second
Btu's = Watts × 3.4
or
Watts = Btu/3.4

FIGURE 2-4   Equivalent values of work.

PROPERTIES OF AIR

Qualities of climate-controlled air are important if the effects of any heating or air-conditioning system are to be maximized for the greatest efficiency and comfort. These qualities deal with temperature, amounts of moisture in the air, and air purity.

Wet- and Dry-Bulb Temperature Measurements

Temperature measurements for climate control purposes are made in one of two ways. A *dry-bulb temperature* reading is taken with a standard thermometer in a conventional manner. A *wet-bulb temperature* reading is taken by putting a wet cloth or wick over the bulb of the thermometer. Usually the thermometer is placed in the middle of a stream of circulating air; the moisture in the cloth surrounding the thermometer bulb evaporates in the moving air stream, lowering the temperature reading on the thermometer. The amount of moisture that evaporates from the bulb wick depends on the relative humidity in the air stream and will have a direct effect on the temperature reading. The difference between the wet- and dry-bulb temperature readings is an indication of the relative humidity of the surrounding air. Relative humidity is a key factor in determining human comfort levels in a heated or cooled environment. For example, an indoor temperature of 78°F with a low relative humidity will be far more comfortable than the same temperature with a high relative humidity. Conversely, raising the relative humidity in the home during the winter months will allow for the achievement of comfortable living conditions at lower temperature levels than with lower relative humidity levels.

Measures of Moisture and Humidity

Two classifications of moisture and humidity are of interest in climate control technology.

Relative humidity: This refers to the amount of moisture in the air relative to the amount of moisture that the air can hold at a particular temperature. This principle is illustrated in the operation of a clothes dryer or hair dryer. These devices raise the temperature of the air, allowing it to absorb increased amounts of moisture (humidity).

Dew-point temperature: This is the air temperature at which moisture condenses on a surface. Dew-point temperature is a function of both air temperature and relative humidity. One illustration of dew point is a phenomenon known as radiation fog. On a clear, cloudless night, as the air cools, heat is radiated from the earth's surface to the night air. The cool air can quickly reach 100% relative humidity and produces a phenomenon known as "radiation fog."

PRINCIPLES OF HEAT TRANSFER

Heat can move in either one of two directions. Consider the illustration of the "thermal hill" in figure 2-5.

Heat will normally flow from a warmer to a cooler object in any isolated system. Eventually, the two objects will reach the same temperature. The movement of heat from a hotter to a cooler substance is referred to as moving heat "downhill," where no external energy is required to accomplish this movement. Conversely, in the air-conditioning process, where heat is extracted from a home with an indoor temperature of 78°F and is moved to the outdoors where the temperature is 85°F, the movement is "uphill"—heat is moved from a warm area to another area in which the temperature is greater. In any application in which heat is moved, three distinct processes of heat transfer are involved—convection, conduction, and radiation. For example, in a conventional oil-fired hot-water boiler, heat is transferred from the combustion of the fuel oil to the surrounding water in the boiler. Water circulates through the boiler, removing heat from the combustion process and transferring it throughout the baseboard radiation to the room air. Eventually, all of the heat within the house will be lost to the surrounding atmosphere, completing the entropy cycle of the second law. Although methods of heat transfer differ, most heating and cooling applications employ at least one, and sometimes all three methods simultaneously.

Conduction

Conduction refers to the transfer of heat between two objects that are in direct contact with each other. Conductive transfer involves the movement of heat from the hotter to the cooler object. In this process, the principle of thermal conductivity, the ability of a substance to either conduct or retard the flow of heat, is most important. The amount of heat exchanged in the process is a function of several basic factors: the thermal conductivity, K, of the substances; the temperature difference between the two objects; the amount of surface area of the two objects in contact; and the length of time that the two objects remain in contact. For example, in a typical hot-water heating system, water in the boiler is heated by conductive heat transfer from the hot metal of the boiler. Likewise, hot water transfers its heat via conductive contact to the finned radiators within the baseboard cabinet.

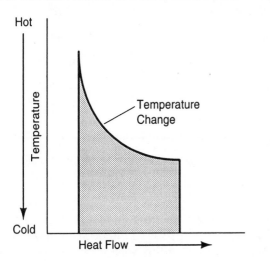

FIGURE 2-5 A "thermal hill" of temperature change. Heat flows from a warmer to a cooler substance. As the two substances approach the same temperature, the rate of heat transfer flows and eventually stops when both objects are at the same temperature.

In addition to specifying the K value of a substance, it has also become common practice to refer to most building materials based upon their ability to resist the flow of heat. This value is known as an R value. R and K values are mathematical reciprocals of each other. These reciprocal values are akin to describing a glass of water as either half full or half empty, and can be used interchangeably, although R values are the more common descriptor. Figure 2-6 lists the R values of selected common building materials.

Material	Standard Thickness	*R* Value
Brick	Inch	0.20
Cement	Inch	0.20
Concrete building block	Inch	0.20
Cellulose insulation	Inch	3.70
Fiberglass insulation	Inch	3.70
Gypsum board (Sheetrock)	1/2 inch	0.45
Plastic (polystyrene)	Inch	4.00
Plywood sheathing	3/4 inch	0.82
Plaster (wall and ceiling)	Inch	0.20
Sheathing board insulation	Inch	2.60
Shingles (asphalt roofing)	Standard thickness	0.40
Wood shingles	Variable standard thickness	0.94
Wood siding	3/4 inch	0.85
Storm windows	Variable, common	1.75±
Wood door (solid entrance)	1-3/4 inches	1.75±

FIGURE 2-6 *R* values of selected building materials. The greater the *R* value, the geater will be the insulating characteristic of the material.

Convection

Convective heat transfer involves the transfer of heat aided by moving fluids. For purposes of our definitions, fluids may be thought of as either gases or liquids. For example, in a typical hot-water heater, water that is heated by the combustion of a burner mechanism located at the bottom of the heater rises as it is heated. Cooler water, which is denser than the lighter, heated water, drops to the bottom of the water heater vessel. In this way, a convective flow is established in the water heater that aids in the heating process. In virtually all types of heating systems, hot air rises and cooler air drops to take its place, setting up convective air currents within the heated room. It is still common practice in central heating system installations to place radiators directly in front of windows and along outside walls of the home. This is done in order to maximize convective heat transfer that is enhanced due to outside air leaking into the home from cracks in the windows and infiltration through the outside walls. With newer building materials and tighter residential construction techniques, this practice is no longer necessary, since air leakage around windows and through outside building walls has been largely eliminated.

In order to maximize convection efficiency, most heating and cooling systems employ fans and circulating pumps to move the air and water throughout the distribution system. When pumps and fans are used, we refer to them as forced-convection systems.

Radiation

Radiation involves the transfer of heat through air, as one might experience heat when standing in front of a fireplace. In this instance, heat is radiated from the wood fire as waves of electromagnetic energy. It is interesting to note that radiant energy does not require a medium through which to travel. For example, radiant energy from the sun travels through the vacuum of space in order to heat the earth's surface. Like other methods of heat transfer, raising the temperature of the radiating object increases the efficiency of heat transfer: the hotter the water temperature in a heating system, the warmer a person will feel while standing in front of the radiator or baseboard convectors. Also, color plays an important part in personal comfort when dealing with radiant heating systems. Dark colors absorb radiant heat, while lighter colors reflect radiant energy. A person wearing dark clothing will be warmer than a person wearing light-colored clothing when standing in front of a fireplace. Likewise, the home with dark-colored roof shingles will absorb a greater amont of radiant solar energy during the winter than one with light-colored shingles. Other applications of this principle are evident in the use of tinted window treatments for both homes and automobiles in order to lower

interior temperatures. Figure 2-7 illustrates the various methods of heat transfer at work in a typical heating situation.

EFFECTS OF HEAT ON GASES AND LIQUIDS

There are three states of existence of matter. In the *solid state*, substances have a fixed shape and volume. In the *liquid state*, a substance will have a fixed volume with no particular shape. In the *gaseous state*, substances have neither a fixed shape nor volume. Gases, liquids, and solids react in predictable ways to changes in temperature. Solids expand when they are heated and contract when they cool. (The only exception to this rule is water, which contracts as it is cooled to a temperature of 39°F. As it is cooled below this temperature, water expands. The results of this process are known to anyone who has ever had to replace plumbing pipes or water fixtures after they have burst from water freezing within.) The predictable nature of expansion and contraction is important when designing heating and air-conditioning systems and associated control devices. With a known volume of water in a boiler, it is important to be able to determine how much the water will expand when it is heated from room temperature to an operating temperature of 200°F. Since liquids, practically speaking, are impossible to compress, an allowance must be made for the expansion of the water within the heating system as it is heated. This is done by incorporating an expansion tank into the piping system. Sizing the expansion tank properly is, therefore, a function of knowing how much expansion is likely to occur as the boiler goes through its normal cycling characteristics during the heating season.

Expansion and Contraction

Different materials expand and contract at different rates as they are heated. The amount that a particular substance will either increase or decrease in length as it is heated or cooled is known as its *linear coefficient of expansion*. The

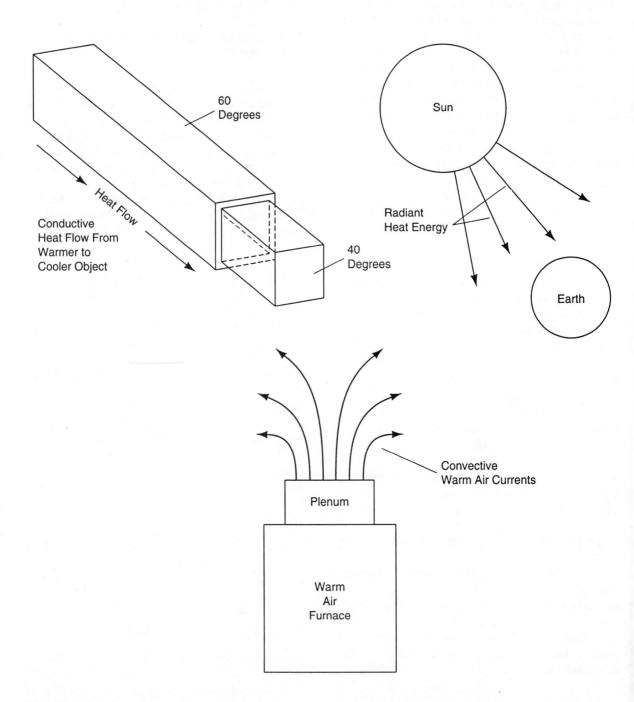

FIGURE 2-7 Heat transfer via convection, conduction and radiation. Most heating and air conditioning systems employ one or more methods of heat transfer during their operation.

amount that the surface area of a material changes during either heating or cooling is known as its *area coefficient of expansion*. *Volumetric expansion* refers to the increase or decrease in the volume of a liquid during heating or cooling.

All gases conform to basic laws of physics that describe changes in their characteristics as they are heated. These laws also specify conditions that govern their temperature and pressure. The two laws of physics that describe the reactions of gases to changes in temperature, pressure, and volume are known as Boyle's law and Charles' law.

Boyle's Law

Boyle's law states that if the temperature of a gas is held constant, then its pressure is inversely proportional to its volume. This law is used in situations where the volume of a given weight of gas at a specific temperature is known, and the investigator wants to determine what the volume of the gas will be at a different pressure. Boyle's law is given by following formula:

$$\frac{P_1}{P_2} = \frac{V_2}{V_1}$$

Where: P_1 equals initial pressure of the known weight of the gas at a specific temperature

P_2 equals final pressure of the known weight of the gas at a specific temperature

V_1 equals intial volume of the gas

V_2 equals final volume of the gas

Whereas Boyle's law addresses changes in the pressure and the volume of gas at a constant temperature, Charles' law deals with the characteristics of gases as the temperature is varied.

Charles' Law

Charles' law illustrates that gases behave similarly to solids and liquids with regard to expansion and contraction. Thus, if the volume

of a specified weight of gas is held constant, then the pressure of the gas will vary directly as its temperature: If you increase the pressure of a specific volume of gas, its temperature will increase proportionally. The formula for Charles' law is stated as follows:

$$\frac{P_1}{P_2} = \frac{T_1}{T_2}$$

Where: P_1 equals initial pressure of the gas

P_2 equals final pressure of the gas

T_1 equals initial temperature of the gas

T_2 equals final temperature of the gas

HEAT LOSS AND HEAT GAIN

Sizing the Heating and Cooling System

Given an understanding of the basic physical principles of energy and heat, the practitioner requires a reference as to the proper sizing of the heating or cooling system. By size, we refer to the heating or cooling capacity of the system in Btu/hr. The size of the heating and cooling system will depend upon how much heat the dwelling loses during the winter and gains during the summer. Detailed calculations are usually performed by both architects and engineers when designing new buildings or retrofitting an existing building with a new environmental control system. Heat loss calculations can be performed either manually or by using specially written computer programs. For purposes of this text, we will present a simple procedure for determining heat loss and heat gain calculations for a variety of structures based upon the construction characteristics of the building. It should be noted that these calculations are not designed to replace detailed heat loss analysis, but rather to quickly approximate heat loss and gain for a variety of different types of structures.

Heat loss: Conventional heat loss calculations are based on a particular design temperature that is usually chosen by either the builder or architect. Thus, if a building is analyzed to have a heat loss of 75,000 Btu/hr, this loss has been calculated to occur when the outside temperature is, say, 10°F. Design temperatures selected for heat loss analysis are usually the coldest temperatures encountered in a specific geographical area during the average winter season. Although the heat loss chart in figure 2-8 is not based on any particular design temperature, the analysis will be accurate enough to place the heating and cooling system in realistic perspective. More detailed analyses must be computed to ensure accuracy in system sizing.

Using the following method, buildings are divided into three categories of construction:—older/uninsulated buildings, modern construction with conventional insulation and double-glazed storm windows, and superinsulated buldings (typical of passive solar homes). For each category, a heat loss figure is assigned, which is then multiplied by the square footage of the structure to yield an approximate heat loss figure. Please note the column showing the spread in the heat loss per square foot. The technician should assign a figure within this spread that is felt to be appropriate given the characteristics of the building in question. For example, an older, uninsulated home that has had attic insulation and storm windows installed would be closer to the 60 Btu/ft² heat loss figure than 75 Btu/ft².

Heat gain: Heat gain can be approximated from the heat loss analysis and is generally calculated to be approximately one third of the heat loss figure. Thus, if a building, heat loss is calculated at 100 Btu/hr, then the heat gain during the summer will be approximately 30,000 Btu/hr. This figure indicates that the cooling system should be sized at approximately 2-1/2 tons of cooling capacity. For a more detailed explanation of the derivation of refrigeration capacity in Btu's, see Chapter 15.

Type of Structure	Heat Loss, Btu/Ft²/Hr	Category
Older, uninsulated buildings Few or no storm windows No fireplace dampers Usually pre-1950s	50–75	1
Conventional, modern construction Storm windows and doors Standard wall insulation (3-1/2 inch fiberglass) Standard ceiling insulation (6 inch fiberglass) Fireplace chimney dampers Usually post-1950s or renovated older homes	30–50	2
Well insulated and passive solar structure Wall insulation = 6+ inches Ceiling insulation = 12+ inches Double and triple insulated windows and doors	25–30	3

Notes: 1. All figures are approximations due to varying building characteristics.
2. Design temperature = 0°F.

FIGURE 2-8 Approximating heat loss in buildings with varying construction characteristics. Multiply the heat loss in Btu/hour by the total square footage of the building to arrive at an approximate heat loss figure. Calculations are based on a design temperature of 0 to 5°F.

SUMMARY

This chapter has focused on the principles of energy conversion and heat transfer, along with the terminology associated with these concepts. While the first and second laws of thermodynamics govern the states and quality of energy, the concepts derived from these laws—namely, the measurement of heat, work, and the properties of gases, liquids, and solids—define the environment within which the HVAC technician works. Given this background, Chapter 3 examines the basic principles of electric power, and is directed toward gaining a broad-based insight into and understanding of the operating characteristics of control devices used in heating and air-conditioning systems.

PROBLEM-SOLVING ACTIVITIES

Review Questions

1. Convert the following temperatures to either the Fahrenheit or Celsius Scale as required.

    a. 51°F
    b. 17°C
    c. 45°C
    d. 82°F
    e. 15°F
    f. 95°C

2. How many Btu's are required to heat 57 gallons of water from a temperature of 47°F to 135°F?

3. Perform the necessary calculations to determine whether propane or electricity is the most efficient method of fueling a domestic hot-water heater, based on an electricity cost of at 0.15 cent per kilowatt-hour and a propane cost of $1.15 per gallon. Assume a recovery rate of 33 gallons per hour for each water heater, with an electrical input of 4500 watts per hour and a gas input of 1 gallon per hour.

4. Explain the differences between the first and second laws of thermodynamics. In this discussion, explain how the actions of each of these laws can be illustrated using a heating or cooling device.

5. Identify four different measurements applied to the concept of heat.

6. Calculate the number of joules of work performed in each of the following two examples.

    a. A person lifting a weight of 8-1/2 ounces 4 feet high.
    b. A person lifting a weight of 5 pounds a distance of 15 feet.

7. How much power is consumed when a 200-watt light bulb burns for 18 hours?

8. How many horsepower can be developed by a 110-volt electric motor that has a current draw of 22 amps?

9. Differentiate between wet-bulb and dry-bulb temperature readings.

10. Describe the differences between heat transfer that takes place by conduction, by convection and by radiation. Identify one type of heating or cooling system process for which each type of heat transfer would be applicable.

11. Using the heat loss chart in figure 2-8, calculate the heat loss of a 1750-square-foot building that has no insulation. In addition, what would the approximate heat gain of the building be?

CHAPTER 3

What Is Electricity?

Most of us have, at one time or another, thought about electricity. Perhaps we were faced by a frustrating and mysterious problem, like an air conditioner that stopped working on hot days, but worked perfectly when taken apart for repair, or the car that refused to turn over only when no help was available. And some of us are simply curious about this widely used technology that makes motors spin, provides lights for our homes, and at times surprises us unpleasantly with dangerous shocks and sparks. One way or another, most of us at some point have asked the question, "What is electricity?"

Electricity can be so complex that those who enjoy mathematics begin to mistake formulas and numbers for a common-sense knowledge of what electrical technology is really about. Formulas and numbers do not travel through wires or cause lamps to glow. To be sure, mathematics, or reasoning with numbers, enables us to calculate the amounts of volts, amps, and ohms in an electrical circuit. But usually repairing faulty equipment involves locating a circuit or component that does not work and no mathematics are involved. How the parts function and connect together is more important than the specific amounts of volts, amps, or ohms. It is a mistake to think that an understanding of electricity depends strictly on mathematics.

Another mistake that people often make is to think that electricity is complicated and hard to understand. Imagine a diagram of all the pipes supplying water for a small city. This diagram could easily be mistaken for a road map of a state or the wiring of a large building. It is difficult to think about all the pipes or roads or wires at the same time. Reasoning and thinking about electricity help to locate the faulty part of a circuit and to fix the problem. A traffic jam may be caused by one accident on the road, just as a lack of water is probably due to one ruptured water main. Understanding electricity is no more complicated than understanding a plumbing system or the traffic flow on a highway.

If understanding electricity is neither dependent on mathematics nor complicated, why do so many people seem to fear it and find it such a mystery? Perhaps one answer is that electricity is invisible, and can be so dangerous enough to cause serious injury and death. In the first instance, our eyes need the help of meters and test instruments to visualize electricity. In the second, our lives depend on exercising caution when working with electricity.

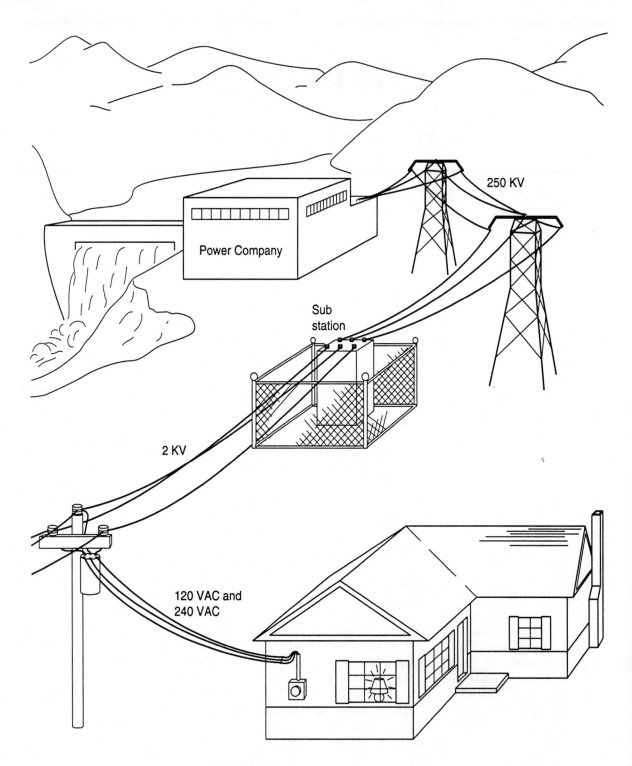

FIGURE 3-1 Power plant to house lighting.

WHY WE USE ELECTRICITY

Throughout the ages, people have banded together in families, tribes, and countries for mutual benefit. Small contributions by each individual have enabled such spectacular achievements as traveling to the moon, as well as more ordinary but still important accomplishments such as the highway system. Cooperation has enabled us to do many things we could not do as individuals. To accomplish anything requires the expenditure of energy.

Just a few generations ago, homes and factories were purposefully located close to a source of energy. Many towns were built near waterfalls. As other sources of energy such as steam, gas, and diesel engines gradually became available, people could choose between living near the source of energy or maintaining their own energy supply. Eventually a solution was found: move energy to the individual.

The inventor Nikola Tesla and the entrepreneur George Westinghouse worked together in 1895 at Niagara Falls to develop a system that produced large amounts of electric energy (figure 3-1). The falling water was used to rotate large water turbines that were connected to electric generators. In this way, electricity was produced efficiently and in great quantities. It could then be transported by transmission wires to customers at remote locations. In homes and factories, the electricity was converted into motion (motors), light (lamps), and a variety of other directly useful forms of energy. It should be noted that Edison developed a direct-current (DC) system of electric power distribution, but it was impractical and far less efficient than Tesla's alternating-current (AC) system. Direct and alternating current will be explained later.

With few exceptions, electricity was, and still is, of very little use until it is converted into motion, light, or heat. We use electricity because it is a practical way to move energy from efficient central generating plants to remote locations. But electricity, the great mover of energy, has little practical value until it is converted to perform some other task.

Over the past 100 years, society has become so skilled in the use of electricity to move things that an associated field, electronics, has developed. With electronics, we move sound (telephone and radio), pictures (television and VCR), and data (computers); see figure 3-2.

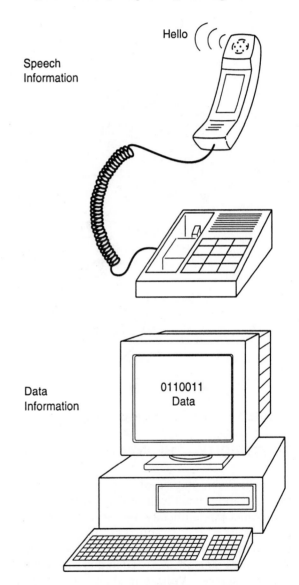

Speech Information

Hello

Data Information

0110011 Data

FIGURE 3-2 Phone and computer communications.

Thus electricity is used to move energy while electronics is used to move information. The air conditioner in figure 3-3 gets energy for cooling from the power station and the information to control room temperature from the electronics of the thermostat. The electricity that flows in electric and electronic circuits is of little use in itself. Rather, it transports the energy and information upon which society depends.

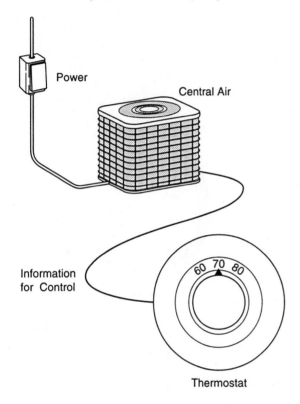

FIGURE 3-3   Air conditioner to plug to thermostat.

ELECTRICITY—THE INVISIBLE MOVER

The invisible nature of electricity can make thinking about it difficult. Electrical technicians overcome this difficulty by developing mind pictures, or models of actual electrical phenomena. Mathematical formulas are models of what is actually going on. Pictured in this way, Ohm's law (Resistance = Volts/Amps) and stepping on a garden hose, which restricts the water flow and mechanical friction, are similar to the resistive nature of electrical circuits (figure 3-4).

Electrical resistance is actually a result of the structure and nature of the universe. George

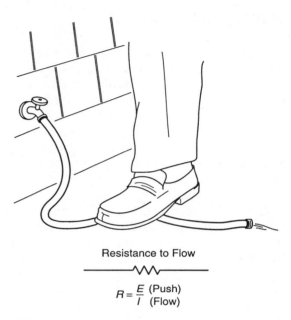

Resistance to Flow

$$R = \frac{E}{I} \frac{\text{(Push)}}{\text{(Flow)}}$$

FIGURE 3-4   Garden hose and resistance.

Simon Ohm described it correctly with his "law," but it should be clear that with or without Ohm, electrical circuits have a resistive nature. Clearly, well-chosen mental pictures, likenesses, and mathematical formulas are necessary to enable the technician to think about electricity and to "see" what is going on inside an electrical circuit. The following likenesses and models have been very carefully selected to provide an accurate, clear, and solid foundation for electrical concepts.

WHAT FLOWS IN ELECTRICITY?

Electricity is the flow of an electron gas. To understand this idea, two points need to be examined closely: the nature of the electron and the physical behavior of a gas.

THE ELECTRON AS AN ELEMENTARY PARTICLE

Scientists have developed a model of the subatomic universe called "the standard model." A basic version of the standard model assumes the existence of three families or groups of vanishingly small particles: leptons, quarks, and messenger particles. Each of these families has six members, for a total of 18 elementary or primary particles (see figure 3-5). Besides being elementary, or not further divisible, these particles also have a quantum nature.

Leptons	Quarks	Bosons That Carry These Forces	
Electron (Electricity)	Up	Photons	Electromagnetic Radiation
Muon	Down	Gravitons	Gravity
Tauon	Top	Gluons	Strong Atomic
Electron-Neutrino	Bottom	W+	Weak Atomic
Muom-Neutrino	Charm	W–	
Tauon-Neutrino	Strange	Zo	

All of the above "matter" particles have "anti-matter" counterparts. For example the anti-matter electron is called a positron, which has a + charge. Protons are made of two UP quarks and one DOWN quark, while neutrons are made of one UP quark and two DOWN quarks.

FIGURE 3-5 Families of particles

Quantum particles taken individually are rather vague smears of energy that lack a definite nature. However, when large numbers of particles such as electrons are considered, the individual changeable natures of single electrons are averaged out. For our purposes, it could be said that we need deal only with large numbers of electrons that, taken together, have a definite nature. Our understanding of electricity is, therefore, "causally disconnected" from the standard model of subatomic particles. For the technician, time is best spent on developing a clear picture of what happens to collections of electrons as they move within circuits. Not surprisingly, many physicists who study elementary particles are often unable to assemble simple electrical circuits.

THE NATURE OF ELECTRONS IN GROUPS

Electrons are so small that they are generally dealt with as a unit of charge called a coulomb, a microscopic bucket of electrons. A coulomb is 6,280,000,000,000,000,000 electrons. Indeed, they are so small that it takes great numbers of them to have any practical effect in electronic circuits. Electrons are vanishingly small.

The negative charge of electrons (–) causes them to be attracted to positive charges and repelled from negative charges, including themselves. They are atomically antisocial and stay as far away from one another as conditions allow. If electrons were collected within a jar, they would act much like a gas and spread out evenly to avoid one another. However, electrons in a jar are not very useful. In most applications, the movement or flow of electrons is necessary.

Unlike heat radiation (the warming rays of the sun), which can travel through empty space, electrons ordinarily require a substance or medium through which to travel. The medium through which electricity flows is called a conductor. One way to think of a conductor is to imagine it to be transparent to an electron gas. The copper in a wire, which appears solid to the human eye, is the hollow of a pipe to electrons. Silver, copper, and gold are all good conductors of electricity. They are all transparent to an electron gas. If electrons could see, they would find it strange indeed that humans cannot simply enter a copper wire. To be sure, stylish electrons would have limited choices in jewelry since silver, copper, and gold would literally fall through their fingers.

Thus electrons taken singly are strange, tiny quantum particles that repel one another. In the large groups that are necessary for practical application, they act like a gas that can be forced through conductors. The conductors in figure 3-6 are substances that are transparent to electrons.

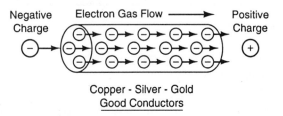

FIGURE 3-6   Bar of copper, silver, or gold.

WIRES ARE PIPES

The technology of both living and nonliving things depends on piping. Human cells require supplies and produce wastes that are carried by the circulatory system. Senses and ideas are piped through a complex array of nerves. Even traffic flow in large buildings is dependent on elevators, hallways, and stairs that, in a more abstract sense, can be viewed as pipes. In a very real way, much of the study of heating and

air conditioning depends on an understanding of pipes that carry water, fuels, refrigerants, and air. As illustrated in figure 3-7, electricity is carried in pipes called wires.

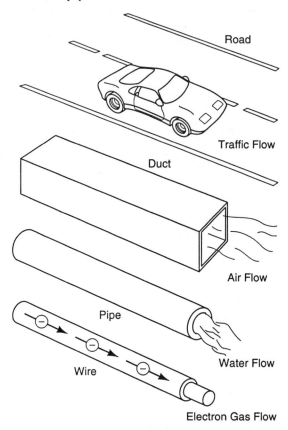

FIGURE 3-7   Water, air (ducts) highway, and wire.

FLOW: THE AMP

Many of the great old theaters and churches have organs with large pipes displayed in a grandiose fashion. The thunderous sounds they emit shake the buildings that contain them and amaze us, yet we are not entirely surprised. There is an expectation of big things from big pipes. Large water mains are meant to carry vast amounts of water. Six-lane highways carry a great deal of traffic. Wires with thick copper centers are intended to carry large amounts of electricity. Recall that since copper is transparent to an electron gas, the copper part of a wire is like the hole in a water pipe. Wires with small amounts of copper (even copper foil on a circuit board) carry small amounts of electron gas. Wires with a great deal of copper carry large amounts of electricity. Different sizes and types of wire are shown in figure 3-8.

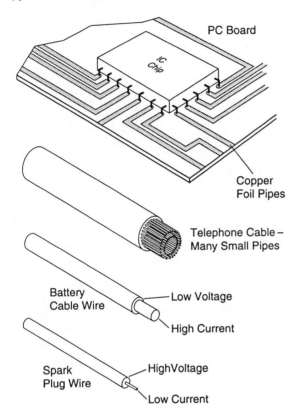

FIGURE 3-8 Big and little wires (note use).

Large and small are useful terms to remember in developing the idea of electricity flow and wire size. More information is needed if electricity is to be understood and used to its full potential. That gigantic number 6.28×10^{18} (move the decimal to the right 18 places), the microscopic bucket of electrons called the coulomb, is a good place to start. Generally, when considering the idea of flow, two things must be specified: the amount (units) and the time it takes to pass. Highways can carry cars (units) per hour (time) or water pipes can carry gallons (units) per minute (time). Technically,

cars per hour and gallons per minute are called the rate of flow. The copper part of a wire has the capacity to carry coulombs (units), and the

APPROXIMATE AMPACITIES OF COPPER WIRE		
Wire Gauge	**Ampacity**	**Ohms Per 1000 Foot**
14	15	2.525
12	20	1.588
10	30	.999
8	40	.628
6	55	.395
4	70	.2485
3	80	.1970
2	95	.1563
1	110	.1239
0	125	.0983
00	145	.078
000	165	.0618
0000	195	.049

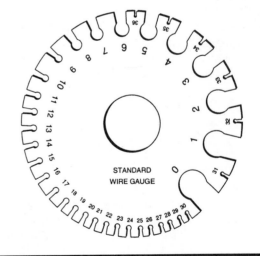

FIGURE 3-9 Wire size and wire gauge.

rate (time) that is used is seconds. Coulombs per second, or the rate of flow of electricity, is used so often that it is referred to as an amp. The amp is a measure of the intensity of flow; hence the letter I is used in electrical formulas to represent the concept of flow and the amp is the unit of measure. Many people use the term amp without realizing that it is coulombs per second, akin to electrical gallons per hour. The copper part of a wire has a capacity to carry amps (coulombs per second). Figure 3-9 illustrates wire size and its capacity to carry amps.

INSULATION: THE PIPE'S WALL

Leaks are common problems in all piping systems. A skinned knuckle, or a leak in the circulatory system heals because blood has the capacity to clot and form its own patch. Unfortunately, most heating and air-conditioning systems continue to leak until they are repaired. What about electrical piping, the wires?

The outer covering of the wire, called insulation, is intended to contain the electron gas within the copper part of the wire. Copper used in water pipes makes a good wall since it is opaque to water. However, since copper is transparent to an electron gas, in electrical pipes (wires), it makes a good hole.

Some materials, such as glass, porcelain, and plastic, are, to varying degrees, opaque to an electron gas. The stylish electron that could not wear jewelry made from silver, copper, or gold would have to settle for plastic rings and chains. Insulators are materials that are opaque to an electron gas (figure 3-10).

The same force or pressure that pushes an electron gas through the copper part of a wire may cause a leak in the insulation. This is especially true if the force or pressure of the gas is very high, as in the ignition system of an automobile. If the spark plug wires are old and cracked, electricity may leak, as in figure 3-11. In damp weather, this may make the car hard to start.

FORCE: THE VOLT

The force in electricity that moves or pushes the electron gas is called electromotive force or EMF. In electrical formulas, the E from EMF is used to represent this electron force or push. Just as the concept of distance is measured in units such as the mile, the concept of EMF is

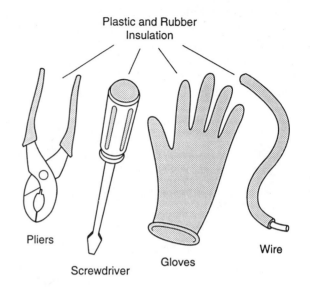

FIGURE 3-10 Insulated tools, gloves, and wire.

measured in a unit referred to as the volt. A relative idea of various voltage levels is useful. The 1.5 volts (V) of an ordinary flashlight battery (E = 1.5 Volts) is a small amount of electrical force that is useful for little devices like toys and portable lights. Flashlights are notoriously unreliable since 1.5 volts is not enough force to push electricity through the oxidized battery, bulb, and switch contacts. Cleaning the contacts

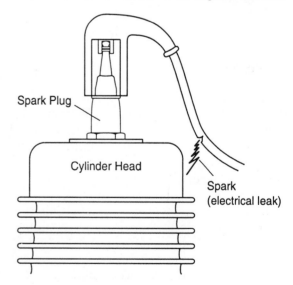

FIGURE 3-11 Spark plug wire spark leak.

by shaking the flashlight can often make it brighter. Car batteries also require clean, firm connections since the EMF is only 12 volts. Note the large cables (pipes) that are connected to the car battery. Most electric power used in the home is supplied by 120 volts. The 120 volts is enough electrical force to push an electron gas through a person, resulting in unpleasant shocks and possible death; extreme caution is always required. Why use such a high voltage for residential service? Lower voltages would require much heavier wire. Imagine a house wired with car battery cables. Television picture tubes, car ignitions and long-distance power distribution (high-tension lines) use voltages from 20,000 (20 kV) to 500,000 (500 kV). It is not uncommon to hear electricity sizzle off electric power lines into the damp night air and TV sets crackle with high voltage when they are turned on. This leaking of electricity into the air is a result of the high voltages used in these applications. The volt is the unit of electromotive force or push.

ELECTRON PUMPS

Thunderstorms result when hot moist air close to the ground becomes lighter and rises in a column. As the air rises, it cools and becomes less capable of holding water vapor. The relative humidity of the cooled air reaches 100% and the water vapor turns to liquid rain. The heavier cool air at the top of the column spills over the side of the rising hot air, reheats, and rises again. Pilots are taught to avoid thunderstorms since the falling cold air may cause the plane to experience a rapid loss of altitude, resulting in a crash (figure 3-12). Lightning is a sure sign of a thunderstorm, but why?

Electrons can be removed from many substances by rubbing or friction, for example, the kind of friction that occurs between the rising and falling air in a thunderstorm. Tremendous numbers of electrons are removed from one air mass and collect on the other. The crowding of electrons together on one air mass produces an electron pressure due to the repulsion of elec-

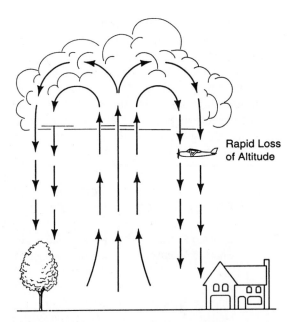

FIGURE 3-12   Thunderstorm and airplane.

trons from one another. Their negative charges repel and they want to escape. On the other air mass, electrons are scarce. When, as in figure 3-13, the electron pressure difference between the two air masses is great enough (millions of volts), the air is forced to conduct a giant spark

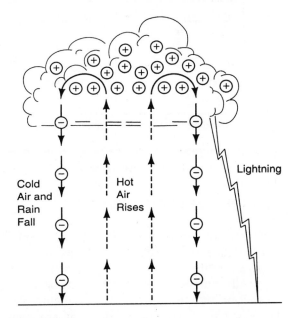

FIGURE 3-13   Lightning discharge during storm.

of electricity. Lightning and thunder, the explosive sound of air being forced to conduct, are the result of this electric discharge. Thunderstorms and all weather systems get their energy from the uneven heating of the earth by the sun.

Thunderstorms can be thought of as solar-powered electron pumps. Heat is supplied by the sun. Rising and falling air produces an electron imbalance by friction. The resulting imbalance causes lightning, the discharge of electricity. Unfortunately, thunderstorms are impossible to control, and the electricity they produce occurs in unpredictable spurts. Practical electric power generation requires that the electron pump be controllable, consistent, and dependable.

Energy Source	Science Principle	Devices	Nature of Application
Bending Force, Shock & Vibration	Piezo Electric Crystal (Molecular Distortion)	Strain Gauge Phono Pick-up	Sensor
Linear & Rotary Motion	Electro-Mag Induction (Coil, Magnet & Motion)	Alternator (Dynamo) Computer Disk	Power* Sensor
Heat	Thermocouple (Unequal Release of Electrons from Dissimilar Heated Metals)	Pyrometer Oven Controllers	Sensor
Chemical	Electrolyte & Electrode Ion Transport	Cell Battery Chemical Fuel Cell	Power Chemical Sensor
Light	Photon Impact	Solar Cell	Sensor Low Power

* Alternators are the primary source of electrical power.

FIGURE 3-14   Sources of electricity and common applications.

SOURCES OF ELECTRICITY

Sources of electricity are electron gas pumps that produce consistent electromotive force or voltage. In effect, they build an electron pressure. All sources of electricity convert some other form of energy to electricity. Figure 3-14 illustrates the forms of energy that are commonly converted to electricity.

Electron pumps, or sources, can be separated into two categories. The first category is made up of devices whose intended use is to produce large amounts of electricity to perform work—to light homes, turn motors, and so on. Dynamos, batteries, and automobile alternators fall within this first category in that the electricity they produce is meant to perform some type of work. The second category comprises sources that are used to convert other forms of energy to electricity for the purpose of sensing or measuring heat, movement, pressure, acceleration, viscosity, and many other things. A dynamic microphone uses the same scientific principles to produce electricity as does the car alternator; however, the microphone is a sensor and the alternator is intended to produce power.

In addition to sources of electricity that differ from one another as to their intended use and the force or voltage they produce, these sources also differ as to the amount of current flow or amps they can supply. AA, C, and D cells all produce 1.5 volts of electrical pressure, but the larger D cell can pump greater amounts of current flow or amps. Large, portable cassette radios require high current, and so generally use D cells. All up-to-date service entrance panels are designed to provide 240 and 120 volts. These panels differ in the amount of amps or ampacity available for use by the circuits in the dwelling. Generally, the technology and design of the source of electricity define its purpose, power or sensor, and voltage. The physical size of the source of electricity has much to do with its ability to pump amps.

THE ELECTRON PUMP: A SCIENTIFIC VIEW

It is now clear that electricity is the flow of an electron gas, and that an electron pump is required to push the gas. Some troubling questions arise. Can we run out of electrons just as we can out of gas in a car? Is there anything left behind when the electrons are transported away, perhaps like the realization of keys left behind on the hall table? If there is an electron pressure, is there an electron vacuum, a place that draws electrons like a shop vacuum draws sawdust?

Answers to these questions depend on an understanding of atomic structure. Atoms are small; this page is hundreds of thousands of atoms thick. They are also very regular structures, and so remarkably few ideas are necessary to understand the atomic model. Just as very few phonetic sounds enable human speech

to communicate all the knowledge civilization has acquired, a few ideas describe the nature of the atom. Atoms combine with one another to make molecules. Molecules are the stuff of which our environment is constructed. Atomic structure is the foundation and language upon which all of technology—and, in fact, the material world--rests. Knowledge of the atomic language of things enables us to control the whims of nature and to unleash the power of technology.

THE ATOM: NATURE'S BUILDING CODE

Building, electrical, and plumbing codes specify what is and what is not permitted in the construction of a building. Codes, complex as they may be, are interpreted by builders to erect durable, safe, and economical structures. Constructing a building requires the fastening together of various items with nails, screws, adhesives, and a great variety of special devices. Fastening is an important part of building. Modularity in distance, like studs placed 16 inches on center, and regularity of tasks, like the wiring of receptacle boxes over and over and in the same fashion, provide uniformity of material, techniques, and skilled labor resources. The results are structurally predictable and economical. How things are held together and the modularity of standard parts (2-by-4 studs, duplex receptacles, 1/2-inch pipe) are woven into the codes. Although a lot of human effort and cooperation are devoted to designing and enforcing building codes, even greater efforts have been devoted over the past few centuries to discovering the codes or rules of the natural universe and the structure of matter. Woven through books on physics, chemistry, electricity, and even molecular biology is a recurrent theme of repetitive standard parts, how they are held together, and the modularity of the resulting structures.

STANDARD PARTS: PROTONS, NEUTRONS, AND ELECTRONS

Imagine a shopping trip to nature's warehouse building supply. The label on the parts bin says, "Protons incredibly small, about 10^{-16} inches or one ten thousand million millionths of an inch across, one positive (+) charge." So as not to be greedy, we place just one in our micro-cart and move on down the shopping aisle. "Neutrons same size as protons, no charge. "What luck, they must be on sale today. We grab a few neutrons and move on. The next sign reads, "Electrons (1/1800 the size of protons), one negative charge (−)." Despite the fact that the electrons look a bit vague and puffy, we toss one in the cart and move to the checkout counter. Charles Coulomb (1736–1806) the sales clerk, explains that we are rather mixed up about this idea of charges. We need not pay for these particles, as nature supplies them in great abundance. The charge has to do with how the particles react with one another. Like charges, negative (−) to negative (−) or electron to electron, repel one another or push apart. Unlike charges, positive (+) to negative (−) or proton to electron, attract or pull together. Our standard parts are bagged, and we are given the receipt shown in figure 3-15.

Coulomb's Nature Corner

QT	ITEM	CHARGE
1	Proton	Positive (+)
3	Neutron	No Charge
1	Electron	Negative (−)

20 Billion Years of Experience

FIGURE 3-15 Receipt particles list and charge.

Upon arriving at the matter construction site, we open the bag to find that we still have the neutrons, but the electron and proton seem to have combined into a single atom of hydrogen. This atom has at its core the heavier proton, surrounded at a great distance by the electron. Strangely, the electron has become smeared into a bubblelike shell. The total or net charge of the hydrogen atom is zero; the positive proton and negative electron balance each other. Let us now leave the imaginary sub-atomic world to build a more specific understanding of the atom.

THE ELEMENTS

Quite without human intervention, nature has constructed 90 or so different types of atoms or elements. Each atom is composed of a core or nucleus of protons surrounded by shells of electrons. Most atoms contain some neutrons in their core or nucleus that add to their weight. Atoms are described by two important characteristics. The first characteristic is atomic number, the number of protons in the nucleus, and the second is atomic weight, the number of heavy particles, protons plus neutrons, in the nucleus. For example, carbon has six protons and six neutrons in its nucleus surrounded by six electrons in shells. Carbon's atomic number is 6 and its atomic weight is 12. Figure 3-16, the periodic table of elements, displays the atomic number above each element and the atomic weight below. Remarkably, just three standard parts—protons, neutrons, and electrons—make up all the 90 elements in the physical environment.

MODULARITY: PERIODICITY AND SHELL

Demolition, or the tearing-down/space-making part of the construction process, reveals how skilled artisans of the past constructed things. Chopping through the discrete layers of an old wall may uncover various coats of plaster, wire mesh, and thin wood lath, materials that are very different from the Sheetrock (plaster board) walls commonly used today. For centuries, scientists have used the same demolition techniques to understand nature, the greatest builder of all. Atomic demolition may

SIMPLIFIED PERIODIC TABLE

ELEMENT	He
ATOMIC NUMBER	2
ATOMIC WEIGHT	4

SHELL

COLUMNS ARE THE PERIODS
OF ELEMENTS THAT ARE ALIKE

SHELL	1A	2A										3A	4A	5A	6A	7A	0
K	H 1 1																He 2 4
L	Li 3 7	Be 4 9										B 5 11	C 6 12	N 7 14	O 8 16	F 9 19	Ne 10 20
M	Na 11 23	Mg 12 24	3B	4B	5B	6B	7B	--------8--------		1B	2B	Al 13 27	Si 14 28	P 15 31	S 16 32	Cl 17 35	Ar 18 40
N	K 19 39	Ca 20 40	Sc 21 45	Ti 22 48	V 23 51	Cr 24 52	Mn 25 55	Fe 26 56	Co 27 59	Ni 28 59	Cu 29 64	Zn 30 65	Ga 31 70	Ge 32 73	As 33 75	Se 34 79	Br 35 80	Kr 36 84
O	Rb 37 85	Sr 38 88	Y 39 89	Zr 40 91	Nb 41 93	Mo 42 96	Tc 43 98	Ru 44 101	Rh 45 103	Pd 46 106	Ag 47 108	Cd 48 112	In 49 115	Sn 50 119	Sb 51 122	Te 52 128	I 53 127	Xe 54 131
P	Cs 55 133	Ba 56 137	La 57 139	Hf 72 178	Ta 73 181	W 74 184	Re 75 186	Os 76 190	Ir 77 192	Pt 78 195	Au 79 197	Hg 80 201	Ti 81 204	Pb 82 207	Bi 83 209	Po 84 209	At 85 210	Rn 86 222
Q	Fr 87 223	Ra 88 226	Ac 89 227	Unq 104 261	Unp 105 262	Unh 106 263												

THESE FIT BETWEEN 57 AND 72

Ce 58 140	Pr 59 141	Nd 60 144	Pm 61 145	Sm 62 150	Eu 63 152	Gd 64 157	Tb 65 159	Dy 66 162	Ho 67 165	Er 68 167	Tm 69 169	Yb 70 173	Lu 71 175

THESE FIT BETWEEN 89 AND 104

Th 90 232	Pa 91 231	U 92 238	Np 93 237	Pu 94 244	Am 95 243	Cm 96 247	Bk 97 247	Cf 98 251	Es 99 252	Fm 100 257	Md 101 258	No 102 259	Lr 103 260

FIGURE 3-16 Periodic table of elements numbers.

be separated into two general areas. First, within the past 100 years, nuclear scientists have constructed great machines that chisel and chip away at the atomic core, or nucleus. This reveals its structure and enables the splitting of heavy atoms such as uranium, which contains 92 protons, 92 electrons, and 146 neutrons, with an atomic number of 92 and an atomic weight of 238. Nuclear fission or splitting has yielded great information about the structure of the atom's nucleus, and has provided vast amounts of energy. Second, scientists have found ways to squeeze together hydrogen, the lightest atomic nucleus, to make heavier atoms. By this process of nuclear fusion, even greater amounts of energy are released. The sun is a controlled fusion reactor producing helium from hydrogen. Hans A. Bethe, the physicist who is credited with describing how the sun works, also contributed to the development of the fusion or thermonuclear bomb, which is thousands of times more powerful than the atomic bomb. Nuclear science deals with such matters as the core of the atom, its nucleus, giant machines, vast amounts of energy, invisible radiation, and rather abstract complicated theory.

Fortunately, our concern is relegated to the outside of the atom. Chemists, material scientists, and electrical engineers are most concerned with the outer part of the atom, the electrons and the shells of which it is composed. This second approach to atomic demolition has been pursued for a few hundred years, and has furnished a clear picture of the electron shells or walls that nature has constructed around the atomic core. This knowledge is the basis for the structure of the periodic table in figure 3-16. Each period or horizontal row ends when a shell or wall layer of electrons is completed. It should be remembered that as nature adds electrons to complete shells, protons must also be added to the nucleus to keep the atom in electrical balance. The electrons and their arrangement in shells govern how atoms combine together, as well as their electrical characteristics. The nature of the material, whether it be ductile or brittle, its boiling and freezing points, and almost all the qualities upon which technologists have capitalized to build modern technical society depend on the arrangements of electron shells.

VALENCE: THE ATOM'S SHEATHING

The durability of a home depends on its outside covering. Quality roofing and sheathing properly installed resist the destructive effects of the elements. High wind will locate any loose external materials and remove them. In addition, partially completed roofing and siding are vulnerable to weather damage. The vulnerability of the atom's outer electron shells is much the same.

The outermost electron shell of the atom is referred to as the valence shell. It is the structure of the valence shell that defines which atoms may join together to form molecules and determines how tightly the atom holds its electrons. The valence, or outer shell, is of such great importance that atoms with similar outer shell structures are stacked above one another in the periodic table. For example, copper (Cu), silver (Ag), and gold (Au) are all in the same vertical row because they all have a single outer-shell electron. Note that silver has more

inner-shell electrons than copper and gold has still more than silver; however, it is the valence or outer shell composed of one electron that makes these elements similar.

Imagine the copper atomic house with one shingle (outer-shell electron). Along comes an electron wind (gas), and there goes the electron, and along with it, one negative (–) charge. The freed electron joins the electron gas and the atom left behind is now short one electron or negative (–) charge. Since the number of protons in the core of the above atom has not been affected, the atom that was formerly neutral now has a positive (+) charge. Copper with 29 protons and 29 electrons will, upon losing an electron, have one more proton (29) than electron (28), or one positive (+) charge.

METALLIC BONDING AND THE CONDUCTOR

Interestingly, copper atoms can bond to one another by the same process as described above. Large numbers of copper atoms in a penny, for example, can lose their outer-shell electrons to an electron gas that blows around within the penny. Note that the total numbers of electrons and protons within a penny are equal. The outer-shell electrons are simply free to move about within the penny. Electron cloud or metallic bonding, as the process is most commonly called, may occur with many metallic elements. It may even take place under special extreme conditions with hydrogen and its lone electron. Copper, silver, and gold are good conductors due to their loosely attached single outer-shell electrons. An electron gas can easily blow through these metals.

INSULATORS

Glass is produced by melting sand, which is composed primarily of the element silicon. Refer to the periodic table in figure 3-16 and count over three rows to the right from copper (Cu), silver (Ag), and gold (Au). Locate the row that contains carbon (C), silicon (Si), germanium

(Ge), tin (Sn), and lead (Pb). All of these elements have four valence electrons. Lead and tin are moderately good conductors of electricity because they have so many electrons that their outer-shell electrons are located a great distance from the nucleus or core. Their siding is loosely adhered. Silicon in the form of glass is a particularly good insulator because its electrons are all well adhered, not only to each individual silicon atom, but between silicon atoms as well.

COVALENT OR SHARED BONDING

The sharing of outer-shell electrons, or covalent bonding as it is called, comes about when valence electrons are shared in an orderly way between neighboring atoms. Glass, for example, is composed of many (poly) crystals, each of which is made of silicon atoms sharing their four outer-shell electrons with four neighbors (figure 3-17). Glass is a good insulator because its electrons are tightly attached and covalently bonded. They cannot be easily blown free.

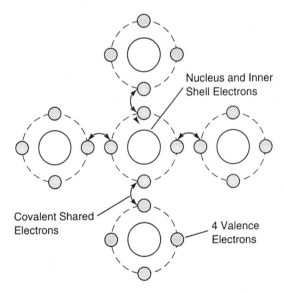

FIGURE 3-17　Silicon atoms.

Covalent or shared bonding holds together most of the important molecules. Living molecules, plastics (polymers), foods, fuels, and even the products or remnants of combustion are covalent molecules. The periodic table can also be viewed as a listing of the hooks or shareable electrons that each atom extends to make covalent bonds. Carbon dioxide (CO_2), a waste product of combustion and respiration, is composed of one carbon atom sharing two electrons with each of two neighboring oxygen atoms. The sharing of two electrons or a double covalent bond, makes for a particularly strong molecule (figure 3-18).

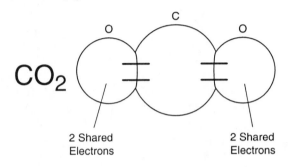

FIGURE 3-18　Carbon dioxide molecule.

IONIC BONDING

Molecules may also be formed when atoms join together by electric attraction. Recall that unlike charges attract. An atom that loses an electron may satisfy its loss by attracting another atom that has an extra electron. It is important to recognize that the atom missing the electron, for example, sodium (Na+), does not regain the electron, but rather satisfies its need for a negative (−) charge by attracting an atom with an extra electron such as chlorine (Cl−). Chlorine retains its extra electron, and hence the ionic bond holds the ion Na+ to Cl−. Figure 3-19 is a diagram of Na+Cl− or common table salt. Ionic bonds are weak and elastic when only a few atoms are involved, but they may, when there are billions upon billions of atoms, form ionic stone mountains.

FIGURE 3-19　NaCl ionic bond.

A SCIENTIFIC VIEW OF THE ELECTRON PUMP CONCLUDED

Is anything left when electrons are removed from a substance? If so, in this instance, what is left behind? Is there a shop vacuum-like electron vacuum? In a manner of speaking, the answer to these questions is Yes. All electron pumps, no matter what type of energy they are converting into electricity, remove electrons from the positive (+) terminal and squeeze them toward the negative (–) terminal. Left behind at the positive (+) terminal are atoms that have lost electrons. They are, in a sense, in an electron vacuum. It would be more accurate to think of these atoms as missing electrons and so exhibiting a positive charge. Atoms at the positive terminal of a battery are positively charged due to a loss of electrons as a result of the pumping action. The atoms in the positive battery terminal cannot move. These atoms make up the matter from which the battery is made. Electrons repelling one another on the negative battery terminal wish to move away. Electromotive force measured in volts is the difference between the lack of electrons on the positive terminal and the surplus of electrons on the negative terminal. Imagine what happens when a wire is connected.

THE ELECTRICAL CIRCUIT

With an understanding of the electron pump, it is now possible to return to the electrical circuit to put together the ideas of the electron pump as a source of push (volts), electron gas flow (amps), and wire as piping.

Circuits are circular looplike things, similar in function to chain drives on motorcycles and bicycles. The chain, like the electrical circuit, has no beginning or end. However, both the chain and the circuit are powered, the former by pedal or gas engine and the latter by an electron pump; (see figure 3-20).

Motorcycles have a purpose. Power from the engine is transferred by the chain to the rear wheel, which turns and moves the rider from place to place. The electrical circuit in figure

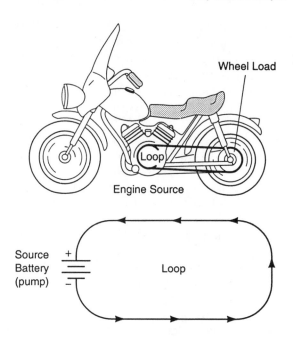

FIGURE 3-20   MC chain and battery wire circuit diagram.

3-20 has no purpose. It lacks the equivalent of a rear wheel, which, in electrical technology, would be called a load. An electrical technician would refer to the circuit in figure 3-20 as a short circuit, in the sense that it is short or missing a load. The distance the electrons travel is too short. In this instance, large amounts of current would flow for no real purpose. Also, motorcycles have clutches that can disconnect the engine power from the chain and rear wheel. The circuit in figure 3-20 has no such device. It seems that two devices need to be added to this circuit in order to make it practical.

CONTROL: A SWITCH, A VALVE, A CLUTCH

Electrical circuits require a control device or component, the simplest of which is the switch. Switches act like valves that can be used to turn a circuit on or off. Thermostats, humidistats, photoelectric relays, and automobile oil-pressure sending units are all switches that turn electrical circuits on and off. Some control devices, such as lamp dimmers on cars and transistors, are capable of varying the amount of electricity between fully on and fully off. Figure 3-21 illustrates a variety of control devices.

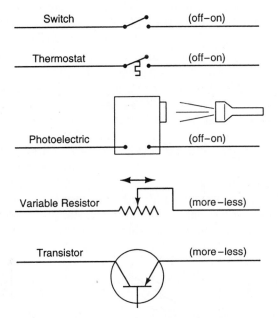

FIGURE 3-21 On-off switches and variable controls.

THE ANATOMY OF A LOAD

Basic biology courses often involve dissecting or taking apart the machinery of life to see how it works. In a similar fashion, much can be learned by taking apart cast-aside machinery. The unknown can be extremely dangerous, so exercise great caution. Refer to the preface of this book for safety information. Seeing first hand what makes things tick is both educational and fun. Do you know any real experts who don't have boxes or cans of used parts?

There is another way of taking things apart that is just as valuable as the actual disassembly of real machinery. Dissecting their inherent ideas and concepts, though harder and more dependent on imagination, usually reveals the multiconcept nature of things. For example, an electric motor is designed to convert electricity to rotary mechanical power. Physically, it comprises a fixed set of coils within a case and a set of coils supported between bearings. The interaction of magnetic fields causes it to spin. Motors have all kinds of arrangements of coils, bearings, and circuits. They differ from one another in many physical ways. Even though the primary intent of the motor is to convert electricity to motion through the magnetic effect, other electrical phenomena come into play. For example, the motor produces unintended and useless heat due to the resistance of its coils. In the following chapter, a variety of electrical loads will be examined from a physical as well as a conceptual perspective.

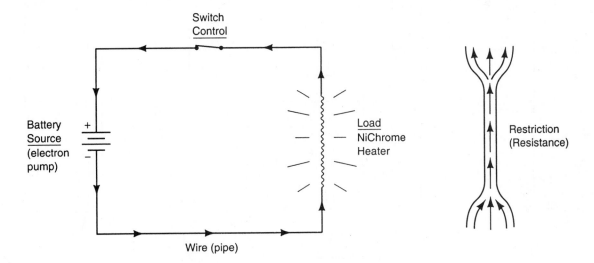

FIGURE 3-22 Source, wire (pipe), and load (restriction).

THE HEATER AS A LOAD

Often, electric heaters are physically composed of thin NiChrome wire strung between heat-resistant ceramic supports. Because the wire is thin and made of a material, NiChrome, that is only a moderately good conductor, the electric current does not flow easily. A heating element of this sort is like the traffic on a four lane high-way merging into two lanes. The electron gas is squeezed through the restrictive NiChrome in such a way that the molecules of the metal heating element are agitated. The shaking of atoms and molecules is heat. Restricting the flow of electrons causes electric energy into be con-verted to heat. Figure 3-22 illustrates the electric heater.

RESISTANCE

An important question may be asked relating to the electric heating element as a load. With a simple load like a heater, how does the amount of force (volts) affect the amount of flow (amps)? To help answer this question, a comparison to the stretching of a rubber band can be made (figure 3-23). As the force stretching a rubber band is increased, the band gets longer. However, as the stretching process continues, it becomes harder and harder to stretch the elastic band. For example, the first pound increases the rubber band's length 3/4 inch, the next pound 1/2 inch more, and the next pound only 1/4 inch more. As figure 3-23 indicates, the relationship between force and

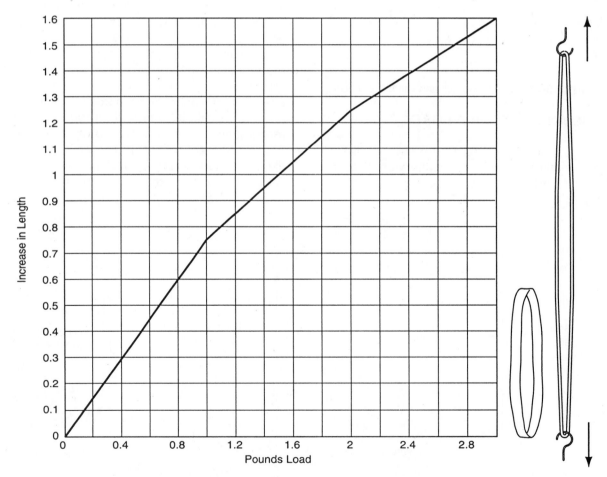

FIGURE 3-23 Rubber band stretch resistance.

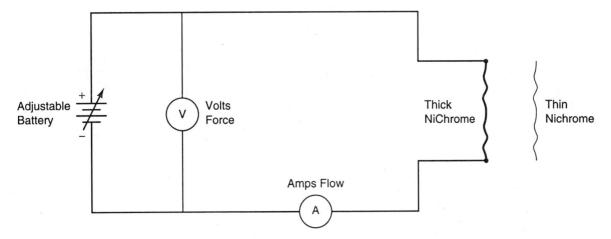

FIGURE 3-24   Test set-up (Ohm's law).

stretch with this rubber band may be presented as an X-Y graph. Note that the results of this simple experiment form a curve on the graph in figure 3-23. Given the electrical question, "What is the relationship between force (volts) and flow (amps) with an electrical load like a heating element?," does the heating element form a curve on a volt-amp chart like the elastic band on a force-stretch chart?

In 1826, George Simon Ohm performed an experiment to describe the force/flow relationship in electricity. To duplicate this experiment requires the following.

1. An adjustable battery with a voltmeter across it to measure force.

2. Heavy connecting wire with an ammeter in series to measure flow.

3. A piece of NiChrome wire for the simple load.

Figure 3-24 shows, in symbolic form, how the foregoing items would be connected to one another.

To test the relationship of force (volts) and flow (amps), the voltage has to be raised in even steps. The resulting increase in amps shown on the ammeter is then recorded. The results are displayed in figure 3-25, which illustrates a table of the voltages applied to the circuit and the resulting flow of amps. In addition, figure 3-25 shows an X (volts) and Y (amps) graph of the results.

Interesting, but what does it mean? It seems that every doubling of voltage results in a doubling of the amps. The X-Y graph shows a straight line. Mathematicians would say that the relationship of force (volts) and flow (amps) with a simple electrical load is linear (straight line), and not at all like the rubber band. More can be learned if the experiment is redone with a NiChrome wire that is thinner, resulting in a greater restriction of electrical flow. Figure 3-26 records the results from the first experiment (figure 3-25), as well as the results of the second experiment with the thinner NiChrome wire.

In comparing figures 3-25 and 3-26, how do the results of the first NiChrome wire differ from those of the second? It seems that both loads form a straight-line relationship of force and flow. That is, the amount the line moves to the right (run) as compared with the amount the line moves up (rise) is consistent from end to end for each individual line. However, the slope or angle of the line representing the thicker NiChrome wire indicates a greater increase in amps for each voltage step than does the line representing the thinner wire. The thinner wire permits a smaller increase in amps for each voltage step than does the thicker wire (figure 3-27).

On each of the triangles in figure 3-27 the movement to the right (run) represents increasing voltage, and the movement upward (rise)

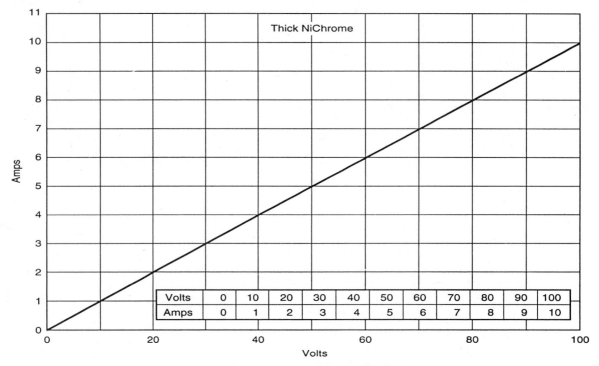

Volts	0	10	20	30	40	50	60	70	80	90	100
Amps	0	1	2	3	4	5	6	7	8	9	10

FIGURE 3-25   Volts 0–100, amps 0–10.

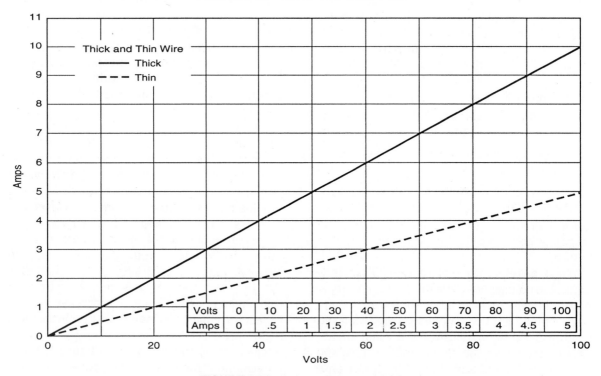

Volts	0	10	20	30	40	50	60	70	80	90	100
Amps	0	.5	1	1.5	2	2.5	3	3.5	4	4.5	5

FIGURE 3-26   Second data set of 3-25.

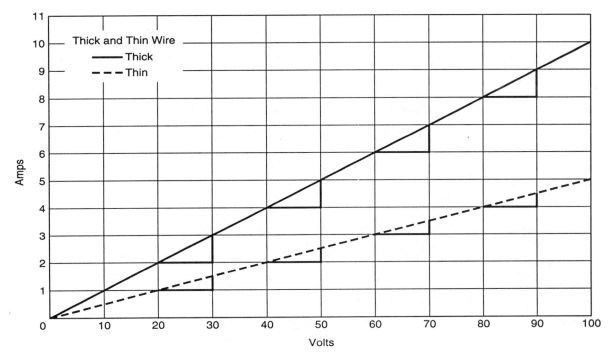

FIGURE 3-27 Rate and rise triangles.

represents the resulting increase in amps. The same voltage pushes more amps with the thicker NiChrome wire load and fewer amps with the thinner NiChrome wire load. It seems that the slope (pitch) of the straight lines in figure 3-27 describes the varying ability of thick and thin wires to conduct or resist the flow of electricity.

George Simon Ohm realized that the ability of a load to resist the flow of electricity could be represented by a ratio. Refer to figure 3-28 and take note of what happens when amperage is divided into voltage E/I. For both the thick and thin wire, the ratio is consistent for any pair of the volts/amps. It is important to notice that the thick wire gives a lower ratio of E/I than does the thin wire. Study figure 3-28 and consider this idea carefully.

Ohm called the ratio of volts (E) divided by amps (I) resistance (R). Ohm's law R=E/I is the formula that defines the nature of simple loads in electrical circuits. Resistance is a lot like electrical friction, and even more like the R

factor of insulation (its ability to resist the flow of heat). Figure 3-29 depicts Ohm's law. It should be understood that resistance is a result of the nature of electricity. Ohm did not invent it; he did, however, describe it well.

RESISTANCE AND OHMS: A REVIEW

The nature of a simple electrical load like a heater was first described by Ohm. Two important ideas need to be understood.

1. In simple electrical circuits, force (volts) and flow (amps) form a straight line; they are linear.

2. A ratio can be made of volts (E) divided by amps (I) that describes the resistance of a load.

Zero (0) ohms is the resistance of a perfect conductor like a very heavy copper wire. Infinitely high ohms is like a disconnected circuit or an open switch. Most electrical loads are somewhere in between.

Volts	0	10	20	30	40	50	60	70	80	90	100
Amps	0	1	2	3	4	5	6	7	8	9	10

Volts/Amps 10/1=10, 30/3=10, 50/5=10, 70/7=10, 90/9=10 AND 100/10=10

All the volts/amps of the thick wire equals a ratio of 10

This ratio of 10 is called *10 ohms*

Volts	0	10	20	30	40	50	60	70	80	90	100
Amps	0	.5	1	1.5	2	2.5	3	3.5	4	4.5	5

Volts/Amps 10/.5=20, 30/1.5=20, 50/2.5=20, 70/3.5=20, 90/4.5=20 AND 100/5=20

ALL the volts/amps of the thin wire equals a ratio of 20

This ratio of 20 is called *20 ohms*

Resistance (ohms) = EMF (volts)/Intensity of Flow (amps)

A poor conductor has high resistance

A good conductor has low resistance

FIGURE 3-28 Volts/amps and thick vs. thin wire loads.

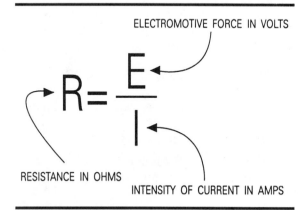

ELECTROMOTIVE FORCE IN VOLTS

$$R = \frac{E}{I}$$

RESISTANCE IN OHMS

INTENSITY OF CURRENT IN AMPS

FIGURE 3-29 Illustration of Ohm's law.

FINALLY, A REAL CIRCUIT!

A practical electrical circuit is shown in figure 3-30. Recall that a circuit is a loop that moves power from the source to the load.

Source: This part of the circuit acts as an electron gas pump. Internally, it uses some other form of energy, chemical in the case of the battery, to remove electrons from the positive (+) terminal and squeeze them on the negative (–) terminal. The source produces an electron pressure called electromotive force, or E, measured in units called volts.

Wires: The electron gas is carried through the circuit by copper that is transparent to it. The inside part of the wire is like the hole in an ordinary pipe, and its size governs how many coulombs per second or amps it can carry. Insulation is used to prevent pressure (voltage) from bursting the pipe and causing electrons to flow in unintended directions.

Control: Often a simple switch is used like a valve to turn the flow in a circuit on or off. Some control devices can regulate the amount of flow between on and off like a water faucet.

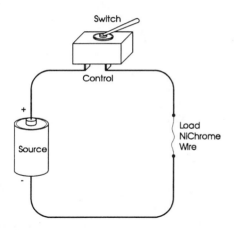

FIGURE 3-30   Battery, switch, and wire heater.

Load: The whole purpose of the circuit is to supply electricity to the load, so that the load may convert it to heat, light, motion, cooling, or whatever is desired. One effect common to all real loads is resistance, a ratio of volts divided by amps E/I. In future chapters, other effects— for example, magnetism and reactance—will be explained.

SYMBOLS AND FORMULAS

Electrical technicians often use symbols to replace actual drawings of electrical components. When thinking about the theoretical nature of circuits, symbols help as they focus attention on the important characteristic of the component. Real heating elements differ physically from each other; however, they may all be represented in a schematic diagram as a zigzag

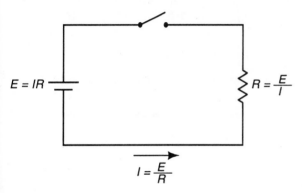

FIGURE 3-31   Schematic of figure 3-30.

line, the symbol of resistance. See figure 3-31, a schematic diagram of figure 3-30. Figure 3-31 replaces components with symbols, showing the same circuit in schematic form.

Numbers and formulas may also be used to represent circuits, especially when it is important to be able to count and measure volts, amps, and ohms. Alongside the schematic in figure 3-31 is the formula for Ohm's law with values taken from the schematic. At the bottom of figure 3-31 are three ways in which Ohm's law can be used to calculate unknown volts, amps, or ohms.

If you have finished this chapter and understand it, you know more about electricity than 98% of the population of the earth. Since a solid foundation is important for thinking about electricity and will support additional material, be certain you also understand the ideas in the first chapter. If not, reread and think about those ideas that are not clear. Ask your instructor for further help.

SUMMARY

In this chapter, electricity was presented as the invisible flow of an electron gas that is used to carry power or information. This flow, called amps, is pushed by a force, referred to as volts, through wires that are like pipes. The inside of the wire, which is usually made of copper, is transparent to the electron gas, and its size controls the maximum amount of amps it may carry. Insulation, which surrounds a wire, is opaque to electrons and acts to contain, or hold in, the pressure or voltage. An electrical circuit comprises a source that acts like an electron pump, a control device that is like a valve, and a load that converts the electricity to useful work. Many loads, such as the electric heater, allow an increase in amps that is proportional to volts. The proportional relationship of volts divided by amps is called resistance and is measured in units referred to as ohms. Resistance, the linear relationship of volts over amps, describes the restricting nature of electrical loads. An understanding of these ideas depends on basic science concepts related to atoms and molecules.

The atom is composed of a nucleus made of positively charged protons surrounded by shell-like arrangements of electrons. Certain substances, such as metals, share electrons between atoms in a rather loose way. Metallic bonding permits the easy movement of electrons and results in conductors. Other materials are made up of atoms that share outer-shell or valence electrons in an orderly way that results in few free electrons. This covalent bonding method occupies most electrons, and often results in a material that is referred to as an insulator. Ionic bonded materials are held together by electric attraction due to a surplus of electrons on one atom and a scarcity on another.

PROBLEM-SOLVING ACTIVITIES

A. A customer complains that a window air conditioner cools poorly. On the service call, the technician finds nothing wrong with the unit. The air conditioner is connected to the electrical outlet with a lamp-cord type of extension. What is the problem and how might it be explained to the customer?

B. A technician sets up toy trains for her son. The engine stops at various points on the track, but restarts when pushed. What is the most likely cause of this problem?

Review Questions

1. Why is gas a better image for electricity than water?

2. Compare copper water pipes and copper wires. How are they similar and how do they differ?

3. Both batteries and plugs have two terminals. Why?

4. The statement "ten volts is flowing" is incorrect. Explain why and define volts and amps in your own words.

5. How is the table of elements periodic and modular?

6. Electricity is used in some chemical reactions to pull apart molecules. What type of bonding depends on charge attraction?

7. Where do the free electrons come from in most conductors?

8. A car battery measures 12 volts with a meter when no circuits are energized, but drops to 10 volts when the starter is cranking. Explain this drop in voltage in terms of an electric pump.

9. What is the resistance of an open switch and a closed switch.

10. A toaster power cord burns near the plug when the toaster is turned on. Describe the two possible causes.

CHAPTER 4

The Dynamics of Energy and Power

One of the most frequently heard customer complaints is, "I just can't get it fixed right the first time." Repeated visits to solve a heating or air-conditioning problem usually involve replacing one part after another until, by chance and at great expense, the faulty component is found. The difference between a "parts jockey" and a professional HVAC technician is analytical reasoning. Analytical reasoning means understanding the scientific logic of how a system functions and what happens when individual components of the system fail. A large part of this book covers the technical details of how individual components work. This chapter is devoted to an examination of the few very important scientific ideas that form a foundation for all physical systems and machines, including HVAC systems.

The following pages explain the ratiolike or leverlike relationships between various energy and power concepts. Mechanical and electric devices will be compared with these concepts so that a clear mental image of abstract ideas, such as energy, can be formed. In this way, the HVAC technician can develop reasoning skills by focusing on the fundamental idea of a lever. The simple lever is a powerful reasoning tool that forms a foundation for much of the specific detail presented in later chapters. The technician who cannot reason is more a laborer than a professional.

ENERGY

Work to Be Done

The grandfather of one of the authors would often say, "Everything in life is work." His intent was to help the family children recognize that a full life requires effort, and he meant work in the popular sense—that is, work as opposed to play. However, from a scientific point of view, his use of the word "work" was incorrect. Scientifically, all of life is work. It cannot be otherwise. Playing football, taking pictures, collecting stamps, dancing, and even daydreaming are examples of work. Every action taken in life is work. Birth itself is work.

Work is defined as force times distance. For example, lifting a 50-pound window air conditioner 3 feet is 150 foot-pounds (50 pounds $\times$ 3 feet) of work. Further, when the air conditioner is installed and working, it moves heat from inside the home to outside the home. Again, a force is exerted to move (in this case) heat, so work is achieved. Though it is not as clear, even thinking is work. Thinking involves chemical and electrical changes in the brain, the moving of molecules, atoms, and electrons. Like the window air conditioner, thinking moves things, and to do that, a force is necessary. If you asked how a friend was doing and the friend responded, "I'm tired, I moved a 2000-

pound car this morning, and the force I used was my car," we would have a scientific description of an automobile accident. Even an accident is work. Again, as that grandfather said, "everything in life is work."

Most technicians and scientists define work as being equal to force times displacement. In this instance, displacement is the distance an object is moved, for example, the 3-foot movement of the window air conditioner mentioned above. As can be seen in figure 4-1, any combination of force and distance, measured either by metric or English units, is a legitimate measure of work.

Special note should be taken of the Btu, kilocalorie, and electrical joule (volt × coulomb). All three of these measures seem to differ from the others in that a distance is not stated. Rather, they displace heat (Btu and kilocalorie) or electricity (joule = volt × coulomb) from one location to another. For example, the air conditioner moves heat from the inside to the outside and does Btu's of work. When charging a capacitor, coulombs (6.28×10^{18} electrons) are forced by volts into the capacitor. Simply stated, electrons are moved from one place to another and joules of work are performed.

Units of Work and Energy

	Btu	Kcal	Ft-Lb	Joule
1 Btu British Thermal Unit	1	.252	778	1055
1 Kilo Calorie	3.97	1	3090	4190
1 Foot-Pound	.001285	.000324	1	1.356
1 Joule (Newton-Meter) also (Volt-Coulomb)	.000948	.000239	.7376	1

FIGURE 4-1   English and metric work equivalents.

"Jean, have you got enough energy left to go to the game after work?" This is a question that implies a common-sense understanding of the idea of energy. On closer examination, the question is, "Do you have enough stored energy remaining to do the work (scientific definition) of going to a game?" Most people realize that energy is the ability to do work. The scientific definition is identical. Technically, however, energy is divided into two forms. To answer the above question, Jean will have to estimate how much energy, called potential energy, she has stored. If her potential energy is enough, she will attend the game and use that energy. Moving energy, actually attending the game, is

energy "doing" work and is called kinetic energy. Figure 4-2 presents examples of potential energy and its use, kinetic energy. Since energy is the ability to do work, it is measured by the units referred to in figure 4-1.

Perhaps one of the greatest accomplishments of Western civilization was the realization that things can be understood and manipulated by taking them apart. For example, in psychology it is common to analyze people's behavior the better to understand what "makes up" personality. The study of genetics enables us to identify what genes affect hair color or cause birth defects. A real benefit of this taking-apart method is that we may then put things back

Energy

Potential   Kinetic

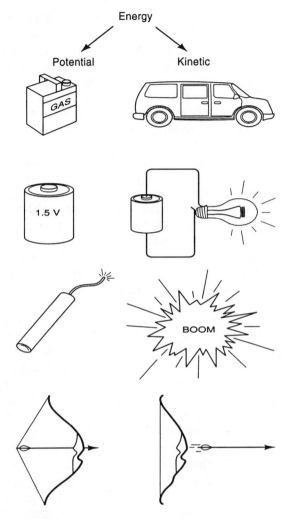

FIGURE 4-2   Potential and kinetic energy.

together in a new or modified way. Through gene splicing, we may replace defective genes, or through counseling, it is possible to ease personality problems. Taking things apart has proved especially useful in science and technology. For example, through chemistry, molecules are taken apart and put back together in new and useful forms. Even things as small as atoms or as large as the universe are understood by physicists through imaginary taking apart. However, in certain instances, the taking-apart method can lead to confusion. To clarify the connection of energy and work, let us look at the example in figure 4-3.

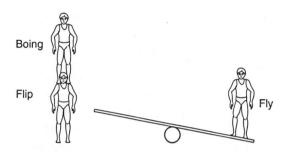

FIGURE 4-3   Potential energy.

In figure 4-3, *Boing* is standing on *Flip's* shoulders, about to jump on a springboard. Since *Boing* had to climb up on *Flip's* shoulders, he represents stored, or potential, energy.

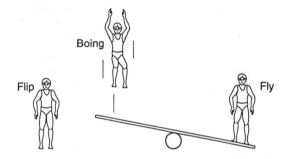

FIGURE 4-4   Kinetic energy.

Boing is forced downward by gravity, and his potential energy becomes kinetic or moving energy (figure 4-4). In figure 4-5, *Boing* hits the board (it's obvious where he got his name), and *Fly*, who is on the opposite end of the board, is lifted upward. Scientifically, *Boing's* falling force is kinetic energy and *Fly's* being lifted

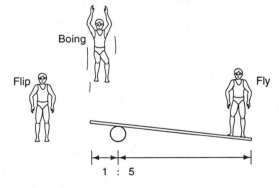

FIGURE 4-5   Ratios of energy and work.

upward is work. Recall that energy and work are measured by the same units. Confusion may result when we separate energy and work, because, as this illustration shows, they are really two halves of one event. If *Boing* doesn't boing, then *Fly* doesn't fly.

Understanding that energy doing work is one event is important because it means that the energy applied to the event is always equal to the resulting work. If *Boing* is heavier, *Fly* will go higher. Technically speaking, the springboard in the illustration is a lever—a lever of energy and work.

The Lever's Fulcrum

The pipe upon which a seesaw rests is technically called its fulcrum. In figures 4-3 and 4-4, the pivot or fulcrum of the lever (springboard) is placed at its center. An important and fundamental idea related to energy and work is

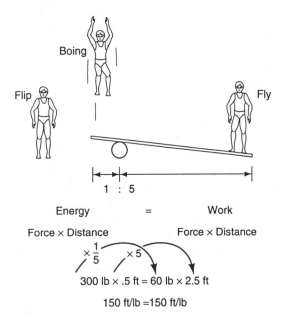

FIGURE 4-6 Energy equals work.

revealed by a thorough understanding of what happens when the fulcrum of a lever is moved off center. Suppose, as in figure 4-5, that the fulcrum of the springboard is set close to *Boing's* end. Imagination tells us that *Fly* will be

lifted a greater distance and *Boing* will fall a shorter distance. This seems like a better arrangement since *Fly* gets pushed more while *Boing* falls less. But wait a minute! If energy and work are one event, then this apparent increase in distance must come at some cost. Thinking back to the beginning of this chapter, remember that both energy and work are equal to force times distance.

Refer to figure 4-6 to see that the loss of movement (falling) on *Boing's* side of the lever and the gain or increase in movement on *Fly's* side are accompanied by an inverse or opposite effect on force. That is, the movement on *Boing's* side of the lever, often called the effort end, is multiplied or increased five times on the resistance or *Fly's* side of the lever. However, the force on *Boing's* side of the lever is reduced by one fifth on Fly's side of the lever. Whatever has been gained in movement or distance has been lost in force. *Fly* is lifted further, but with less force. *Fly* does not gain any great increase in the time he is in the air for his gymnastic exercises. Therefore, by moving the fulcrum back and forth various distances, different force combinations are made available to *Boing* and *Fly*. During rehearsal, the springboard gymnastic team will, by trial and error, find the best position for the fulcrum in order to push *Fly* as high into the air as possible. Technically, the team will find the fulcrum position that gives the best mechanical advantage. It is important to understand that mechanical advantage enables the best matching of energy with the work to be done. It does not, however, give any increase in the total amount of energy available. To do that *Boing* would either have to wear a lead belt or eat more pizza.

It has been said that "simplicity of character is the natural result of true wisdom." Translated into the technical world, this means that a "rock solid" understanding of a few scientific principles explains most of the modern devices used by society, including those in the field of HVAC. First, let us be absolutely clear about the principles revealed by the springboard example above. We will then illustrate how these principles explain other mechanical and electric devices.

Boing's falling on the lever (springboard) exerts a force (how much he weighs) over a distance (how far he pushed down on the board). Energy equals force times distance. When *Boing* is on *Flip's* shoulders, he is potential or stored energy. When he steps off of *Flip's* shoulders, he is moving or kinetic energy. *Boing's* kinetic energy is transferred to *Fly* by the lever. *Fly* has work performed on him. While on the lever, he is pushed upward by a force times the distance the lever moves up. Since this is one event, energy in equals work out. The springboard or lever transfers energy from *Boing*

Ratio 5:1

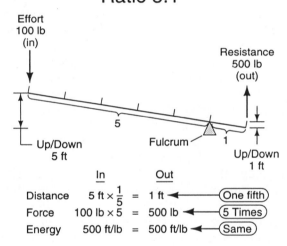

	In	Out	
Distance	$5 \text{ ft} \times \frac{1}{5}$	= 1 ft	One fifth
Force	$100 \text{ lb} \times 5$	= 500 lb	5 Times
Energy	500 ft/lb	= 500 ft/lb	Same

Ratio 1:3

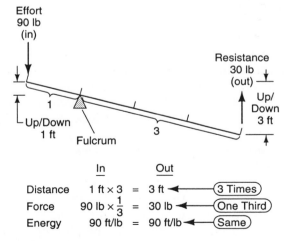

	In	Out	
Distance	$1 \text{ ft} \times 3$	= 3 ft	3 Times
Force	$90 \text{ lb} \times \frac{1}{3}$	= 30 lb	One Third
Energy	90 ft/lb	= 90 ft/lb	Same

FIGURE 4-7 Energy Levers.

to work done on *Fly*. By moving the fulcrum, the best mechanical advantage can be selected (the best combination of force times distance), but we can get no more work out than the amount of energy we have put in. Figure 4-7 shows how the lever of energy and work may be adjusted for mechanical advantage. Note how the ratio of the effort end of the lever to the resistance end establishes how force can be given up for distance or vice versa. The lever illustrates wisdom that technologists use to explain many things.

Useful machinery generally has a source of energy, like *Boing* in figure 4-7: a place in which work is being performed (the lifting of *Fly*), and mechanisms that couple the source of energy to the work to be done. The springboard gymnastic team used a lever both to transfer or connect the energy to the work and to modify the ratio of force and movement. All machinery, mechanical as well as electric, requires the connecting of energy to the work to be done. Many machines also use leverlike parts to adjust the ratio of force and movement to yield the best mechanical advantage. Recall that energy doing work is one event in time. The relationship of energy and time will be examined shortly. It is appropriate now to look at a few devices that connect the energy source to the work to be done. The examples that immediately follow were selected because they are usually described in terms of force and motion. Other devices will be introduced later that are leverlike, but in common practice generally are described in terms that include time.

Figure 4-8 illustrates three energy (force times distance) lever ideas that represent this important concept. At the top of this illustration is a lever that is three times longer on the left side of the fulcrum, the effort end, than it is on the right side, the resistance end. A ratio of 3 to 1 is obvious, but how does this affect force and distance? Imagine this lever in motion; as the left, effort end is depressed, the right, resistance end moves up a lesser amount. We know that the ratio of 3 to 1 supplies the necessary information that for every three units of distance moved down on the left side, the right side will move up one unit of distance. A 6-inch movement down on the left will move the right side up one third that distance, or 2 inches. If distance (displacement) has been lost, then force must be gained or increased, but by how much? Again, the ratio of 3 to 1 supplies the information for the solution of this problem. Recall that we lost distance by one third, so we must gain force by the opposite, or three times. A downward force of 2 pounds on the left will lift a weight of 6 pounds on the right. Whatever

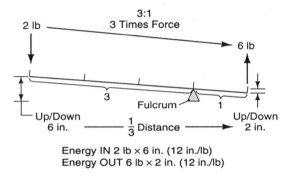

Energy IN 2 lb × 6 in. (12 in./lb)
Energy OUT 6 lb × 2 in. (12 in./lb)

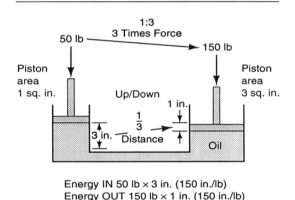

Energy IN 50 lb × 3 in. (150 in./lb)
Energy OUT 150 lb × 1 in. (150 in./lb)

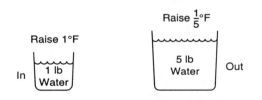

Energy IN 1 lb Water Raised 1°F = 1 BTU
Energy OUT 5 lb Water Raised $\frac{1}{5}$°F = 1 BTU

FIGURE 4-8 Three energy/work ratios.

was lost in distance or displacement is gained in force. To double-check this calculation, remember that energy equals work. The left, effort end of the lever is moved downward 6 inches with a force of 2 pounds or 12 inch-pounds of energy. The right or resistance end of the lever is moved upward 2 inches with a force of 6 pounds or 12 inch-pounds of work. Energy must always equal work. Pulling nails with a claw hammer is a clear example of the lever.

The second example in figure 4-8 is a bit less obvious, but no more complicated than the first. Hydraulic as well as pneumatic cylinders depend on fluids for the transmission of energy. Hydraulics generally use oil, a noncompressible liquid, while pneumatics make use of air, a compressible gas. Look at the small cylinder on the left and the larger cylinder on the right. What is it that makes one cylinder large and the other small, and what will affect the amount of fluid pushed by the pistons of the cylinders? It is the end or face of the piston that pushes against the fluid. How about a ratio of piston area? Let us assume that the surface of the piston in the cylinder on the left is 1 square inch and the surface of the piston on the right is 3 square inches, a ratio of 1 to 3. At this point a mental picture is helpful. Imagine pushing the left piston down 3 inches. If its area is 1 square inch and it is moved 3 inches, it will displace or move 3 cubic inches of fluid. The right cylinder will be moved upward, but, since it has three times more surface area (3 square inches) it is moved up only one inch by the displaced 3 cubic inches of fluid. A loss of distance by one third must bring an increase in force of three times. A force of 50 pounds on the left will lift 150 pounds on the right. In this way, a hydraulic jack can exert enough force to lift a car. Often pneumatic devices are used as actuators in HVAC systems to provide an appropriate force to operate valves that control the flow of water and air.

At the bottom of figure 4-8 is an example of the leverlike idea that is, on the surface, not at all like a lever. At the left of the illustration is a small gas flame, much like the pilot light on a kitchen stove. Let us assume that the flame is a heat energy source of 1 Btu. In Chapter 2, the Btu was defined as the amount of heat that raises the temperature of one pound of water 1°F. If we were foolish enough to place a finger in the flame, the result would be a painful burn. The flame is perhaps 1200 degrees at its hottest point, but the volume that it can heat is small. Kitchens are often a bit warmer than the rest of the house owing, in part, to a stove pilot. The high temperature and small volume of the pilot

are transferred to an increase of a few degrees in the temperature of the larger volume of the kitchen. One pound of water can be heated 1 degree by 1 Btu of heat energy or 5 pounds of water heated 1/5 degree. In this instance, the leverlike idea is that a given amount of heat energy can make small things very hot or larger things just warm. The trade-off is between temperature increase and mass. How is a heating system like a lever?

$$\frac{\text{The Doer}}{\text{(Force} \times \text{Distance)}} = \text{Power} = \frac{\text{The Done}}{\text{(Force} \times \text{Distance)}}$$
$$\frac{\text{Energy}}{\text{Time}} = \text{Power} = \frac{\text{Work}}{\text{Time}}$$

UNITS OF POWER

	BTU/h	Kcal/s	Foot-Pound/s	Horsepower	Watt
1 BTU/hr	1	.00007	.2161	.000393	.2930
1 Kcal/s	14290	1	3087	5.613	4184
1 Foot-Pound/s	4.628	.000324	1	.00182	1.356
1 Horsepower	2545	.1782	550	1	746
1 Watt (Joule/s) Amps × Volts	3.413	.000239	.7376	.001341	1

FIGURE 4-9 Power equivalents.

POWER

Power and Time

In the previous section on energy and work, nothing was said about the concept of time, but time is important to us all. "How long will it be before our energy supply is used up?" Or, on a more personal basis, "How long will it take to get the job done?" These are both important questions. Energy, the reader will recall, is force times distance. There is no mention of time.

Work is also force times distance, and again there is no mention of time. Technically, when "how long" (time) is added to the concept of energy or work, a new idea, "rate," is introduced. The rate at which energy is being used or work is being performed is power. A simple example will clarify the difference between energy and power. One gallon of gasoline contains a specific amount of energy. The high-powered engine in a race car would consume this gallon of gas in a few minutes and travel only a few miles. The more frugal, low-powered moped driver may travel for hours and go more than 100 miles on the same gallon of gas. The rate of fuel consumption (energy) is much higher for the race car than for the moped.

The formula at the top of figure 4-9 is a mathematical representation of power. Note that both energy and work equal force times distance—all the same idea. That idea becomes power when it is divided by time. Also shown in figure 4-9 are the basic English and metric units of power. Special units of power that are often used to describe technological processes are displayed at the bottom of the figure. For example, horsepower (550 ft x lb/sec) is used to compare a rotating power engine with the number of horses it would replace. Joule per second has two uses as a measure of power. The first use is mechanical; a joule per second is newtons (force) times meters (distance) divided by seconds. The watt, an electrical measure of power, is also a joule per second. Electrical watts are equal to volts (force) times coulombs (displacement) divided by seconds. Another, more common way of looking at the watt is that it is equal to volts times amps. The reader should refer to figure 4-9 and think about how all the units of power represent the same basic idea. A common 1500-watt portable electric heater operates on 120 volts and draws 12.5 amps. A 1500-watt electric heater for an automobile would have to operate on 12 volts. How many amps would it require? Why are electric car heaters not practical?

Power Levers

The idea that energy must equal work was described in relationship to the lever. The lever and leverlike devices portrayed in figure 4-8 showed how force and distance (displacement) may be rearranged by a ratio. The following devices act in the same way as those presented in the section on energy levers and conceptually are identical to them. However, these devices are more easily understood from the perspective of power than from that of energy.

Figure 4-10 once again shows the springboard gymnastic team, but in this case, the

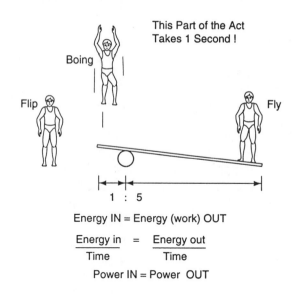

Energy IN = Energy (work) OUT

$$\frac{\text{Energy in}}{\text{Time}} = \frac{\text{Energy out}}{\text{Time}}$$

Power IN = Power OUT

FIGURE 4-10   Power in equals power out.

concept is based on power. Note that in this illustration the time is 1 second for both *Boing* and *Fly*. Many technical mechanisms are similar. Gears like those at the top of figure 4-11 are often used to exchange or adjust the ratio of torque (rotational force) for RPM (revolutions per minute). Revolutions per minute include both movement (revolutions) and time (minutes). Thinking in terms of torque and RPM involves distance (revolutions), force (torque), and time (minutes), which is a power idea. Most rotary power-producing machines (electric motors and gas engines) produce high RPM and low torque. To take efficient mechanical advantage of these devices often requires a transmission comprising gears, or their less costly close relatives, pulleys and belts or sprockets and chains. HVAC systems make extensive use of these mechanisms, especially with regard to blowers and actuating devices.

Note in figure 4-11 that the driving gear on the left has ten teeth while the gear on the right, called the driven gear, has 50 teeth. Rotate the gears in your mind. Imagine how many times the driver has to turn in order to turn the driven gear once (five turns of the driver to one turn of the driven gear, or a ratio of 5 to 1). Another way to look at this ratio is to note that the driver gear has ten teeth to the driven gear's 50 teeth, which yields a ratio of 1 to 5. These ratios seem opposite to each other; however, the main point is that there is a ratio of 1 and 5. Which direction the ratio goes is established by formulas for the scientist, but the technologist will find common-sense reasoning more useful. If the left driver gear turns five times for each single turn of the driven gear, there must be a reduction in RPM by that ratio. If RPMs (distance) are given up, then torque (the twisting force) must be gained by the opposite ratio. Recall that power in equals work out. Reason through this same idea while looking at the pulleys and belt in the middle of figure 4-11. A good feel for this can be developed by riding a ten-speed bike and taking note of the effect of the sprocket chain ratio changes (gear changes).

This power lever concept is the basis of many technical devices; it even explains the basic ratio of the electrical transformer. Referring to the bottom of figure 4-11, assume that a transformer has 500 turns on the input coil (called the primary coil) and 50 turns on the

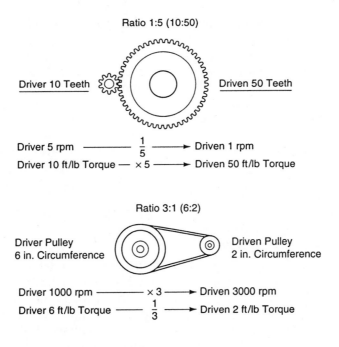

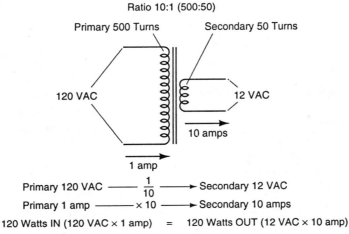

FIGURE 4-11 Gear, pulley, and turn ratios.

output coil (called the secondary coil). More details about how the transformer works will be presented in the next chapter, but for now the ratio of these coils is important. This ratio of 500 to 50 provides a turns ratio of 10 to 1. In this case, the voltage (force) will be reduced by the ratio of 10 to 1, so that 120 volts at the input results in 12 volts at the output. This drop in voltage (force) must be compensated for by movement, and in electricity movement is coulombs per second (or amps). The resulting amps at the output will be ten times higher than the amps at the input. This transformer drops the voltage by one tenth and raises the amps by a factor of 10 to 1. The electric power of volts times amps (watts) in must equal volts times amps (watts) out. How are the electrical transformer and the automotive transmission similar?

None of the power-leverlike ideas explained above changed time. In each case, the time in was the same as the time out. If we ignore the difference between potential (stored) and kinetic (moving) energy, there is a leverlike effect that can be related to time. For example, a small electric resistance heater may take two hours to heat 50 gallons of water, while taking a hot shower or using a clothes washer may exhaust this supply of hot water in one half hour. The low power of the water heater accumulated energy in the form of hot water (Btu's) for over two hours, which was used up in a high-power mode in only one half hour.

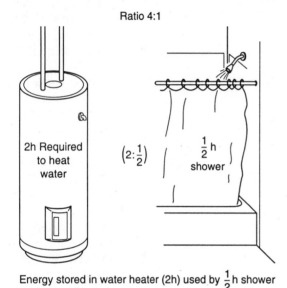

Ratio 4:1

Energy stored in water heater (2h) used by $\frac{1}{2}$ h shower

FIGURE 4-12 Energy/time levers.

Figure 4-12 shows a lever whose length represents the time ratio of two hours to one half hour. This is equivalent to a low-powered electric water heater operating for two hours equaling 50 gallons of hot water that may be

used up in one half hour. In a similar way, the amount of time a heating system is on during a 24-hour period (high power times short time) equals the heat loss of the building over the same 24-hour period (low power times long time). This leverlike time ratio can be adjusted by changing the capacity (power) of the heating system or reducing heat loss through additional insulation. The HVAC technician needs to be concerned with both heat gain and loss, as well as the leverlike connection between the two.

EFFICIENCY, OR GETTING WHAT WE WANT

Energy is indestructible, so if 10 Btu of energy is released by a gas flame, it must do 10 Btu of work. All energy and work leverlike phenomena produce work that is equal to the energy consumed. Unfortunately, we often perform unintended work while consuming energy. A good example of unintended work is the automobile. It uses much of the energy in the gasoline to heat the atmosphere, whereas what we want is to get from one place to another. Since gasoline is wasted in performing work that we don't want, the automobile is not very efficient.

Efficiencies of some common devices are listed in figure 4-13. Of these, the automobile, at approximately 25%, is the least efficient; 75% of the energy in the gasoline is used to do work that is not desired. Figure 4-13 shows that efficiency is a number that compares input (denominator of the fraction) with output (numerator of the fraction). For example, if a 100-

$$\text{Efficiency} = \frac{\text{Energy OUT (performing desired work)}}{\text{Energy IN}}$$

$$\text{Efficiency} = \frac{250 \text{ Joules (car motion out)}}{1000 \text{ Joules (gas line energy in)}} = .25 \ (25\%)$$

$$\frac{250 \text{ Joules of work}}{1000 \text{ Joules of gasoline}} \longrightarrow$$

Lost energy (undesired work) = 750 Joules of exhaust heat, friction, etc.

FIGURE 4-13 Efficiency.

horsepower motor performs only 90 horsepower of desired work and 10 horsepower of undesired work, the efficiency is calculated by putting the output 90 horsepower over the input 100 horsepower. The resulting .9 would be changed to a more understandable percentage rating by multiplying by 100. Remember, since energy is indestructible, efficiency is really a percentage rating of how much of the energy is performing work that we want. Efficiency is a very important concern in the selection and operation of HVAC systems.

INERTIA AND MOMENTUM

The leverlike connections of force, movement (displacement), and time have been the primary focuses of this chapter. However, the machinery of HVAC systems and the heat they supply or remove from buildings is also affected by another important scientific concept. Central to this concept is the idea of change: change in speed, direction, or temperature.

The automobile pictured at the top of figure 4-14 has been accelerated by the application of energy supplied by gasoline. In a general way, it could be said that energy has been installed in the vehicle to overcome its inertia. The car now has an impetus to continue in a straight line. Scientifically, the car has a quality of motion called linear (straight-line) momentum. It wants to continue going in a forward direction. The screech heard while braking for a stop sign is a result of the tires' resisting momentum. Both tire and brake wear are a direct result of momentum, which, in turn, is a result of acceleration from the energy supplied by the gasoline. Low-mileage Mike overcomes inertia by consuming fuel, but then throws away momentum through tire and brake wear. Inertia is a tendency to resist motion, while momentum is a resistance to stopping. The hammering of water pipes is another good example of inertia, in that water accelerates when valves are opened and stops short (momentum) when they are closed.

Rotary inertia can be readily observed in the labored slow starting of blower motors. Gener-

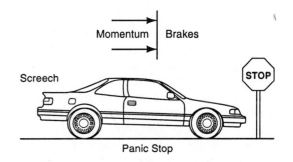

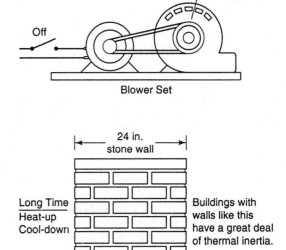

FIGURE 4-14 Flywheel effects.

ally, the larger the motor and blower, the more inertia the motor has. Angular momentum is the tendency of a motor to keep going even after its power has been disconnected. Note that the saw in figure 4-14 is still turning even though it has been disconnected from power. One must exercise caution while working with rotating machinery and make certain that the equipment has actually stopped. Many serious injuries have been caused by the angular momentum of saws and grinders.

At the bottom of figure 4-14 is an old building with thick masonry walls. Due to thermal inertia, many old buildings take a long time to heat up from a cold start. And due to thermal momentum, these same buildings cool down slowly. The tendency of large structures made of high-specific-heat materials that heat and cool slowly is an important consideration in heating and cooling system design and adjustment. Thermal inertia and the momentum of water with its high specific heat often make hydronic heating systems more stable than hot-air systems. The thermal mass of the water stabilizes the system temperature, resulting in more even heat.

SUMMARY

Energy is applying a force and moving a distance. So is work. Energy and work are two ideas that often describe one event, like a radio that is consuming electric energy and doing the work of producing music. With the appropriate equipment, any combination of force times distance can be adjusted to yield any other combination of force times distance, as long as they equal each other. This idea can be easily understood by the comparison of a lever.

Mechanical advantage means using leverlike devices to match the energy source to the work to be done. When time is divided into energy, a new idea, power, is derived. Power is the rate at which (how fast) energy is used. A lever with high power on one end and low power on the other is like a heating system, in that the furnace is on for only a short time while the house loses heat at a slow rate for a long time. The on versus off time of heating and cooling systems has a great deal to do with comfort and economy. Although all of the energy consumed produces work, not all of the work that is done is useful (for example, wasted heat going up the chimney). Efficiency is a measure of useful work over input energy times 100 (to make it a percentage). Finally, moving cars tend to keep moving and hot buildings tend to stay hot as a result of momentum. Machines may be slow to start-up and cold buildings may take time to heat up owing to inertia. Large, heavy things, like old masonry buildings, heat and cool slowly because of their inertia and momentum.

PROBLEM-SOLVING ACTIVITIES

Troubleshooting Case Studies

A. A customer complains that the new gas-fired, hot-air furnace that had recently been installed makes a few popping noises when it turns on and again when it shuts down. There are no problems with the amount of heat delivered or the temperature control of the unit. The homeowner also says that the furnace turns on and off, or cycles, more often than would be expected. (*Important note*: The new furnace was retrofitted into an existing duct system.) During an on-site inspection, the technician observes the following.

   1. The popping noise is due to air pressure ballooning ducts upon blower startup and shutdown.

   2. Though the furnace is properly sized, the system does indeed cycle frequently.

   What practical solution or adjustment may solve these two problems? (Hint: The solution involves a mechanical leverlike ratio adjustment that will have an effect on the leverlike relationship of furnace heat to building heat loss.) Identify the solution and explain the mechanical ratio adjustment and its effect on the duct noise, as well as on the rapid furnace cycling.

Review Questions

1. Describe the leverlike relationship of force and motion involved in the mechanical movement of a common water valve.

2. What form of energy does a flashlight battery contain when the flashlight is off? How does that form of energy change when the flashlight is turned on?

3. Explain the leverlike idea that describes why a truck must shift to low gear to climb a steep hill.

4. From a scientific point of view, the statement that "life should be a balance of work and play" is both true in that it illustrates an important idea and false in that it uses a word incorrectly. Explain how it is both true and false.

5. If it is assumed that money is energy, how is saving $1 every day for a year and buying something for $365 similar to the concept of power?

6. Ancient peoples heated water for cooking by placing fire-heated rocks in the water. Describe this leverlike process as it relates to furnaces and buildings.

7. Why do hydraulic jacks require so much pumping to lift a car just a few inches?

8. Lowering the temperature while no one is home saves energy, but if it is lowered too much, energy is actually lost. What does this idea have to do with inertia and momentum?

CHAPTER 5

Sources of Electricity
and
Effects of Loads

The practical use of electricity depends on the availability of a wide variety of sources that produce an electromotive force or voltage. These sources convert other energy forms into electricity, either to perform such work as running motors, or to sense and measure other phenomena, such as heat and light. Similar devices that change resistance rather than produce electricity from other energy forms will also be described. Electric energy provided by such sources is piped through wires to loads that make use of the electricity by doing work or acting upon the information furnished by sensors. In this chapter, these ideas will be explained in greater detail, as will the relationship between magnetism and electricity as it relates to both sources and loads. Furthermore, three effects—resistive, inductive, and capacitive—will be discussed in connection with a variety of loads.

SOURCES OF ELECTRICITY

The first law of thermodynamics states that energy cannot be created or destroyed. In ordinary everyday life, people often seem to believe otherwise. Government programs have to be paid for with tax dollars, car loans come due, and high salaries are the result of a great deal of work. Just as there is no free lunch in life, there is no source that creates electricity

from nothing. Sources of electricity are really converters that, through the application of scientific principles, change other forms of energy into electricity. Also, a variety of devices are available that vary resistance in response to other energy forms, such as heat and light. Since these resistance devices are similar to, and often confused with, sources of electricity, they also will be described. Usually, small devices, whether they actually produce electricity or simply change resistance, are intended to measure the other energy form and to provide an electrical output that represents the amount or intensity of the other energy (sensors). Larger devices, such as car alternators and power station dynamos, are intended to produce large amounts of electricity to perform work.

Heat

HVAC technology has as its central purpose the manipulation of thermal energy. To accomplish this purpose, frequent measurements of temperature are necessary. Gas furnaces require devices to ensure that pilot flame ignition and room temperature are controlled through information gathered by the thermostat. Devices that convert heat into either electricity or changes in resistance depend on the scientific principles presented below.

Thermocouples: Over 150 years ago, Thomas Seebeck discovered an effect that is still widely used to convert heat (temperature) to electricity. His discovery was based on the fact that different metals "free up" different numbers of electrons when they are heated. Metals with loosely attached outer-shell electrons, upon

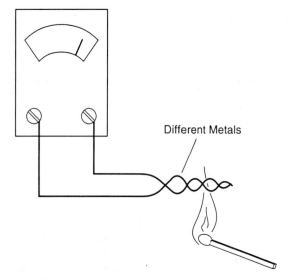

FIGURE 5-1   Thermocouple.

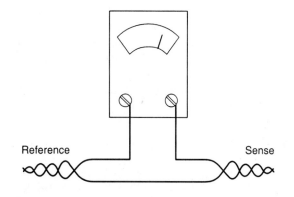

FIGURE 5-2   Reference and sensing thermocouple.

being heated, will release more electrons than will metals with few outer-shell electrons. If these two different (dissimilar) metals are fused together, a junction is formed that, when it is heated, provides a small voltage. Figure 5-1 shows the simple circuit used to measure this

heat-derived voltage. It is important to understand that the circuit in figure 5-1 will not permit accurate temperature measurements. To perform precise measurements, it is necessary to compare two thermocouple junctions, as shown in figure 5-2. In this illustration, one thermocouple is heated, while the other, the reference junction, is placed in an ice bath. In this way, the heated junction is referred to or compared with the junction held at 0°C and accurate temperature measurement is made possible.

One of the most common uses of a thermocouple in HVAC is to sense the presence of a gas pilot flame. Single thermocouples provide enough electricity to hold the relay points closed that allow an electric solenoid main burner gas valve to be energized and held open. If the pilot light goes out and there is no heat to the thermocouple, the relay that holds the gas valve open is released and shuts down the main burner.

Thermocouples may also be connected in groups called thermopiles. Thermopiles provide enough electric power from a pilot flame heat to operate a main burner gas solenoid valve directly without the need for any external source of electricity. Power outages do not affect the operation of these burners.

Bimetal switches: In HVAC appliances, dissimilar or different metals are also used in another way that is very different from the way in which thermocouples are used. Recall that the thermocouple produces electricity when it is heated. Bimetal switches are devices that open or close circuits; they do not produce any electricity. To understand bimetal switches, it is necessary to take a brief look at the physics of expansion and contraction.

When most materials are heated, their atoms and molecules vibrate more vigorously. These vibrations cause the atoms and molecules to push against one another, which results in an increase in the physical size of the material. In this way, things get bigger when they are hot and smaller when they are cooled. Bridges, buildings, and even HVAC ducts and pipes are designed to permit movement caused by expansion and contraction. Long runs of radiator pipe are bound to expand when they are heated. If some room for movement is not designed into the system, buckling, noise, and eventually leaks in the piping may result.

Due to internal molecular differences, materials expand and contract at different rates. When exposed to identical amounts of heat,

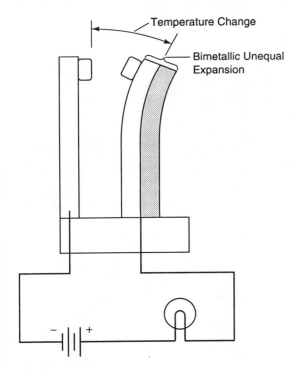

FIGURE 5-3 Bimetallic switch.

aluminum expands twice as much as iron does, and copper is midway between the two. Each material has a coefficient of linear expansion that describes how its length changes with temperature. If two different metals with different coefficients of linear expansion are fastened together as shown in figure 5-3, heating this device will cause unequal expansion of the metallic sandwich and result in its bending or buckling. By adding a set of electrical contacts, the bimetal device in figure 5-3 becomes a switch.

Many ordinary heating and cooling thermostats are bimetal switches. To enhance the

bimetal action, a long bimetal strip is rolled into a compact coil as detailed in figure 5-4. To prevent problems due to contact point dirt and oxidation, a small glass tube with contacts fused into the end and a drop of mercury within is fastened to the bimetal coil. As the coil winds and unwinds with changing temperatures, the mercury rolls from one end of the tube to the other, making and breaking the heating or cooling circuit. Bimetal mercury thermostats are very common and very durable.

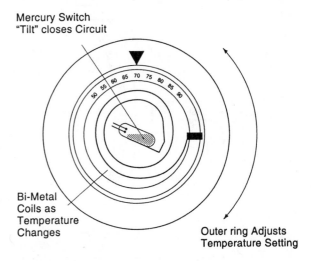

FIGURE 5-4 Room thermostat.

Window air-conditioner units utilize bimetal switches in two places. First, the thermostat that controls room temperature is bimetal, often with conventional contact points. Second, a circuit breaker that "trips" from an overcurrent or overheated condition is fastened to the compressor pod. Both of these bimetal switches control relatively large currents and are subject to contact point deterioration. To minimize contact point arcing, switches are generally designed to snap or make and break circuits rapidly. The click of a wall switch is caused by a toggle action that rapidly snaps the contacts together and apart. This same snap action can be added to bimetal switches.

Figure 5-5 shows a bimetal switch that has the ends of the bimetal strip trapped. When the temperature increases, tension is built up in the strip until enough is accumulated to cause a rapid buckling in the opposite direction. This action can be compared to attempting to fit an oversized wall stud in place. The stud buckles left or right, but will not stand straight up within the wall framing . In this way, a snapping action provides long contact point life. Circuit breakers pass electric current through the bimetal switch, thereby heating it. Overcurrent conditions cause the breaker contacts to snap open.

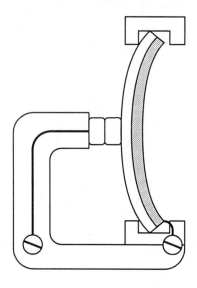

FIGURE 5-5   Trapped bimetallic switch.

Resistors temperature-dependent (RTDs): Temperature-dependent resistors are constructed of metal resistive elements, which, when heated, increase in electrical resistance. The temperature and resistance of these resistors rise together, resulting in precision devices that are used for accurate temperature measurement. Ordinarily, they are employed with electronic circuits, and at present are not in common use in the HVAC field. Again, it should be made clear that RTDs do not produce electricity.

Thermistors: Although most materials increase in electrical resistance with an increase in temperature, thermistors are an exception. Thermistors are widely used in solar energy applications and automated environmental control system, because of their unique negative tem-

perature coefficient; that is, their ability to decrease in resistance as the temperature increases. Electronic control units that employ digital temperature readouts and combustion and exhaust gas analyzers often employ thermistors as the temperature-sensing devices.

Thermistors are made of semiconductor materials that, given their special manufacturing process, conduct electricity better at higher temperatures. One application related to combustion is exhaust gas analysis.

Light

Heating, ventilating and air-conditioning systems make use of devices that respond to light in a variety of ways. Light is an indicator of both pilot and main burner combustion flame. Solar heating units may utilize components that sense light to track the movement of the sun or to measure the intensity of solar radiation. To understand how light-sensitive devices function, a brief description of the nature of light is necessary.

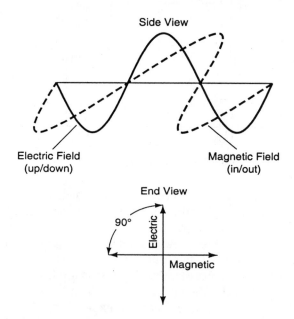

FIGURE 5-6   Electromagnetic waves.

From a human perspective, light is what the human eye sees. Scientifically, however, light is a narrow part of an energy spectrum called electromagnetic radiation. As the name implies, electromagnetic radiation results when an electric wave and a magnetic wave travel through space in a zigzag fashion. At the speed of light (300 million meters per second), these two waves become one; at slower speeds, electricity and magnetism act in a more separate manner, although they are closely connected. Figure 5-6 shows how the electric wave and magnetic wave relate to each other as they travel through space.

The electromagnetic radiation spectrum varies from low frequencies to very high frequencies, that is, how many zigs and zags the wave makes in 1 second. The effect might be compared to the sound produced by a piano keyboard. If an electromagnetic radiation keyboard were available, we could play from radio waves through microwaves, light, x-rays, and gamma rays, and finally to cosmic rays. Figure 5-7 illustrates the frequency spectrum of electromagnetic waves from low frequencies to high frequencies. Light, a narrow band of the entire spectrum, is also shown in figure 5-7. The human eye sees different frequencies of light as different colors. Although this description of radiation refers to frequency, light is most often described by wavelength. Wavelength is how long a wave is. Higher frequencies have more waves within the same distance, so each wave is shorter. As frequency increases, wavelength becomes shorter. The "nm" units in figure 5-7 represent nanometers (10^{-9}), or thousand millionths of a meter.

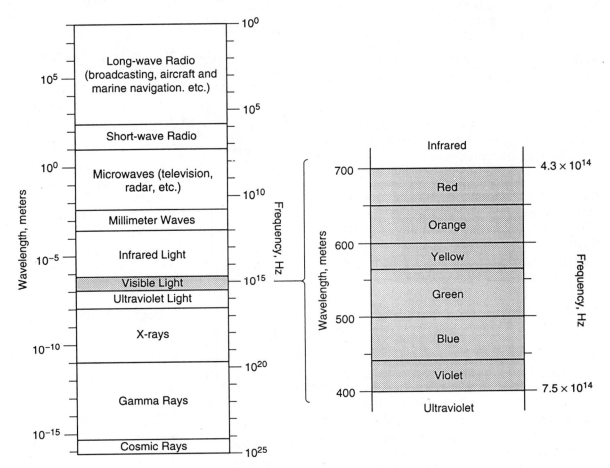

FIGURE 5-7 Electromagnetic spectrum.

Photovoltaic cell: The zigzag energy in light can be converted to electric energy with a photovoltaic or solar cell. These devices are composed of specially manufactured layers of silicon, either as a single-crystal structure (monolithic) or as a polycrystal structure (amorphous). This manufacturing process is referred to as "doping." The monolithic types are more costly and more efficient than are the amorphous types. Generally, the efficiencies of these devices are less than 20% and the amount of electricity they produce is small. Typical outputs for single solar cells are under a volt at .1 amp. Even though these devices are costly and relatively inefficient, they are used to produce power in remote locations, as well as in industrial applications where their relatively high cost of installation is a less important factor when compared with the high costs of utility-purchased electricity.

Photoresistive cell: There are a number of devices that change resistance when exposed to light. These photoresistive devices are included in this section because they are similar and are often confused with sources of electricity; however, they do not produce electricity. As the name implies, photoresistive devices vary their resistance with the intensity of the light.

Cadmium sulfide cell: By far the most common light-sensitive device used in the HVAC appliance is the cadmium sulfide (cad) cell. In the dark, the cad cell can have a resistance as high as 1 megohm, whereas in bright light, its resistance drops to a few hundred ohms. Since this cell can handle relatively large currents and varies its resistance over a wide range, it can be used to control relays directly. The cad cell is sensitive to the wavelength of light from a gas or oil combustion flame. It is also inexpensive and durable. It is composed of a layer of cadmium sulfide upon which is deposited two closely placed grids of metal. The grids do not touch, however, and in the presence of light, cadmium sulfide frees up electrons, becomes conductive, and completes the circuit between the two grids. Perhaps the cad cell's only drawback is the fact that it is slow to "turn off"

after the light source is removed. A cad cell may take as long as a tenth of a second to return to its high-resistance state. Oil burners and gas power burners commonly incorporate cad cells to sense ignition in their safety circuits.

Phototransistors: Phototransistors are usually employed in conjunction with electronic controls, and at present find little application in HVAC. Due to their increased use in optical electronic circuits their cost has dropped. They can be constructed to change resistance to selected wavelengths of light, making flame-color sensing possible, and they interface well with electronic controllers. Phototransistors lack the durability of cad cells and require electronic "know-how" in servicing. Chapters 9 and 10 cover these topics in greater detail.

Chemical

The conversion of chemical energy to electric energy typically is what we consider the type of energy conversion that takes place in the ordinary battery. The technician needs to be familiar with the batteries that are used in HVAC equipment, as well as with the tools of the trade.

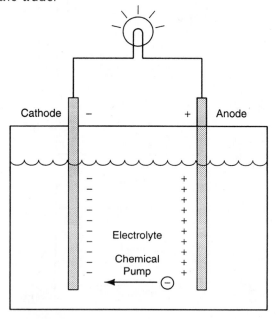

FIGURE 5-8 Voltaic cell.

Recall from Chapter 2 that the second law of thermodynamics states that energy is indestructible. Batteries are a good example of this law in that the electric energy withdrawn from a battery must come from chemical changes within the battery. When it is discharged, the battery is chemically different from what it was in the charged state.

To help you develop a practical understanding of batteries, we first should clarify two common terms, battery and cell. A cell is two electrodes, commonly called anode (+) and cathode (–), between which is an electrolyte (the chemical structure of the battery, see figure 5-8).

The voltage of a cell depends on the chemical nature of the electrodes and the electrolyte. For example, ordinary flashlight cells have zinc and carbon electrodes with manganese dioxide as the electrolyte. This chemical combination produces an electrical pressure of 1.5 volts. One might think of this cell as a single-stage pump. No matter what we do, this combination of chemistry cannot give us more than 1.5 volts of pressure. Just like any pump, the pressure of this cell will drop if too much flow (amps) is drawn from the cell. For this reason, the testing of a cell requires that a load be placed on the cell. Dead flashlight cells will indicate 1-5 volts when measured with a voltmeter, but only a small portion of a volt when measured with a battery tester. The best test of any battery arrangement is how it, and the device it powers, functions under actual operating conditions.

The physical size of a cell is also important. Common 1-5-volt cells are available in sizes ranging from the tiny AAA to the larger D cell. If these cells are all the same voltage, what is the difference? Again, the electron pump illustration will clarify the idea. Bigger cells can supply more amps (coulombs per second) or flow, just as bigger pumps can supply more gallons per hour. AAA cells work well in digital clocks, which require only millionths of an amp, while flashlights require the greater currents available from a D cell. The cell's chemistry defines the voltage and the cell's size, the amp capacity.

Higher voltages are supplied by grouping cells together so that each cell boosts the voltage (pressure) of the previous cell, just like the multistage air compressor used to fill scuba tanks. Each pump (cell) raises the pressure of the preceding one. Batteries are cells that are grouped together to boost or aid each other. Cells connected end to end (plus terminal to the minus terminal) are connected in series to one another. If they are supplied in a single package, this unit is called a battery. How many 1.5 volt cells are in a 9-volt portable radio battery? The answer is given in figure 5-9.

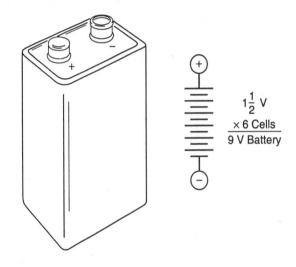

FIGURE 5-9   Nine-volt radio battery.

Batteries, and the cells of which they are composed, have characteristics in addition to voltage that need to be considered in their usage. Figure 5-10 displays the common types of batteries and the main characteristics of each.

Danger lies in the invisible and the unknown, and this is particularly true of batteries since they are chemical in nature and supply electricity. While working with batteries, observe all the manufacturer's cautions and instructions. Following is a partial list of particular dangers.

1. Do not attempt to recharge non-rechargeable batteries, as they will explode.

2. Recharge rechargeable batteries only with chargers intended for use with those batteries, and be certain the polarity is correct. Follow the manufacture's instructions.

BATTERY CHARACTERISTICS

Type	Voltage per Cell	Recharge	Characteristics	Uses
Carbon Zinc	1.5 V	NO	Inexpensive, gradual voltage drop, moderate to low current drain, moderate shelf life and cold performance	Flashlights, toys, and non-critical applications
Alkaline	1.5 V	NO	Moderate cost, gradual voltage drop, moderate to high current drain, long shelf life and good cold performance, most cost effective	Halogen flashlights, high current drain toys, cassette radios, portable tools and test equipment, critical applications
Lithium	1.5 to 3.6 V	NO	Expensive, gradual voltage drop, high energy content, extremely long shelf life	Instrumentation, computer stand-by, clocks, medical, military
Nickel Cadmium	1.25 V	Up to 1000 Times	Expensive, voltage stable till cell exhausted, very high currents possible, poor to moderate charge retention, should be fully discharged or loses capacity	Rechargeable tools, vacuums, thermostats, RC Toys
Lead Acid	2 V	YES	Expensive, gradual voltage drop, high current possible, can be partially discharged, requires special handling, dangerous	Automobiles, motorcycles, marine

FIGURE 5-10 Battery characteristics.

3. All batteries contain corrosive materials that can burn, poison, and blind; never open batteries.

4. Loose batteries can short-circuit and cause fire, especially nickel cadmium batteries.

5. Lead acid batteries require special precautions, and hence, reference to an automotive text is suggested.

Vibration and Pressure

Crystals have fascinated people through the ages, with all sorts of magic and special meanings being attributed to both the natural and synthetic varieties. Jewelry, microchips, abrasives, and even crystal balls exhibit real technical wonders, as well as being identified with quackery and fraud. As is usually the case, the most magical things are those that are understood the least. Here is a general idea of what crystals are.

Recall that the periodic table of elements in Chapter 3 contains columns of elements that share similar outer-shell electron structures. Atoms can form crystals by sharing outer-shell electrons. Under the proper conditions, atoms will join with each other in a regular repeated fashion, like bricks in a wall. If the bricks (atoms) are of a unique shape, a wall with a unique shape will result. Elements in each of the columns of the periodic table form unique crystal shapes. Often these crystals are irregular and broken up, but under certain conditions, single large monolithic crystals are formed. Remarkably, soot, which is composed primarily of the element carbon, forms a diamond when crystallized. However, rather than romance, it is the electrical nature of the crystal that interests us.

Since the crystal is a structure that is held together by electrical charges (the sharing of electrons), it is understandable that forces that act to bend or distort the crystal will have electrical results. The distortions of each atomic bond in a crystal will add together, due to its regular structure, and produce a negative charge on one side of the crystal and a positive charge on the other. Figure 5-11 is a simplification of this phenomenon.

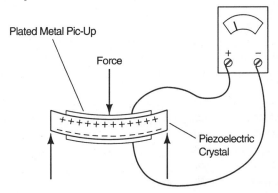

FIGURE 5-11   Electricity from a crystal.

The action picture in figure 5-11 is referred to as the piezoelectric effect. When the crystal is struck with a spring-loaded hammer, this effect can provide enough voltage to jump a gap and ignite a gas flame. Stoves, barbecue grills, and even small cigarette lighters use the piezo-electric effect as a source of ignition. Interestingly, this process is reversible, that is, electricity flowing through a crystal will cause it to bend. Many small electronic devices use components called "chirpers" to generate an alarm. By using the effect in both directions at the same time, a crystal can serve as a micro tuning fork to keep digital clocks and computers running with great accuracy.

Motion and Magnetism

Imagine a world without electricity—walls without receptacles for supplying power for lights, heaters, TV sets, drills, and vacuum cleaners. How would you make a hole or get the dust off the carpet? Ever try to read by candlelight? One hundred years ago, life was very different. People had little time to invent, explore, or just enjoy life. Obtaining the resources needed simply to stay alive was a full-time job. What changed this?

The answer to this question is complicated, but two things come to mind. The first is the millions who contribute skills that provide homes with heating, cooling and light.

By far most of the electricity used today is furnished by a process discovered by both Michael Faraday in England and Joseph Henry in the United States in the 1830s, and known as electromagnetic induction. For practical purposes, all the electric power we use today is generated by large dynamos based on the scientific principle of electromagnetic induction. The automobile alternator is, in general, a small dynamo.

To produce electricity by electromagnetic induction, three things are necessary. First, a magnetic field is required. It could be from a permanent magnet or an electromagnet. Second, a coil of wire is needed—the more turns, the better. Third, there must be relative movement between the two, that is, either the magnet or the coil must be moved. Figure 5-12 reveals the simple setup necessary to produce electricity by electromagnetic induction.

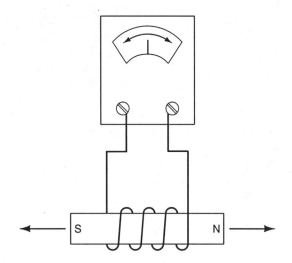

FIGURE 5-12 Electromagnetic induction.

To increase the amount of electricity produced by electromagnetic induction, either stronger magnets can be used, more turns of wire can be wound on the coil, or the speed of the relative motion between the two can be increased. The net effect of any or all of these changes is to cause each wire in the coil to "cut" more lines of magnetism. The nature of magnetism and its relationship to electricity are so important that the next section is devoted to it.

THE NATURE OF MAGNETISM

Both magnetism and electricity are invisible. We know them through their effects, such as the attraction of iron or sparks. It is important to understand that the words used to describe them are not the things themselves. However, words enable us to share ideas about their nature. If you do this, that will happen, and so on. After reading this section, it would be useful to "play around" with some magnets and iron filings to gain direct knowledge about magnets.

Magnetic Fields as Circuits

Permanent magnets are pieces of material that have all their electrons revolving in the same direction. This ordered arrangement of electrons causes the material to concentrate

magnetic forces at two locations, called the north pole and the south pole. The magnetic forces at these poles give rise to the flow of magnetic lines. Magnetic lines flow through the magnet, from pole to pole, and radiate or fanout through space, and then return to the magnet. Figure 5-13 illustrates a bar magnet and the lines, often referred to as flux lines, that travel through space.

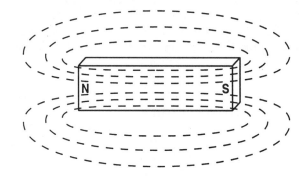

FIGURE 5-13 Magnetic lines.

Take note that in figure 5-13 the flux lines are looplike. Like electrical circuits, the flux lines travel around and around. However, it is important to understand that while electricity and magnetism always appear together, magnetic circuits and electrical circuits are different things. A careful examination of figure 5-13 shows that the magnetic lines crowd together as they travel through the magnet and spread out through space. A more technically accurate description would be that the flux density (lines per square inch) is high at the magnet's poles and low through space. The reason for this is that the magnet is more permeable, or magnetically conductive, than the surrounding space. This could also be described from the opposite perspective—that space is more reluctant than is the magnet. Reluctance is like magnetic resistance. Here you might be wondering if there is a magnetic equivalent of Ohm's law.

Rowland's Law

An American physicist, Henry Rowland (1848–1901), observed that in simple magnetic circuits, the relationship between magnetic force MMF (Magnetomotive force) and magnetic flow phi (number of flux lines) is direct and linear. Remember that with Ohm's law, resistance was a ratio of volts (force) divided by amps (flow). In a like way, Rowland used a ratio of magnetic force divided by magnetic flow to describe the concept of reluctance. As the name implies, reluctance represents how poorly a piece of material conducts magnetism. Reluctance and permeability are opposite ways of describing the ability of material to conduct magnetic lines or flux. Rowland's law is compared with Ohm's law in figure 5-14.

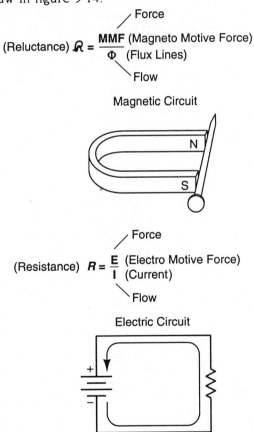

$$\text{(Reluctance)} \; \mathcal{R} = \frac{MMF}{\Phi}$$

Force — MMF (Magneto Motive Force)
Flow — Φ (Flux Lines)

Magnetic Circuit

$$\text{(Resistance)} \; R = \frac{E}{I}$$

Force — E (Electro Motive Force)
Flow — I (Current)

Electric Circuit

FIGURE 5-14 Rowland's and Ohm's laws.

Further details about Rowland's law are beyond the scope of this book. It is important for the HVAC technician to understand the nature of magnetic circuits and the relationship

of force, flow, and reluctance. Why do motors, transformers, and relays have such heavy metal cases and cores? All of these devices have both magnetic and electrical circuits. An incorrectly adjusted relay is actually a magnetic circuit failure.

Magnetic Terminology

To bring together the ideas of the previous section and prepare for the next, read the following brief definitions. A few terms not mentioned above have also been included.

Flux: The movement of magnetic lines.
Flux density: The number of magnetic lines per area. How concentrated a magnetic field is at a location.
Hysteresis: Magnetizing a piece of material north-south then remagnetizing it south-north consumes energy. This would constitute a hysteresis loss. Hysterical behavior is thought to be due to hidden memories. Many materials can be magnetized and can be considered to have magnetic memories. Computer disks and cassette tapes are examples.
North pole: One of two opposite poles of a magnet where the flux lines are concentrated. A north magnetic pole will attract a south pole and repel a north pole. The earth's north pole is actually south magnetically, but is referred to as north because it will attract the north pole of a compass needle.
Magnetic shield: Magnetic field lines cannot be blocked, but are bypassed around or away from other objects. Many stereo speakers use highly permeable material to shunt or bypass the magnetic field away from TV screens. Magnetism can distort the picture.
Magnetomotive force (MMF): The force that causes a magnetic field.
Permeability: How well a material passes magnetic lines.
Reluctance: How much a material resists or opposes the flow of magnetic lines.
Retentivity: How large a magnetic memory a material has; how well it retains magnetism.
Saturation: The point at which a piece of material cannot be magnetized any further. Metal cassette tape has a higher saturation

point than does ordinary tape. Rare-earth magnets have a very high saturation point.
South pole: One of the two opposite poles of a magnet where the flux lines are concentrated.

Electricity and Magnetism

Many effects in science work both ways, so it is no surprise that Michael Faraday, who induced electricity with magnetism in the 1830s, may have been influenced by Hans Christian Oersted. In the 1820s, Oersted noticed that a compass needle was deflected when it was placed near a current-carrying conductor or wire. Figure 5-15, shows a segment of a wire that is part of a complete circuit through which current is flowing. As long as the current flows through the wire, a magnetic field circulates around the wire. The central idea is that every current flow is surrounded by an accompanying magnetic field.

netic field. Electromagnetic fields are very useful. Motors, speakers, relays, and solenoids all depend on electromagnetic principles. Electromagnetic fields can be made very strong, and can be easily controlled by varying the current flow.

A long wire ordinarily would have a long magnetic field, but a long wire is cumbersome to use technically. The solution is to coil the wire as illustrated in figure 5-16. A close look at this illustration reveals that the circulating magnetic field around a straight wire adds together in such a way that the coiled wire provides a north pole and a south pole at opposite ends. The strength of the electromagnet depends primarily on the number of turns and the intensity of current flow. The ampere-turn, which is amps times turns, is a common measure of the strength of an electromagnet. Electromagnetic devices will be explained in later chapters.

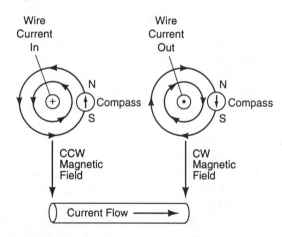

FIGURE 5-15   Magnetic field surrounding a current.

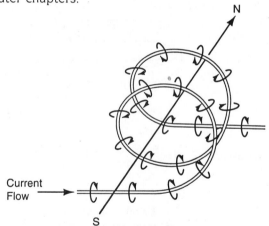

FIGURE 5-16   Electromagnetic coil.

Under certain circumstances, both the magnetic field and the electric field can leave the wire as electromagnetic radiation (radio waves). Ordinarily, the magnetic field continues to surround the wire as long as there is current flowing through it. Magnetic fields around the wire or from a permanent magnet are of the same nature. Because the source of the magnetic field is electricity, the field around the wire is commonly referred to as an electromag-

EFFECTS OF DC LOADS

Books about electricity use the concept of resistors to explain circuit configurations (series and parallel circuits, for example). As a result, students can be left with the impression that electricity is a subject that relies mainly on calculations based on complicated circuits made up entirely of resistors. It is no wonder that when an HVAC technician sees no resistors in a schematic of a heating or air-conditioning system, the question arises: "What were all those complicated resistor circuits for that I struggled with in electricity?" This confusion comes about because two important topics, resistance and circuit configuration, have been mistaken for one. To avoid this confusion, the remainder of this chapter will focus on the effects of DC (direct-current) loads. These effects include resistance, inductance, and capacitance. The next chapter will look at these same loads with AC (alternating current) as the source.

Resistance: The Formula in Action

In Chapter 3, the basic electrical circuit was described as having a source, a control, and a load. Resistance, the nature of the simplest load, was defined as a linear relationship of force (volts) and flow (amps). With a simple resistive load, if we push twice as hard (double the voltage), twice as much flow will result (double the amps). George Simon Ohm recognized this relationship and created a description we now call "Ohm's law." Recall from Chapter 3 that resistance is really a ratio or fraction that results when we divide volts by amps. It is not something tangible. Rather, it is an effect. Also, resistance is a bit backward in that as the ohms increase, the component becomes a poorer conductor. A good, clean, tight connection, for example, has 0 ohms of resistance, while an open circuit has infinitely high resistance. Be-

yond these important concepts lies the quantification or counting of volts, amps, and ohms that can easily be accomplished by applying Ohm's law to simple circuits.

An example is the circuit in figure 5-17, in which the voltage and amperage are both known. The resistance is unknown, but can easily be calculated by dividing the current (I) into the voltage (E). Try it! Dividing 120 volts by .5 amp gives a resistance of 240 ohms. In the field of electronics, a resistance (or resistor) that is more than 10% off value could stop a circuit from functioning, so it is common practice to calculate resistance in this way. More often than not, the electrical components used in HVAC equipment are either good and conduct the right amount of current, or they are bad and have infinitely high resistance and do not conduct any current at all.

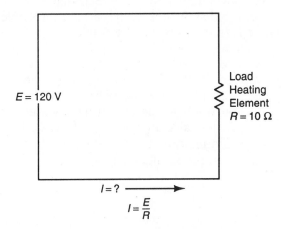

FIGURE 5-18 Calculating current.

Common also are bad circuit components that are internally short-circuited. Short-circuited components have no electrical resistance and conduct so much current that they often trip breakers or blow fuses. Simple continuity testers are available to check components in order to determine whether they conduct any current and/or the proper amount of current under a circuit loading condition. Chapter 8 shows how to use continuity checkers, as well as how to use an ohmmeter, which checks resistance directly.

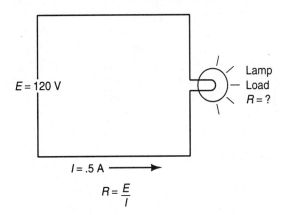

FIGURE 5-17 Calculating resistance.

The circuit in figure 5-18 shows both voltage and resistance. Suppose a fuse keeps blowing and we need to check the current flow. To do this, we divide the resistance into the voltage. Try this yourself. Even though you may own this book, or any book, you can't own the ideas in it unless you think them through for yourself. So let's reason this one out. Dividing ten ohms into 120 volts yields 12 amps. If the fuse that is blowing is an 8-amp fuse, there must be either a short circuit in the heating coil or too little resistance, due perhaps to the fact that the coils are touching each other. Understanding how to measure and calculate current flow is useful knowledge for the HVAC technician. In Chapter 8, ammeters and troubleshooting techniques will be covered in more detail.

Figure 5-19 shows a circuit that is missing—you guessed it—voltage. Take note of how the Ohm's law formula has been rearranged (transposed) so that the unknown is always the single letter (variable) on the left. This can be done with all mathematical formulas and is the main reason for most formulas. If the .25 amp is multiplied by the 48 ohms, we find that the source provides 12 volts. Because the equipment you will use operates from a variety of voltages, the calculation and measurement of voltage are important. Voltmeters will be covered in greater detail in Chapter 8.

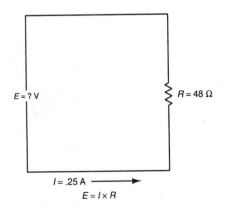

FIGURE 5-19 Calculating voltage.

Now that the concept of resistance has been described and calculated, the question arises,

"What components offer the effect of resistance?" Motors, lamps, relays, switches, and heating elements—in fact, all components—have some electrical resistance. Expect high-quality switches to have very little resistance (good conductors) and small components that require small amounts of electricity such as indicator lamps, to have high resistance. It is important to recognize that some components change resistance due to heat (resistance generally rises) and light (resistance generally drops). Most circuit components often exhibit other effects in addition to resistance. The next sections cover these other effects and how they react to DC.

Coils and Inductive Kick

As a child, one of the authors surprised himself when he received an electrical shock. In his mind, this should not have happened. The source was a D cell, only 1 1/2 volts, and the load was an electromagnet. The electromagnet was a simple coil wound around a large nail, like the coil in figure 5-16. Curious about what would happen, the young experimenter added more and more turns to the coil and took note of the increasing magnetic effect. At one point, while disconnecting the electromagnet, "*zap*"! How could the low-voltage D cell provide such high voltage for a shock? In this section, we will develop an understanding about coils that the child did not have, but first a word about shocks.

Pilots who fly beyond their skill and knowledge crash. HVAC technicians who work with electricity need personal instruction from certified teachers, appropriate licenses from government agencies, experience with competent "masters of the trade," and specific first-hand instruction in safety practices. The environment in which the technician works is full of well-grounded metal and is often wet. These conditions mean that electricity can easily flow through the technician, causing death. The HVAC technician is warned to perform no work that is beyond his or her skill and knowledge. It is a matter of life and death!

In the early 1800s, Heinrich Lenz observed an electrical effect that relates to coils and

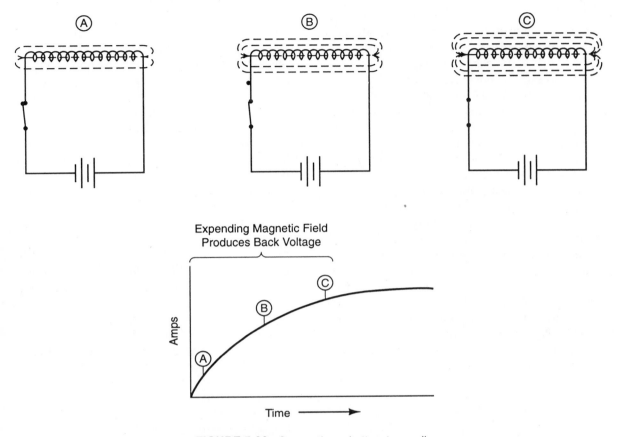

FIGURE 5-20 Connecting a battery to a coil.

magnetism. A general way to think about the effect he discovered is that it is the electrical version of Newton's third law: "For every action there is an equal but opposite reaction." A good way to understand Lenz's law is to examine what happens to a coil when it is connected to and then disconnected from a battery.

Figure 5-20 illustrates a coil that has just been connected to a battery. Figures 20 A, B, and C follow closely in time, usually small parts of a second. As the current begins to flow through the coil, a magnetic field appears in A, then gets stronger or moves outward from the coil in views B and C. Recall that to make electricity, one needs a magnetic field (magnet), a coil, and relative motion between the two. All present in figure 5-20, A-C. The magnetic field is expanding, and hence it is moving. The result is

that the same coil that is carrying current and expanding its magnetic field induces, into itself, a back voltage. The back voltage is a reaction to the original action of the current flow and magnetic field expansion. Looked at in electrical terms, the process is as follows: A voltage (battery) is connected to a coil and current begins to flow. A magnetic field expands around the coil, and that same expanding magnetic field induces a back voltage into the coil. Then the back voltage pushes against the original battery voltage (reaction). The total effect is that when a coil is first connected to a battery, current flow rises slowly because a back voltage is produced (induced) by the expanding magnetic field. Once the magnetic field has expanded completely, the back voltage stops, as there is no longer any relative motion between the coil and the field. At the bottom of figure

5-20 is a graph showing current rise as time goes on. Take note that after a period of time, the inductive effect is completed and the current flow stabilizes. At this point, the only effect limiting current flow is the coil's resistance. The inductive effect is dependent on time, whereas resistance is not. Every time a coil is connected to a DC source, the inductive effect takes place. All coils react to current changes. If the coil is large and has an iron core the effect is greater.

The unit of inductance is the henry. A 1-henry inductor will produce one·volt of back voltage, while the current is increased at a rate of one amp in one second or one volt of forward aiding voltage if the current is decreased at a rate of one amp in one second.

If you think about it, energy is required to overcome this effect (induced back voltage) and to expand the magnetic field. Energy is stored in the fully expanded magnetic field, and will remain stored as long as current flows through the coil. A current-carrying coil can be thought of as a magnetic balloon.

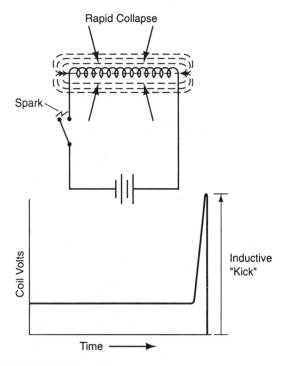

FIGURE 5-21   Inductive kick.

Opening the switch or disconnecting the battery from the coil in figure 5-20 is like sticking a pin into the magnetic balloon. The fully expanded magnetic field around the coil is a result of the current flow through the coil. Imagine what would happen if the current flow were to stop. Figure 5-21 shows the disconnecting of a coil and the resulting rapid collapse of the magnetic field. At the bottom of the illustration is a graph of the voltage applied to the coil from the battery and the voltage resulting from disconnecting the coil. Remember that the magnetic field is a result of current flow, so that if current flow is stopped, the magnetic field must collapse. The magnetic field must also release the energy stored in it. The result, as shown in figure 5-21, is that the rapid collapse of the magnetic field develops a high voltage (force) that arcs across the switch contacts as they are opened. The inductive kick developed when a coil is disconnected is like the hammering of water pipes when a valve is slammed shut. Moving water has momentum, and in a way so does the magnetic field around a coil.

The inductive kick is a demon that can destroy switches, motor brushes and commutators, and electronic components, and can cause lightning-like radio interference. However, it is useful in developing high voltages. Automobile ignition coils convert energy of 12 volts stored in a magnetic field to a brief, fuel-igniting spark of 50 kilovolts (50,000 volts).

Coils exhibit the most observable effects on DC when they are either connected or disconnected. Small increases or decreases in coil current cause the same effects, but to a much lesser degree. An increasing current causes an expanding magnetic field and a resulting back voltage that momentarily holds the current down. A small decrease in current is partly counteracted by a collapse in the magnetic field, which provides a forward voltage. In this way, inductive coils can be used to "choke" out rapid current changes. Many electronic modules that are used in conjunction with electrical appliances have choke coils or filters in their power input lines. Due to the inductance effect, coils oppose a change in current flow. Inductors delay current changes.

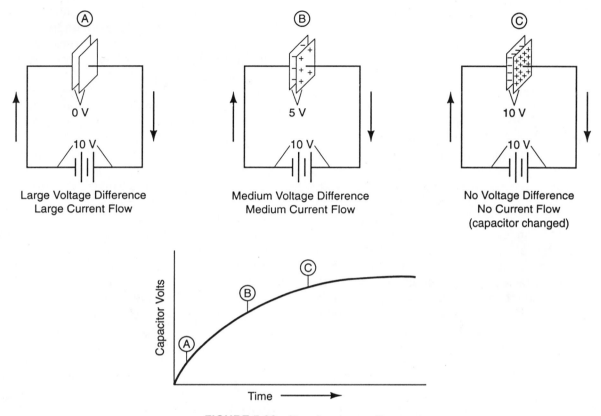

FIGURE 5-22 Charging a capacitor.

Capacitive Charge and Discharge

While the inductive effect depends on the magnetic field nature of electricity, the capacitive effect depends on its electric field nature. By electric field we mean the attraction that positive + (protons) holds for negative − (electrons). The electric field surrounds the positively and negatively charged objects, like the earth and a cloud just before a lightning strike. You can feel that charge when it causes pages of a magazine to stick together and see it holding lint to your clothes or dust to your TV screen. Recall that a magnetic field is a result of current (flow). An electric field is a result of voltage (force). The capacitor is a device whose function depends on the electric field.

There are many different types of capacitors, but they all function in the same way. Their differences lie in their construction and applica-

tions. Every capacitor comprises two conductive surfaces or plates, separated by an insulator. The plates are intended to hold the positive and negative charges, while the insulator between them prevents current flow between the plates.

Figure 5-22 is a simplified drawing of a capacitor when it is first connected to a battery. The events that occur proceed from figure 5-22A to B, then to C. If the battery in view A could talk (remember, it is an electron pump), we might hear it say, "I would like to get rid of some of these electrons at my negative terminal pushing against each other. Look at all that neutral space on the left plate of the capacitor. I'll push some up there. And my positive terminal would like to vacuum up some electrons; a little bit of pull, and I bet I get some from the right neutral plate of the capacitor." And that is just what happens when the capacitor is connected to the circuit. Electrons flow to the left plate and from the right plate. At the beginning of this process, the capacitor has no voltage across its plates, so a large current flows to and from the capacitor. As the electrons are accumulated on the left plate and removed from the right plate, a voltage, or pressure difference, begins to build up between the plates. Figure 5-22B shows the capacitor charged up to 5 volts, or halfway to the battery voltage. Since the difference between the battery and capacitor is now only 5 volts (battery 10 volts less capacitor 5 volts equals a 5-volt difference) the current flow to and from the capacitor begins to drop off. Finally, in figure 5-22C, the capacitor has accumulated an electrical charge of 10 volts, equal to that of the battery. It might be said that the battery has pushed all the electrons it could onto the left plate and removed all the electrons it could from the right plate of the capacitor. Current flow now ceases. As the capacitor charges, the accumulated electrons on the left capacitor plate are surrounded by a negative electric field that expressed itself through the insulator and pushed electrons from the right plate toward the positive battery terminal. Never during the charging of the capacitor did any electrons actually flow through the insulator. If that had happened, the capacitor would have been destroyed.

The quantity of charge that a capacitor can hold depends on a number of factors. First, like any container, the capacitor has a capacity. It can hold a larger charge if the plates have a larger area, if the insulation is made thinner, or if an insulating material that expresses an electric field with greater strength is selected. Second, a higher charging voltage will force the capacitor to accept a greater charge. Too high a voltage will short-circuit the insulator, so one must be careful not to exceed the voltage rating of the capacitor.

The capacity of a capacitor is rated in farads. A one-farad capacitor will hold one unit of charge (remember the coulomb, 6.28×10^{18} electrons) when one volt is placed across it. A one farad capacitor would fill up a small room, so capacitors are ordinarily measured in micro- (10^{-6}) or pica- (10^{-12}) farads.

Let us disconnect the charged capacitor. It holds electricity, but generally not enough for it to be used as a batterylike, long-term source of power. However, it is useful as a power source for special applications. Some digital memory circuits employ capacitors for very-short-term power backup. More often, when the capacitor is used to store energy, it acts like an accumulator. Photo flashes use small batteries that charge up the capacitor over a few seconds. When the picture is taken, the full charge of the capacitor is "dumped" through a xenon gas tube that supplies a bright, but very brief, source of light.

Earlier we discussed how an inductor can be used to smooth, or stabilize, current. In a very real way, the capacitor is the opposite of the inductor. If a capacitor is connected across the terminals of a DC power supply, it will act to smooth the voltage. If the power supply voltage rises, current will flow to the capacitor to charge it up, and in this way act as a load to lower the voltage. Conversely, if the voltage drops, the capacitor will discharge into the supply lines and tend to boost the falling voltage. DC power supplies generally have one or more large capacitors to perform this voltage-stabilizing function. Due to the capacitive effect, capacitors oppose and/or delay a change in voltage.

ACTUAL DC LOADS ILLUSTRATED

There is a very large difference between actual or real electrical circuits and the ideal ones that books like this present. Real or actual items are never as perfect as ideal ones. For example, residential electrical service is 120 volts, but it is not uncommon to measure voltages as low as 110 volts or as high as 128 volts. Measurements of volts, amps or ohms that fall 10% above or below the nominal (what it is specified to be) are usually within acceptable limits. Further, real electric devices generally exhibit more than one effect (resistance, inductance, and/or capacitance). Also, electric devices are often affected by other factors, such as moisture, heat, and vibration. Even measuring instruments and equipment have an effect on what is being measured. Actual experience with "experts" is necessary to develop a sense of confidence, skill, and safety in working with real circuits.

Resistive

Figure 5-23 shows a 60-watt lamp, like a head lamp, connected to a 12-volt car battery. We know the battery is 12 volts and that the lamp consumes 60 watts of power. One could try to follow a set procedure to come up with amps and ohms, but that's not really critical thinking. Amps or ohms cannot be found if all we know is the voltage. Two of the three electrical measures in Ohm's law will permit us to calculate the third. What about power? We studied that in Chapter 4. Do you remember what power is equal to? If not, here is a brief reminder: watts equals amps × volts ($P = I \times E$). How many amps times 12 volts equals 60 watts? By trial and error on scratch paper, you will soon find that 5 amps times 12 volts equals 60 watts. Now that we have both the volts (12) and the amps (5), the resistance can be calculated by using the version of Ohm's law

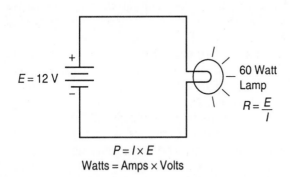

$$P = I \times E$$
$$\text{Watts} = \text{Amps} \times \text{Volts}$$

FIGURE 5-23 A 60-watt lamp connected to a 12-volt source.

shown in figure 5-23. Divide 5 amps into 12 volts. The answer is 2.4 ohms. If we remove the lamp and measure its resistance with an ohmmeter, (see Chapter 8), it will read 1.5 ohms. Have we made an error, or is there some characteristic of the lamp that would explain

this difference? Lamps run very hot, and their hot operating resistance is higher than their cold resistance. Heat affects things, which is a good fact to keep in mind when troubleshooting HVAC equipment. Things change with temperature. Is there any other effect besides resistance that the lamp offers? Look closely at the lamp in figure 5-23. The symbol for it is a small coil that looks like the filament of a real lamp. If it's a coil, it must be magnetic. We can check this by placing a small magnet close to the lamp and watching the filament bend slightly toward or away from the magnet. The lamp must then have some inductance. The amount of inductance is very small, and in this case, it is not relevant to its function.

Magnetic and Inductive

All electric motors are composed of coils that are designed to produce the magnetic fields that cause the motor to rotate. The

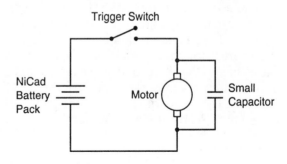

FIGURE 5-24 A cordless electric drill.

schematic diagram in figure 5-24 is of a cordless drill. This type of motor has brushes that make and break electrical circuits in the rotating part called the armature. These brush-type motors are commonly used where high power and low weight are required (see Chapter 7 for an explaination of how this and other types of motors work). For now, all we need to realize is that the brushes are constantly connecting and disconnecting armature coils. The coils are intended to be electromagnets; however, as much as we may not desire it, they also are inductors. Remember what happens when a coil is disconnected from a battery. The magnetic field collapses and a spike of high voltage called an inductive kick is produced. Brush arcing and radio interference are the results. This type of motor often fails as a result of brush and commutator wear. What might be the purpose of the small capacitor connected across the brushes? If you recall, a capacitor can be charged. In this case, the voltage kick finds a place to go by charging the capacitor. The capacitor helps to prevent brush arcing. Small disk capacitors are often found connected across contact points of relays and thermostats for this purpose. Figure 5-24 shows a motor that is primarily magnetic, but, because it is made up of a series of coils, it is also inductive.

Capacitive

Figure 5-25 shows a simple timing circuit schematic. This circuit includes a battery, resistor, and a capacitor. Here, the time it takes to fully charge the capacitor up to the level of the battery voltage is controlled by the resistor. A 10-ohm resistor may permit the capacitor to charge in less than a second, while a resistance of 100,000 ohms may slow the process to 10 minutes. The resistor reduces the flow of current to the capacitor and thus it takes longer

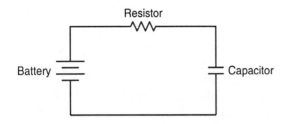

FIGURE 5-25   A timing circuit.

to fill. Oxidized battery terminals will increase the time a photo flash takes to charge and prepare for the next picture. The corrosion on the terminals acts as an electrical resistor within this circuit. This circuit is both resistive and capacitive; however, since no coils are present, it is not inductive.

A Mixed Load

As the previous examples have illustrated, most real electrical loads are mixed. The point is that in the area of electricity, and in the sciences in general, observations of effects in the real world are purified and simplified so that important general rules can be made and laws established. For example, although zone valves, blower motors, and ignition transformers all exhibit inductance, they are very different from one another. If you have studied this chapter well, with the help of your instructor and experience, you will learn to move between the practical and theoretical parts of HVAC. Many people in the field limit themselves by getting stuck either in practice or in theory. True understanding requires both.

SUMMARY

In this chapter, electricity was described as coming from a source. All devices that are considered sources of electricity are actually energy converters. Heat is converted to electricity by the thermocouple and light to electricity by the photovoltaic cell. However, due to their similarity, a number of heat- and light-sensitive devices that are often confused with sources of electricity were explained. Bimetal switches (ordinary thermostats) respond to temperature changes by the unequal expansion of two metals, and in this way can be used to open and close electrical circuits. Also, temperature-dependent resistor devices (RDTs) and thermistors change resistance as a result of temperature change. The commonly used cad cell changes resistance as light intensity changes, and the photoelectric transistor, in a general way, also changes resistance.

Batteries, the most common chemical sources of electricity, have been explained from a practical application perspective. Cells grouped together in series can be seen as a multiple-stage pump, progressively raising the voltage. Alkaline batteries are a good source of disposable energy, while NiCads have the advantage of being rechargeable.

Spark igniters, which convert mechanical force to electricity are piezoelectric crystals. These crystals will also work backward, in that they will bend in response to an applied voltage. Many of the sources of electricity, like the piezoelectric crystal, are reversible.

The importance of magnetism as a producer of electricity was described. To produce electricity, a coil, a magnetic field, and relative motion between the two are necessary. Because of the extensive application of magnetic devices in the HVAC area, magnetic fields were discussed at some length. Magnetic fields were presented as circuitlike in nature with concepts similar to elec-trical circuit concepts, for example, Rowland's Law. Also, impor-tant connections between magnetism and elec-tricity were enumerated, with the technician cau-tioned not to mistake electrical and magnetic circuits as being the same, but rather to recognize their similarities and connections.

The long and difficult road to understanding electrical circuits was begun in this chapter by separating the ideas related to the effects of electrical loads from those related to circuit configuration. Resistance was further examined through calculations based on Ohm's law. The inductor, and the effect it exhibits on DC, was also detailed. Opposition to current rise and fall, as well as inductive kick, was described. In contrast to the inductor, which depends on magnetism for its effect, capacitors are seen to accumulate a charge and to be useful in DC circuits in order to stabilize voltage. The induc-tor delays and opposes a change in current, while the capacitor delays and opposes a change in voltage.

Finally, mixed-load real circuits were de-scribed as an introduction to thinking about actual rather than theoretical effects. Future chapters will further develop real circuit applica-tions in a step-by-step fashion. From a safety-related point of view, all technicians should become aware of the potential hazards of the HVAC environment.

PROBLEM-SOLVING ACTIVITIES

Case Studies

A. Rechargeable tools used by the technician exhaust their charges in progressively shorter periods of time. The technician uses these tools for only a short time, and then recharges them. What is the problem and how might a second battery pack be used to help to solve it? (Hint: See the battery chart.)

B. A new oil burner was unpacked, forgotten, and left exposed near a sand-blasting area. Months later, its exterior was wiped clean (the fuel connections were still capped), and it was put into service. Upon a call for heat, the burner ignited, and then shut down. What might be the problem?

C. A consumer attempted to adjust a blower relay, and in doing so received a shock. After all the power had been disconnected, the technician cleaned and adjusted the points. Draw a simple diagram of the relay, showing its electrical contacts and it's magnetic circuit. Refer to relays pictured in later chapters. Why, considering safety, economics, and good customer relations, might it be wiser to replace the relay than to try to fix it?

D. During a house remodeling project, a thermostat was removed and then replaced crooked. The customer complained that the indicated temperature setting and the actual temperature differ by 7%. Explain what has happened and how it may be easily corrected.

Review Questions

1. A 1500-watt portable room heater is designed to operate from 120 volts. How many amps must it draw?

   a. What would be the hot resistance of the heater?

2. Draw a simple circuit using a meter, a cad cell, and a battery to measure light.

3. Batteries are chemical in nature. With the instructor and other class members, develop a list of all the major hazards of working with batteries.

4. Speakers are stationary magnets with movable coils attached to a flexible diaphragm. How can the speaker convert electricity to sound?

5. The speaker of question 4 will also work as a microphone. Explain this.

6. A common dimmer switch is rated for 300 watts of incandescent lighting and the instructions state that it is not to be used with motors. What is the reason for this?

7. Even after they are unplugged, TV sets can "hold a charge." The picture tube is made of glass, with a metal coating on both the inside and outside of the tube. Besides serving as a display device, what else is the tube acting like?

8. What are the characteristics of the HVAC working environment that make electricity especially dangerous?

9. What two methods are used to reduce point deterioration on bimetal thermostats and switches?

10. Use Ohm's law to fill in the missing values in the following.

Volts (E)	Amps (I)	Ohms (R)
25	10	?
120	2	?
240	?	120
5	?	500
?	50	2
?	0.1	5000

CHAPTER 6

Electric Power Distribution and AC Effects

At this point in your understanding of electricity, it is necessary to move toward those devices (both sources and loads) that are used with alternating current (AC). The efficient and practical distribution of electric power depends on current that flows first in one direction and then in the other. At first, the idea of AC seems unnecessarily complicated, and indeed complications are added. New concepts relating to theory have to be developed and new devices require explanation. Perhaps it was because of these complications that Thomas Edison relied on DC for the first commercial power generating station. However, Nikola Tesla realized the great advantages, both as to the durability of equipment and the economies of power transmission, that were available only with AC. In the late 1800's an ongoing battle raged between the well-known Edison and the eccentric Tesla over the two concepts. On Edison's side was the record of his many successful inventions and a great deal of wealth. Tesla, on the other hand, had the support of the entrepreneur George Westinghouse. Together with Westinghouse, Tesla built the Niagara Falls power station based on AC principles. The resolution of the dispute came shortly after the opening of the Niagara station, when Edison's popularity no longer could overcome the clear advantages of AC.

Today, all power stations use ideas developed by Tesla.

In this chapter, we will describe the generation of electric power, dealing first with the DC generator and then moving on to the AC generator, more commonly referred to as the dynamo or alternator. Terminology and descriptions of the AC wave will follow. Our introduction to AC will continue with the transformer, an electrical gearbox-like device, and with the AC effects of inductance and capacitance. The discussion then moves on to the idea of power factor, an important way of rating motors and other AC devices. The chapter ends with a common-sense look at the nature and advantages of three-phase AC.

ROTATING GENERATORS

In the previous chapter, it was seen that electricity could be produced by moving a magnet back and forth through a coil of wire. However, back-and-forth motion is not a desirable mechanical arrangement. Hand saws, piston engines, and even the old butter churn all require that the device first be moved in one direction, and then stopped before being moved in the other direction. All this stopping and then starting in the opposite direction wastes energy and time. Sawing in one direction only (rotary motion), as with the circular or band saw, is a better arrangement. Turbine engines produce more power and operate more efficiently than do piston engines (reciprocating motion). It would seem that generating electricity by rotating either the coil or the magnet would make a great deal more sense, and it does.

DC Generators

Until about 1960, automobiles produced electricity to charge the battery with a device called a generator. This device has long since been replaced by the alternator. Let us see how it, and all generators, work and why they are no longer used in the automobile.

The large magnet in figure 6-1 has a strong magnetic field that flows through the magnet, and then from pole to pole through space. The magnetic field is strongest directly in front of each pole. This stationary part of the generator is referred to as the "field." Between the poles is a single-turn coil. Imagine the coil rotating and the sides of the coil passing through the magnetic fields concentrated by the poles of the magnet. The coil will produce the largest amount of electricity when the coil sides are near the poles. What about when the coil is in the up and down position? If you guessed that since the coil sides at that point are away from the magnetic fields, no electricity would be produced, you are correct. When the coil sides are near the poles, they produce the most electricity, and when they are farthest away, they produce no electricity. The result of rotating the coil in this way is that the electricity produced goes up and down, up and down, and so on.

There is also another effect that occurs as the coil is rotated. The coil side on the left moves up through the south magnetic field, and then in one half turn moves down through the north magnetic field. The reversal of magnetic poles causes a reversal of electricity. It seems that the electricity not only is rising and falling, but also is reversing direction. The current goes up and down, as well as back and forth. Recall that this device needs to provide current flow in one direction since DC is required to charge the battery. One solution to the reversal of current flow (AC) is to switch the connections to the generator at the same speed at which the generator turns. That is not practical, however, so a better solution is to design a rotary switch that is connected to the generator shaft. Look closely at the end of the coil and see how it is connected to two half circles. These two copper half circles rotate with the coil and are called the commutator. Electrical connections to the generator coil are made through carbon brushes that are held against the commutator by spring tension. Carbon is a material that is both conductive and self-lubricating. The spinning coil and commutator assembly are called the "armature." As the armature spins, producing constantly reversing current, the commutator also constantly reverses the connections to the brushes. In this way, it switches the current from AC to DC. The output of the generator is shown in the graph in figure 6-1.

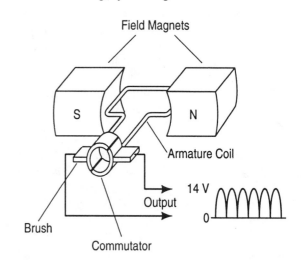

FIGURE 6-1 The DC generator.

Many improvements can be made in the design of the simple DC generator illustrated in figure 6-1. For example, the armature can be constructed of many coils, with many segments on the commutator. An electromagnet may be used to increase the strength of the permanent magnet field. However, even with these improvements, a major fault remains. Inductive kick of the armature coils, as well as high current load, results in sparking and premature failure of the commutator-brush arrangement. The AC-producing alternator does not suffer from these problems.

Alternators

Many problems can often be solved by reversing things. Alternators are perfect examples of this process in that the spinning current-producing part of the generator, called the armature, becomes the stationary part of the alternator. Further, the stationary magnetic field part of the generator called the field becomes the spinning part of the alternator. Figure 6-2 shows a simplified diagram of an alternator in which the spinning part that provides the magnetism is referred to as the rotor and the fixed or stationary current-producing coil is called the stator. As the rotor is turned within the stator, electric current is induced in the stator coils. As in the generator, the current rises and falls, depending on how close the rotor's magnetic poles are to the stator coils. Realize also that with every half turn, the poles of the rotor are reversed, and thus the direction of current reverses. Electric current produced by the alternator, as by the generator, rises and falls and periodically reverses direction. The generator produces electricity in the spinning armature, while the alternator produces electricity in the stationary stator. Improvements in the alternator are provided by increasing the number of coils in the stator and the number of magnetic poles in the rotor. Tesla realized that the output of the alternator was AC. However, rather than changing the AC to DC with brushes and a commutator, he designed the induction motor, which requires AC for operation (motors are explained in Chapter 7). Tesla further realized that lighting and heating devices work equally well on power from either DC or AC. Why then use the generator and its frail commutator?

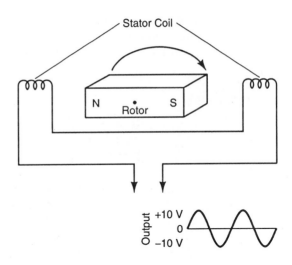

FIGURE 6-2 The alternator.

Mention was made earlier of a battle between DC, which Edison supported, and AC, developed by Tesla and Westinghouse. Edison's primary arguments against AC generation were that it was complicated and dangerous. Experience quickly proved that AC was far superior to DC, and that Edison's arguments were unfounded. Today, all electric power systems rely on AC, and thus it is important that the HVAC technician develop a clear understanding of alternating current.

AC THEORY AND NOMENCLATURE

Have you ever watched the pendulum of a grandfather clock? It swings to the right, then goes slower and slower, and for just an instant stops. It begins its trip to the left slowly at first, then it gains speed as it travels through the center, and it slows to a stop at the far left. Again, it gains speed on moving to the right, and the process continues. A string on a musical instrument follows the same pattern of motion, that is, rapid travel through the center and slowing to a stop at each end of its vibration. Most vibrating or swinging things follow that motion pattern. The pistons in a gasoline engine move rapidly through the middle of their stroke and then slow to a stop at each end, and so do your feet when peddling a bicycle. All of these are examples of nature's most common back-and-forth, or up-and-down, motion pattern. Mathematicians call this pattern the sine wave.

The Sine Wave

The alternator produces electricity that follows the sine wave pattern. As the magnetic rotor moves past the stator coil, a great deal of electricity is produced. Then the rotor magnet moves away and the electricity becomes less and less, until the approaching opposite magnetic end of the rotor causes the stator coil again to produce more electricity. However, since the rotor has made one half turn and now moves the opposite magnetic pole past the stator coil, the electricity flows in the opposite direction. Every half turn of the simple alternator in figure 6-2 reverses the polarity (+−, −+) of electricity to the output connections. The electrical output rises to its maximum for just an instant as the pole passes the stator coil (just like the clock pendulum reaching far right), and then rapidly goes through zero and reaches maximum output in the opposite direction (like the pendulum reaching far left). In this way, the spinning alternator produces electricity that follows nature's most common pattern, the sine wave, shown in figure 6-3.

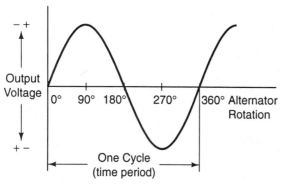

Frequency = Number of Cycles in One Second
Hertz (H$_z$) is Cycles per Second

FIGURE 6-3 The sine wave.

To grasp this idea, all you need to do is see the connection between figure 6-2 (imagine the rotor turning) and figure 6-3. Take note of the electricity rising to maximum in one direction, and then falling through zero to a maximum in the other direction. It may seem that the electricity (electrons) in the wires are moving forward and backward, but getting nowhere. Even though that is true, AC electricity, just like DC, can provide heat, light, and, with Tesla's induction motor, motion. Recall that electricity is a transporter of power, and that it is almost never used without being converted into something else.

Periodic Wave Terminology

If you and I were the smartest people on the earth but spoke different languages, we could accomplish very little. Therefore, it is necessary to learn the commonly used terms when describing the AC sine wave. If the alternator is spun fast, lots of sine waves are produced, but "lots" is too general a term. The frequency is "how many" sine waves are produced in 1 second. By very accurate control of alternator speed, the power company supplies electricity at a frequency of 60 hertz (Hz). The hertz was named for Heinrich Hertz, who discovered radio waves; it really means cycles per second. The grandfather clock depends on the time period, or frequency, of the pendulum for accuracy, just as many electric clocks depend on the frequency of 60 Hz to keep accurate time. Also, some electric motors rotate at speeds that depend on the power line frequency. Common induction (washing machine, blower, etc.) motors generally run at 1725 or 3450 RPM. Another common term used to describe waves, again a bit like the clock pendulum, is *period*, which means how long it takes for one wave or cycle to occur.

Equating AC and DC

"How much work will it do?" is a question that, in one form or another, is asked about cars (how fast is it?), tools (how much power does it have?), and just about every other object that we are interested in acquiring. Although it may not be obvious, the answer is often stated in such a way as to compare the new object with an old one. Humans, it seems, like to hang on to old ideas. This leads to some strange comparisons. Electric motors are rated in horsepower, although very few horses are used today to operate machinery, especially those types of machines that are electric. This same type of thinking applies to AC and DC electricity.

Recall that power in electricity is measured in watts, which are calculated by multiplying volts (force) by amps (flow). With DC, that makes good sense since the volts and amps are constant from second to second. However, AC proves to be a problem in that it is constantly changing. The sine wave in figure 6-4 could be a representation of volts or amps. Ordinarily, as the volts rise and fall, the amps that are caused to flow also rise and fall in step. Note that the

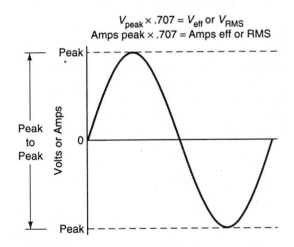

$$V_{peak} \times .707 = V_{eff} \text{ or } V_{RMS}$$
$$\text{Amps peak} \times .707 = \text{Amps eff or RMS}$$

FIGURE 6-4 Peak volts or amps.

maximum voltage in the sine wave is referred to as the peak. If a measure is taken from the maximum voltage in one direction of the wave to the maximum voltage in the other direction of the wave, it is referred to as peak-to-peak voltage. Neither of these measures would seem proper to use when comparing AC and DC

since the AC voltage, or its resulting current, is only at the peak for a short time. This would be like a worker who hustles for one hour a day and slows down for the other seven hours. Equating AC to DC is a little tricky, so let us take a closer look.

A wise person compares things on the basis of what they will actually do. Technically, the "what they will do" is power—power to provide heat, light, and so on: power measured in watts. The common measure of AC volts and AC amps was developed by comparing the work that AC electricity would perform with the work that DC electricity would perform.

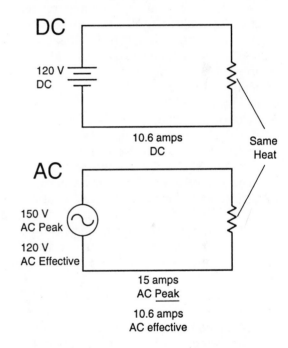

FIGURE 6-5 Effective or RMS volts or amps.

As figure 6-5 illustrates, the amount of power available to provide heat with 150 volts AC peak and 15 amps AC peak can be demonstrated by trial and error to be equal to the amount of heat provided by 120 volts DC and 10.6 amps DC. Since AC volts or amps are at peak for only an instant, they produce less power than do the same continuous DC volts and amps. This may seem like cheating, but for practical purposes, all AC measures of volts and amps are "effective" AC volts and amps. "Effective" in this instance means that 170 volts AC peak has the same effect as 120 volts DC. For example, suppose your instructor tells you to measure the voltage of a circuit that is described as 100 volts AC peak. When you connect the AC voltmeter to the circuit, you note a reading of 70.7 volts. Did you make a mistake? You check the meter settings and take a second look at the reading. Again the meter needle rests at 70.7 volts AC. No error has been made. Almost all AC voltmeters are designed to read "effective" AC volts, not peak volts. Also, almost all AC ammeters read "effective" AC amps, not peak amps. As a matter of fact, the common residential service at 120 AC volts is actually 120 AC volts effective and about 170 AC volts peak.

This idea of effective volts or amps is often referred to as RMS volts or amps. The root of the mean of the squares (or RMS) is simply a mathematical way of arriving at the same value as the trial-and-error derived effective value. Effective and RMS volts are the same, as are effective and RMS amps. For practical purposes, all AC volts and amps are measured, labeled, and discussed in this system as effective or RMS volts and amps. For example, a 240-volt, 200-amp residential service entrance is actually 240 volts AC effective or RMS and 200 amps AC effective or RMS.

Many electronic components, as well as electrical insulation materials, can be damaged by instantaneous high voltage. For example, insulation rated for 120 volts would not be acceptable for use on 120 volts AC effective or

RMS because the voltage rises to a peak of 170 volts. A more specific understanding of this idea can be developed by looking at figure 6-6 and taking note of how conversions between effective and peak voltage can be made.

Although most of the technician's work in the area of HVAC does not involve the measure of peak AC volts or amps, it is important to realize that the meters you use and the equipment you install and maintain are rated in effective or RMS AC volts and amps. This somewhat confusing issue of effective versus peak volts and amps is a good example of how advertising and marketing people can use complex ideas to confuse the consumer.

One of the important ratings of stereo systems is wattage—the AC volts times the AC amps that drive the speakers. Home stereo systems are usually rated in true watts or in watts based on effective or RMS volts times effective or RMS amps. A 120-watt amplifier is a true 120 watts. Take note that sometimes true watts are referred to as RMS watts because they are based on RMS or effective volts and amps. Car stereo systems seem to be rather surprising devices in that they contain high-wattage amplifiers in packages that are much smaller than home stereos. A careful reading of the ratings of car stereos reveals that they are often rated in peak watts which are calculated by multiplying peak AC volts by peak AC amps. The rating here is obviously misleading.

TRANSFORMERS

The word transform means to change or transport from one form to another. For example, a sawmill may be used to transform a tree into lumber and carpenters may transform the lumber into a house. In each case, a conversion or transformation has occurred in which little was lost and the new form is more desirable than the old. Electrical transformers are devices that transform some combination of volts and amps to some new and more desirable combination of volts and amps. Generally, little electric power is lost as transformers are usually 95% or more efficient (certainly

---

Peak AC volts or amps X .707
=
Effective or RMS AC volts or amps

Effective or RMS AC volts or amps x 1.414
=
Peak AC volts or amps

FIGURE 6-6 Effective or RMS to peak conversion.

no power is gained, as we cannot get something for nothing). An important point to remember is that transformers change the combination of volts and amps at the input to a different, more useful combination of volts and amps at the output. They are very efficient at doing this, and suffer little power loss. Usually, the total amount of watts applied at the input of the transformer is the same as the total watts at the output, less a very small amount lost primarily as heat. Another important idea is that transformers ordinarily will function only with AC.

The Lever, Gears, and Transformers

Remember the springboard gymnastic team in Chapter 4? The team spent time moving the pivot or fulcrum of the lever. In doing so, they were able to give up distance traveled and to gain force, or to lose force and gain distance. The energy of the falling gymnast was transferred through the lever to the rising gymnast. Automobile transmissions perform in a similar way in that selecting gear ratios is like adjusting the fulcrum of a lever. For example, the motion (RPM) of a small input gear called the driver is reduced when it is in mesh with a larger, slowly moving output gear called the driven. Revolutions per minute are reduced, but twisting force or torque is increased. In the case of the transmission, the energy is transferred through the pushing of the driver gear teeth against the driven gear teeth. Electrical transformers have a ratio that is like the length of the lever arms and the ratio of driver to driven gear teeth in transmissions. However, unlike the lever and gear transmission, the transformer transfers energy from the input to the output without physical contact or moving parts. The transformer moves energy from the input to the output by magnetism in the form of a complete magnetic circuit.

A close look at the top drawing in figure 6-7 reveals that the transformer is composed of two physically independent coils called the primary and secondary. These independent coils are wound close to each other on a strong

magnetic core, or frame. To understand how a transformer functions, it is necessary to keep the electrical, as well as the magnetic circuit in mind.

Recall that a rising, falling, then reversing AC voltage will produce a current that increases in one direction, drops to zero, then reverses and rises to a peak in the opposite direction, and falls back to zero again. If this AC current is flowing in the primary coil of the transformer in figure 6-7, what will the magnetic field surrounding it be like? The magnetic field will first expand and the flow of flux lines will increase to maximum and then return to zero. The north and south poles of the magnetic field will then reverse, and again the flux lines will increase to maximum, then return to zero. This constantly changing magnetic field caused by current in the primary coil has an effect on the secondary coil. Remember that magnetism crossing a coil produces electricity, and hence electricity will be induced into the secondary coil. It could be said that the primary coil produces a changing

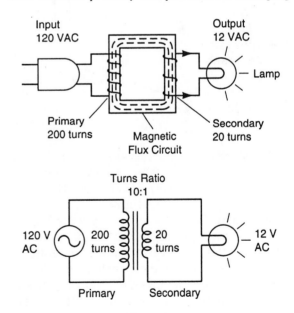

FIGURE 6-7 The transformer.

magnetic field from the AC current within it and the secondary coil converts the changing or moving magnetic field back to electricity. Since the magnetic field is constantly changing in both strength and direction, the current induced in the secondary will be AC—electricity to magnetism to electricity—electrical circuit to magnetic circuit to electrical circuit. This constantly changing alternating current is the source of the magnetic variations; hence AC is required for the transformer to function.

Turns Ratio

Gear transmissions have gear teeth or gear ratios. For example, a gear with 10 teeth spinning a gear with 50 teeth is a ratio of 10 to 50, or it could be simplified to 1 to 5. Likewise, electrical transformers have a similar ratio based on the number of turns on the primary coil compared with the number of turns on the secondary coil. The turns ratio of the transformer shown in figure 6-7 is 200 to 20, or simplified as 10 to 1.

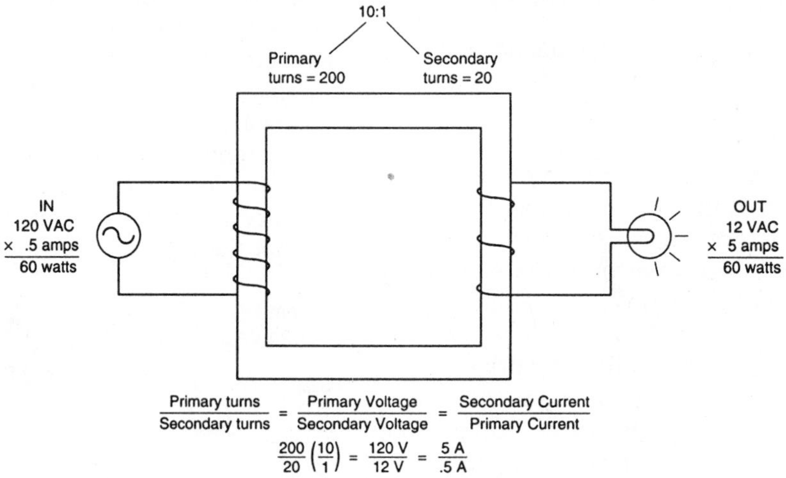

ratio. If the turns ratio varies from 10 to 1, then the 120 volts at the input will drop to 12 volts at the output. The secondary coil has one tenth of the turns of the primary and one tenth of the voltage of the primary. Look at figure 6-8 and take note of the connection between the 10-to-1 turns ratio and 120 volts input to 12 volts output. The lower output voltage seems to be a loss, but that loss is made up for by a gain in amps. The current in a transformer is described as being inversely (opposite) proportional to the turns ratio. In the transformer shown in figure 6-8, the turns ratio is 10 to 1. Since the current is inversely proportional to the turns ratio, it will increase ten times—opposite the voltage decrease of one tenth. The .5 amp in the primary coil will increase to 5 amps in the secondary. Look closely at the formula in figure 6-8. The last section that represents current is upside down, or opposite to the first turns ratio section. Voltage is directly proportional to the turns ratio, whereas current is inversely proportional. The reduction of voltage by one tenth is balanced by an increase in current of ten times. In this way, the transformer, like a lever or gearbox, can reduce volts (force) while increasing amps (movement or flow). Recall again that power in equals power out. The power at the input, 120 volts times .5 amp, or 60 watts, is equal to the power at the output, 12 volts times 5 amps or 60 watts. No gain or loss in power has occurred. Rather, the electrical output of the transformer is a more useful combination of volts and amps.

FIGURE 6-8 Transformer calculations.

Transformer calculations are not complicated since the relationship among volts, amps, and the turns ratio is easily understood. However, during these calculations it should be kept in mind that the watts (amps times volts) at the input of the transformer are, for practical purposes, the same as the watts (amps times volts) at the output of the transformer. No gain in power is possible. Like the lever or gearbox, the transformer can "give up" volts to gain amps or amps to gain volts. Figure 6-8 shows a step-down transformer that drops the voltage. Since the turns ratio is 10 to 1, the voltage will follow this ratio directly. It can be said that the voltage is directly proportional to the turns

$$\frac{\text{Primary turns}}{\text{Secondary turns}} = \frac{\text{Primary Voltage}}{\text{Secondary Voltage}} = \frac{\text{Secondary Current}}{\text{Primary Current}}$$

$$\frac{200}{20}\left(\frac{10}{1}\right) = \frac{120\ V}{12\ V} = \frac{5\ A}{.5\ A}$$

Applications

Transformers are used in many applications throughout the areas of electricity and electronics. Many types of transformers exist. Some are specially designed for a particular purpose, while other, general-purpose types are available with a wide variety of specifications. Ordinarily, each transformer is designed to serve one of a variety of possible purposes. Therefore, what follows is a listing of purposes, with brief discussions of some transformers designed to fulfill those purposes.

Step-down transformers: All of these transformers have fewer turns in the secondary than in the primary. Their main application is to reduce voltage; however, since they also increase current, they may be used to provide high currents. Utility pole transformers drop voltage from about 15,000 VAC (volts AC) to 240 and 120 VAC for residential power. While reducing the voltage to levels that are less dangerous, they also increase the current available. Small step-down transformers that drop 120 VAC to 5-25 VAC are very common. HVAC low-voltage control circuits make extensive use of these transformers, which greatly reduce the dangers of both electric shock and fire. Wiring of these low-voltage control circuits is simplified in that codes are less stringent and techniques less rigorous. Bell transformers are used for these same reasons. A great deal of electronic and electric equipment today is supplied with enlarged plugs that contain small transformers. This reduces the weight and cost of such devices as cordless telephones, NiCd battery chargers, stereo systems, and electrical toys. A further advantage of this type of design is that the device is not subject to the heat that the transformer produces. Some devices that are called by different names are actually step-down transformers. For example, arc welders are primarily big step-down transformers designed to provide the high currents—between 50 and 500 amps—necessary for welding. Even the type of soldering gun that has two copper wires terminating in a heated tip is a step-down transformer. It provides high currents to the tip, which acts as a resistance heater. Often these guns heat poorly due to loose tip clamp screws. Low-voltage/high-current circuits require tight, clean connections.

Step-up transformers: These devices all have more turns in the secondary than in the primary. This results in an increase in voltage and a decrease in current. As a group these transformers require careful handling and respect since they produce voltages that are potentially lethal. In the area of HVAC, the most common example of this type of step-up device is the oil burner or gas power burner ignition transformer. It provides a 10-kv (kilovolt) spark at 23-ma (milliamps) across electrodes to initiate combustion of the oil/air or gas/air mixture in the combustion chamber. The spark either may be terminated after successful ignition or may be continuous during the entire heating sequence. A transformer very much like this is used with neon signs to provide the high voltages necessary for neon to conduct electricity to produce light. Later in this chapter, the extensive use of step-up transformers by the power company will be examined as it relates to economical power distribution. Some of these transformers raise the voltage to .5 million volts. A good example of a step-up transformer is owned by one of the authors. It steps 120 VAC up to 2000 VAC and provides .5 amp. As its output is almost certainly lethal (high voltage and substantial current) and it has no practical use, its present application is as a doorstop. As experienced as we would like to think we are, under no circumstances would we connect a killer like this to power. High voltage is dangerous.

Isolation transformers: One of the biggest hazards for a person working with electricity is accidentally to touch a "hot" wire while being grounded. Of the two current-carrying wires that make up 120-volt residential electrical service, one wire is referred to as the "hot" wire and is color-coded black; the neutral wire is white. These hot and neutral wires power 120-volt circuits. A more complete description of power distribution will be presented in Chapter 7;

however, for now, the hot wire can be thought of as the wire that is hot in reference to ground. This means that if a technician is grounded by a pool of water, appliance chassis, or any grounded device, and touches this hot wire, electrocution is possible. The hot wire is hot to ground. This ground-to-hot reference may be eliminated by a transformer that has an equal number of turns in the primary and secondary and is called an isolation transformer. Recall that a transformer transfers power from the primary to the secondary by a magnetic field. In most transformers, there is no direct electrical connection between the primary coil and the secondary coil. Since the only connection is magnetic, reference to ground is eliminated. The isolation transformer is useful in bench servicing procedures in that it reduces the likelihood of a hot-to-ground shock. All transformers that have no direct connection between the primary and secondary act to remove ground reference. This is useful in electronic controller applications as well.

INDUCTANCE

A 10-foot piece of thin wire connected within a DC circuit offers nothing more than resistance. If the volts and amps are known, the resistance is a simple calculation. If you fully understood the previous section on "Equating AC and DC," then you may realize that the thin 10-foot piece of wire connected to AC can be understood in exactly the same way. If the AC volts and AC amps are known, then resistance can be easily calculated. The idea of resistance, the linear or straight line relationship of force (volts) and flow (amps), is the same in either AC or DC circuits—resistance is resistance. However, if that same piece of wire is wound into a coil, an additional magnetism-related effect comes into play.

The nature of this additional magnetic effect was discussed previously (see Chapter 5 on coils and inductive kick). Review of the DC inductive concept presented in its section will be helpful. The circuit shown in figure 6-9 is a DC source connected through a switch to the

coiled-up 10-foot piece of wire. When the switch is closed, current begins to flow and an expanding magnetic field produces a back voltage (counterelectromotive force, or CEMF). The result, as shown at the lower left in figure 6-9, is that the current, or amps, is held down as long as the magnetic field is expanding. In

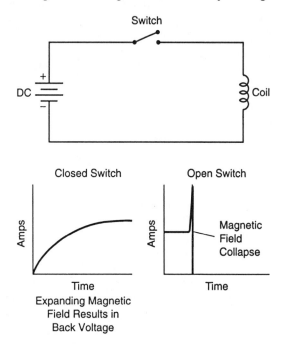

FIGURE 6-9 A coil on DC.

a very short time, the magnetic field fully expands, after which the back voltage drops to zero. After this inductive start-up, the only effect limiting current flow is simple resistance. However, if the current is raised, the magnetic field expands, producing a back voltage. If the current is lowered, the magnetic field contracts, producing an aiding forward voltage that prevents the current from dropping rapidly. The inductive effect opposes or delays changes in current. The faster we try to change the current, the faster the magnetic field expands or collapses, producing still greater back voltages or forward voltages. Recall that the coil stores energy in its magnetic field. For example, if the switch is opened suddenly and quickly, then the magnetic field collapses suddenly. This produces a spike (brief pulse) of high voltage that pushes current through the air gap between the opening switch contacts. This inductive kick produces sparks if any coil is rapidly disconnected. The lower right of figure 6-9 shows the current flow resulting from inductive kick. What has just been explained is the nature or effect of a coil connected to DC. Coils or inductors connected to AC are a complicated extension of this principle.

"Complicated" and "complex" are good words to describe the effects of inductors on AC. Electrical technology is a large, detailed picture, especially the part related to inductors on AC. At first, the detail seems overwhelming and one can easily feel very ignorant. Fitting together the first two pieces of a jigsaw puzzle is really tough. The last few are easy, so keep an open mind, shut off the stereo, and grab as many of these puzzle parts as you can.

The Henry Defined

Figure 6-10 shows the label on a box that contains an inductor coil. Let us get the simple parts of these inductor specifications out of the

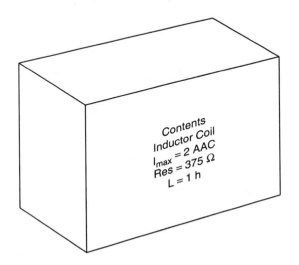

FIGURE 6-10 Inductor specifications.

way. First, Imax or maximum current means that if more than, in this case, 2 amps is allowed to flow through this inductor, it burns

up. The wire thickness and this inductor's ability to get rid of heat will permit a maximum current of 2 amps. It's all right below that and cooks itself above that. Second, the specified resistance of 375 ohms means just what it says. In a DC or AC circuit this inductor will act like a 375 ohm resistor. If it were removed from a circuit and measured with an ohmmeter, the meter would read 375 ohms. All that is left to discuss is the L=1 henry specification.

In Chapter 5, the henry was defined as a unit of measure based on how the inductor reacts to changing currents. Recall that a 1-henry inductor produced 1 volt of back voltage if the current was raised 1 amp in 1 second. Figure 6-11 is the mathematical formula that represents this idea. You will not need to memorize this formula, or even use it for calculations, but it will help you to focus and bring detail to a mental picture of the idea of what a henry is. In figure 6-11, the triangle symbol (delta) is used to represent "change in." For example, if the firing of a pistol cartridge were to be described, it might be said that at time 0 the hammer falls; at delta 10 microseconds, the primer burns; at delta 30 microseconds, the main charge burns, and so on. Referring to the formula in figure 6-11, if a delta T is 1 second and delta I is 1 amp, then 1 over 1 is 1. If the L (inductance concept) on the left is 1 henry (inductance unit), then the

---

Mathematically:

$$L \text{ (in henrys)} = \frac{e}{\Delta i / \Delta t}$$

Where: e = volts of CEMF
Δi = change of current in amperes
Δt = change of time in seconds

FIGURE 6-11 The formula for henrys.

formula says that 1 henry equals E over 1. What number over one equals one? One henry equals 1 Volt of CEMF over 1 amp in 1 second of change. If this were a 2-henry inductor and the current changed 1 amp in 1 second, how many volts of CEMF would be produced? Well, twice as much inductance would provide twice as much back voltage. If you got the idea that more inductance, in henrys, or more rapid change in amps per second produces a greater reaction (more CEMF), you have this puzzle part.

It should now be clear that inductors react to changing currents. In contrast, resistance is not affected by changing currents. Resistance is not reactive whereas inductance is. Resistance is passive and laid back, while inductance reacts. This might be thought of as one friend who just listens while you talk and another who responds. This reactive idea about inductors is called inductive reactance.

The Concept of Inductive Reactance

This puzzle part connects AC electricity to the inductor's nature to react to changing current. Remember the sine wave as it relates to electricity? First, the current rises in one direction to maximum, and then it drops back to zero, only to rise back to maximum in the other direction and again return to zero. Alternating current provides a constantly changing current. This current change is not as simple as 1 amp in 1 second and requires a more complex mathematical expression. The expression depends on a mathematical scheme called angular velocity measured in pi (π) radians. Rather than spend the next ten pages explaining this concept, let us just say that each cycle of the AC sine wave is mathematically represented as 2π.

But wait a minute, 2π describes one sine wave and the power company supplies 60 of them in 1 second. Correct! The frequency of 60 Hz will also need to be considered, since most electrical ideas are based on seconds and 60 sine waves are provided by power companies in the United States (Europe uses 50 Hz). Both the description of one sine wave (2π) and the number of waves in 1 second (frequency) will be required to describe the nature of change of the electricity provided by power companies. Refer to figure 6-12 and find the 2π and f. These two numbers 6.28 (2π) and 60 (fre-

quency) constitute the nature of change of AC electricity. That's why they are placed in the formula for inductive reactance.

Now that the nature of change of AC has been described mathematically by the $2\pi f$, all that is left is to describe how the inductor reacts to change—first the change 2π times f, then the inductor's reaction. Remember the henry, which tells what the inductor does when current is increased or decreased. All that is required to complete the formula for inductive reactance is the henry. Again look at figure 6-12 and ask yourself why the formula contains 2π, f, and L. For the moment, don't worry about the part of the formula to the left of the equals sign. Rather, read the explanations on figure 6-12 of the items to the right of the sign.

About 30 years ago, both of the authors

Inductive Reactance

$$X_L = 2\pi f L$$

Where: 2π = pi radians (describe one sine wave of AC)

f = frequency (number of sine waves in 1 second)

L = induction in henrys (how the inductor reacts to change)

X_L = inductive reactance in ohms (how *this* inductor on sine wave AC of *this* frequency reacts and provides ohms, over and above its simple resistance)

FIGURE 6-12 Inductive reactance.

heard this idea for the first time, and were confused. Our instructors reexplained it and we discussed it between ourselves. Gradually, the idea of inductive reactance became clearer and clearer. Don't be discouraged; few people understand this concept the first time they are exposed to it.

The train of logic on the right side of this equation yields a result that is called inductive reactance. The X_L represents the concept of inductive reactance measured in ohms. X is often used in science to represent new mysterious things or ideas, like X rays. In one sense, the X in this formula represents a new type of ohms. Don't be confused. Ohms are ohms—it's the source of these ohms that's new. Ohms of resistance are due to a restriction in the electrical circuit, while the inductive reactance ohms are caused by magnetic field expansion and contraction around the inductor. Ohms of resistance are present with both DC and AC, but ohms of inductive reactance occur only with AC.

As a brief review let us calculate the X_L inductive reactance of the inductor specified in figure 6-10 and refer to figure 6-12. We multiply 2π (6.28) by f (60) and then by L (1 henry). Use a calculator or a piece of scrap paper.

$$6.28\ (2\pi) \times 60\ (f) \times 1\ (\text{henry}) = 376.8\ \text{ohms}$$

The inductive reactance of a 1-henry coil connected to 60 Hz sine wave AC is 376.8 ohms. If the coil were stretched out into a straight wire, these ohms would disappear. Why?

Now it seems that the inductor from figure 6-10 has two kinds of ohms: the basic resistance of 375 ohms and the 376.8 ohms of inductive reactance. If the total effect of this inductor is to be found so that this idea can be used to represent "real inductors," it would seem that the two types of ohms (resistive and reactive) should be combined. Generally, that is correct; however, it is not that simple.

Real Inductors

Real inductors cause a delay in current. With DC, this delay can actually be measured in seconds, but this is not practical with AC since it is constantly rising and falling. Rather, the idea of phase shift is used. A good example of phase shift is working a night job. Your body tells you it's ready to work at 8 A.M., but the job does not start until 8 P.M. You are out of phase by 12 hours or one half day. You may adjust to this, but inductors don't. Pure inductors (the part that is inductive reactance X_L) delay current so that it flows 90 degrees after the voltage (force) is applied. Because one cycle of the sine wave is produced by one rotation or turn of a simple alternator, it is represented as 360 degrees. From the beginning of the wave through one positive peak, down through zero, then through the negative peak and back up to zero is 360 degrees. This can be likened to 24 hours in one day. If the night worker is out of phase by one half day, then, in a similar way, current in a perfect inductor is out of phase by one quarter of a cycle or 90 degrees.

When the inductor is connected to DC, its current delay can be measured in specific time intervals. However, since AC provides a constantly changing voltage, there is no starting place from which to measure a time delay in current. Rather, it is easier to think of the current in an inductor as playing catch-up with the voltage. The voltage rises to maximum, then 90 degrees later the current rises to maximum. Currents in inductors are always lagging behind the voltage.

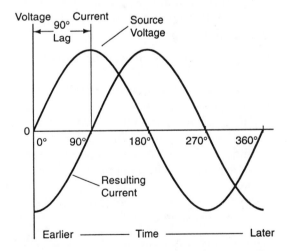

FIGURE 6-13 Current and voltage in a pure inductor.

The graph in figure 6-13 shows that the AC voltage rises first (left, earlier part of the graph) then the current "lags" (comes later) by 90 degrees.

Before all of what has been explained about inductors is put together to represent the real inductor, a brief listing of the separate concept or ideas will be helpful.

1. Inductors are coils of wire that offer ordinary resistance. These ohms of resistance are the same whether the wire is straight or wound into a coil. These ordinary ohms of resistance can be measured with an ohmmeter.

2. Whenever the current that flows through an inductor is raised or lowered, a magnetic field expands or contracts, producing a voltage that opposes the current rise or fall. The inductor reacts.

3. Rapid current termination forces the magnetic field around an inductor to collapse, producing a spike of high voltage and current. The inductive kick is detrimental to electrical contacts, but is useful for spark ignition.

4. The unit of measure of the nature of the inductor to react is the henry. A 1-henry inductor will produce 1 volt of CEMF if the current is raised 1 amp in 1 second.

5. When connected to AC, the inductor reacts to produce additional ohms that are due to inductive reactance. Inductive reactance is based on a mathematical description of the sine wave (2π) times the frequency times the henrys of the inductor. These new inductive reactance (X_L) ohms would disappear if the coil were stretched out to a long wire. Don't mistake these reactive ohms for the wire's resistive ohms.

One of the most useful skills for an HVAC technician to develop is backward reasoning. That skill involves examining a result and trying to reason out what must have happened to get to that end product. This skill is used to make efficient repairs to HVAC systems. Let us

approach real inductors in this way and see what results.

Take a sheet of paper (graph paper is preferable). Examine figure 6-14 closely. Although it looks complicated, start by looking for things that are familiar—things that were discussed in the last few pages.

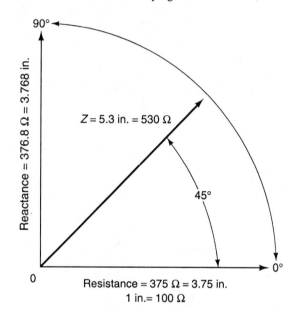

FIGURE 6-14 A real inductor.

Copy the 3.75-inch horizontal line from the bottom of figure 6-14 on the middle of a sheet of paper. This line represents the 375 ohms of ordinary resistance. It is 3.75 inches long to represent 375 ohms on a scale of 1 inch = 100 ohms. This line must be a graphic representation of the ordinary resistance of the wire from which the coil is wound. "O.K.," you might ask, "but why is it horizontal, and further, what is the 0 degrees label next to it for?" Ask yourself, "Does ordinary resistance affect the time relationship of volts and amps or cause any phase shift?" The answer is No. Resistance causes 0 degrees of phase shift. So much for the meaning of the 0 degrees. Note the 0 degrees to the right of the line on your paper.

Now copy the vertical line from figure 6-14 on the paper. It begins at the left end of the resistance line and goes up 3.768 inches. Its length must represent 376.8 ohms, which, you may recall, was the inductive reactance calculated earlier. But above this line that represents inductive reactance is 90 degrees. The last time 90 degrees was cited was in reference to phase shift. A pure inductor causes the current to lag the voltage by 90 degrees. The idea is that since both inductive reactance ohms and the 90-degree phase shift are caused by magnetic field expansion and contraction, then these ohms are also phase shifted. Notice that the line representing inductive reactance ohms is rotated 90 degrees up and to the left. This counterclockwise rotation indicates that, like current, ohms of inductive reactance lag by 90 degrees.

The next step is to make the two lines that have been drawn on your paper into a box. To do this, draw a line straight up from the right end of the resistance line and a line to the right from the upper end of the inductive reactance line. This process is an easy way to find the vector of resistance and inductive reactance. Graphic drawings of vectors are often used to represent two or more forces acting on an object. You already have this idea as a practical concept. For example, if a plane were pointing north but actually moving or flying northwest, it could be assumed that the wind was coming from the east and pushing the plane west. The force of the plane's engine pushes it north while the wind pushes it west. As a result, it travels northwest, a direction that is a combination of these two forces. The actual path of the plane is a vector of the two forces. Though they are not forces, resistance and reactance also can be illustrated by vector combinations.

Drawing the vector is the last step. Draw a line on your paper that connects the lower left corner of the square with the upper right. Measure the line with a ruler. It should be about 5.3 inches long. Recall that the scale is 1 inch = 100 ohms. This line represents 530 ohms. It has to be ohms because it is a combination of ohms of resistance and ohms of inductive reactance. Label the line, as it is in figure 6-14, with a Z. The 45 degrees label next to the line is midway between 0 degrees and 90 degrees.

Let us try to reason out the 45 degrees. A real inductor has both resistance and inductive reactance, so the actual phase lag of current depends on how much resistance and inductive reactance there are. If the coil has a great deal of inductive reactance and very little resistance, the phase lag of current will be close to 90 degrees. If the coil is mostly resistance, then the phase lag is close to 0 degrees, or no phase lag at all. The coil that has been used as an example has about the same amount of resistance and inductive reactance and a resulting phase lag of current of 45 degrees.

At this point, it should be realized that electrical theorists use up the alphabet quickly, so, of those letters left, the Z is rather exotic and mysterious. As this also is true of whatever this last vector line represents, Z seems like a good choice. Z represents impedance. Impedance to traffic flow would include road construction, which, like resistance, restricts flow, as well as a deer, which, like inductive reactance, reacts to moving vehicles by blocking traffic. Ohms of impedance include the collective effect of resistance and reactance. These two types of ohms cannot simply be added as they occur at different times (90 degrees apart).

It is rumored that one of the authors has a 15-inch subwoofer in the car stereo that can blur the image in the outside rearview mirror. This speaker is connected through a crossover coil. The crossover coil's inductive reactance ohms are high at high frequencies and stop the high frequencies from getting to the subwoofer. But low frequencies cause little inductive reactance and the boom passes right through to the speaker.

The previous section on inductance contains many difficult concepts. The more that you understand of this material, the closer you are to becoming a top-notch HVAC technician. Understanding inductance is difficult.

CAPACITANCE

Chapter 5 described how a capacitor, two plates separated by an insulator, can be charged up. If a DC source is connected to a capacitor, current rushes to fill or charge it up. The current at first is high, and then later, as the capacitor charges, it falls off. Another way to describe this is that the capacitor starts out with no voltage and high current. As it charges, the voltage rises closer to the source and the current drops. After a short time, the voltage of the capacitor is equal to the DC source voltage and current flow stops. Viewed from the perspective of the capacitor, current comes first, and then voltage. In this way, capacitors cause a delay in voltage. First current, then voltage—first it fills (amps), then it builds pressure (volts). What might happen if the DC source volts were raised and then lowered?

Suppose a 10-volt battery charged a capacitor to 10 volts. If the 10-volt battery is replaced with a 12-volt battery, more current will flow to the capacitor until it charges to 12 volts. Therefore, if the source voltage is raised, current will flow until the capacitor charges to the new, higher voltage. The capacitor's voltage follows the source, but is delayed due to the flow of charging current. In this way, the capacitor causes a delay in voltage. Do not be confused. The voltage that is delayed is the voltage of the capacitor, not at the source.

Now the 12-volt battery is replaced with the 10 volt battery. Recall that the capacitor has been charged to 12 volts. This time the capacitor pushes (remember, its voltage is higher than the source) current back into the source. Again the voltage drops, first in the source, and then current flows from the capacitor into the source, and it drops from 12 to 10 volts. The capacitor's voltage follows the source, but is delayed by the time it takes to charge, or, in this case, discharge.

The Farad Defined

The reader may recall from Chapter 5 that a 1-farad capacitor was described as being capable of holding one coulomb (6.28 times 10 to 18 electrons) if 1 volt of electrical pressure was placed across its terminals. This simply means that if we squeeze electrons into a 1-farad capacitor with a force of 1 volt, it will hold 1

coulomb. When connected to DC, the capacitor acts like an electron gas storage tank. As the capacitor functions due to electric fields, on DC it might be thought of as an electric field balloon. To get an idea of how the capacitor functions on AC, it is helpful to define this same farad in another way; that is, it is the same farad, but defined from a different perspective.

Since AC is constantly changing and reversing direction, a more useful definition of the farad will be based on change. Again, change will be represented by the delta. Figure 6-15 is the mathematical formula for farads based on change. Look at the C, which represents capacitance (the concept) in farads (the units). Imagine that the C on the left of the formula represents a 1-farad capacitor. Then 1 equals everything on the right side of the equation. The right side is a fraction under a fraction, so let us explain the lower fraction first. If the voltage is changed (delta) 1 volt, over a change in time of 1 second (delta) t then the rate of change is 1 volt per second. Given that rate of change, a 1-farad

---

Mathematically:

$$C = \frac{i}{\Delta v/\Delta t}$$

Where: C is in farads

   i = charging current

   Δv = change in volts supplied by
       the source

   Δt = change in time in seconds

FIGURE 6-15 The formula for farads.

capacitor will have 1 amp of current flowing to it. With a 1-farad capacitor, if we increase the source voltage 1 volt in 1 second, then one amp flows to it. What happens if we decrease the source voltage 1 volt in 1 second? In this case, the capacitor discharges at 1 amp. This definition of the farad simply means that if the source voltage is increasing, the current will flow to a capacitor, and if the source voltage is decreasing, the current will flow from the capacitor. Further, the capacitor's voltage is always playing catch-up to the source voltage.

It should now be clear that capacitors react to changing voltages. Recall that inductors react to changing currents. Therefore, capacitive reactance may be viewed as an effect that is the opposite of inductive reactance. Current and voltage exchange places in the formula for the henry and farad. Compare figures 6-11 and figure 6-15.

The Concept of Capacitive Reactance

The constantly changing AC voltage is always rising or falling relative to the capacitor's voltage. To connect the idea of AC and the capacitor's nature of reacting to changing voltage, it should be recognized that the voltage of the capacitor is always trailing the AC source voltage. As the AC source rises, current flows to the capacitor and the voltage of the capacitor rises. As the AC source voltage falls, current flows from the capacitor to the source, and the voltage of the capacitor falls. This is a game of the capacitor's voltage trying to catch up to the source voltage. In a general way, it could be said that the capacitor "acts" to conduct AC electric current by constantly charging and discharging.

How well a capacitor acts to conduct current depends primarily on the size of the capacitor and the AC frequency. For example, a small capacitor quickly charges to the AC source voltage. This results in a small charging and discharging current. In other words, a small capacitor does not act as if it is conducting as well as a large capacitor does. If the frequency is decreased, the amount of time the capacitor has in which to charge and discharge is longer, so that it will quickly charge and no longer act like it is conducting. A larger capacitor takes longer to charge, so current is constantly flowing. A higher frequency offers less time for the capacitor to charge so current is constantly flowing. Bigger capacitors act to conduct AC better as they never fill up. Higher frequencies seem to pass through capacitors better since

there is not enough time for the capacitor to charge up.

The two ideas discussed above can be used to compare capacitors and inductors. First, inductors cause current to lag or fall behind, while capacitors cause voltage to lag or fall behind. Second, inductors act to conduct less current at high frequencies, while capacitors act to conduct more current at high frequencies. Remember the formula for inductive reactance in figure 6-12? The formula for capacitive reactance in figure 6-16 is made up of parts that are similar to the inductive reactance formula.

On the left of figure 6-16 is X_c, which represents the ohms of capacitive reactance. These ohms are due to the nature of the capacitor to conduct by the process of charging and discharging while it "tracks" the changing AC source voltage. On the right of the formula are 2π (sine wave) times f (number of sine waves in one second) and C (farads, or how the capacitor reacts to voltage changes). The capacitor conducts better if it has more farads (C

$$X_c = \frac{1}{2\pi fC}$$

Where: 2π = pi radians (describe one sine wave)

f = frequency (number of sine waves in 1 second)

C = the capacitor in farads (how the capacitor reacts to change)

X_c = capacitive reactance in ohms (how *this* inductor reacts and provides ohms given these conditions)

$\dfrac{1}{Over}$ = the reciprocal makes $2\pi fC$ vary in the opposite direction

FIGURE 6-16 Capacitive reactance.

in the formula) and if a higher frequency provides less time for it to fill up (f in the formula). As any of these numbers go up, the ohms of reactance (X_c) go down. This is just the opposite of the reaction of the inductor. A 1 is placed over the items on the right side of the equation to reverse the effect they have on the left side. This reciprocal process can be easily understood by example. The number 2 is lower than 4, but their reciprocals (put a 1 over them) are opposite. The number 1/2 is bigger than   1/4; their orders have been reversed. In this way, the effect of X_c capacitive reactance, is the reverse of X_L inductive reactance.

Imagine that you have just removed the case from a window air conditioner and inside you see a capacitor labeled 10 μF and 330 VAC. This capacitor is shown at the top of figure 6-17. As has been done in the past, let us deal with the simple stuff first. The 330 VAC means that if a voltage higher than this is placed across the capacitor, its insulation will short-circuit and it will be destroyed. In some cases, this can cause a dangerous explosion of the capacitor. Some capacitors are polarized + and –; if they are connected backward, a dangerous explosion may also result. Reverse polarity is not a concern with this capacitor since it is labeled VAC (recall that AC is constantly reversing polarity).

The 10 μF is the rating of the capacitor in farads. In this instance the μF (microfarads) label is used. MFD means the same thing (microfarads). Farads are big units and most capacitors are only a tiny fraction of a farad. Micro means millionths. This is a 10 millionths of a farad or 10-μF capacitor. To convert 10 μF into farads, the decimal is moved six places to the left. Thus 10 μF is the same as .000010 farad. Find this value on the formula in figure 6-17. Also note that the values have been filled in for 2 π (6.28) and *f* (60). When these three values are multiplied by each other, the result is .0038. Next, 1 has to be divided by .0038. The capacitive reactance of a 10-μF capacitor at 60 Hz is 265.4 ohms. Since this formula uses a reciprocal, an increase in farads or frequency will result in a drop in ohms of capacitive reactance.

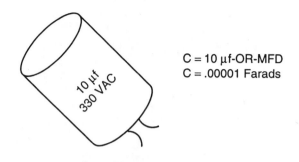

$$C = 10 \text{ μf-OR-MFD}$$
$$C = .00001 \text{ Farads}$$

Capacitive Reactance Formula

$$X_C = \frac{1}{2\pi \, fC}$$

$$X_C = \frac{1}{6.28 \times 60 \times .00001}$$

$$X_C = \frac{1}{.0038}$$

$$X_C = 265.4 \ \Omega$$

FIGURE 6-17 Calculating capacitive reactance.

In a very general way, it could be said that capacitors and inductors are "kind of" reciprocals of each other. Capacitors act to conduct better at high frequencies, while inductors conduct better at low frequencies. Capacitors cause a voltage lag, while inductors cause a current lag. Connected to DC, the inductor passes current limited only by the ordinary resistance of its wire, while the capacitor's insulator permits no current actually to pass through it.

Real Capacitors

Actual capacitors are very like the ideal concept of capacitors described in the previous section. Since capacitors comprise two plates with an insulator between them, no current can actually flow through the capacitor. Though this idea is correct, some capacitors, called "electrolytic capacitors," are manufactured in such a way that they permit a small leakage current between the plates. Leakage current will be discussed in Chapter 9. Ignoring the small leakage current, how much current flows through a capacitor as a result of ordinary resistance? The answer is None. No current actually flows through the capacitor because the resistance of its insulator is infinitely high. In most cases, all that is left to deal with is X_C, or ohms of capacitive reactance.

A drawing representing the effect of the 10-μF capacitor would be a simple line going from zero phase down 90 degrees to show a voltage lag. The line would represent 265.4 ohms of capacitive reactance. But an illustration showing this is not necessary. However, it is necessary to summarize the nature of capacitors.

1. A capacitor connected to DC will draw current for the instant that it is connected, after which the current will flow to or from it only if the source voltage changes. Otherwise, it will act like an insulator.

2. A capacitor may be used to briefly store a charge of electricity.

3. A capacitor connected to AC is constantly charging and discharging, which makes it act as though it is conducting. It acts like a better conductor if it never gets a chance to charge to either polarity of the sine wave. Bigger capacitors and higher frequencies contribute to this ability to act like a conductor.

4. AC causes the capacitor constantly to lag behind in voltage, as it is always charging or discharging and playing catch-up. Capacitors cause a voltage lag of 90 degrees.

Most of the other concepts and ideas in this book are not nearly as difficult as those in the previous section, which you should go back to and review. Also, as you study and gain experience, many of these ideas will fall into place.

POWER AND THE EFFECT OF PHASE SHIFT

Before examining ideas related to power, a brief review of some basic concepts is helpful.

Recall that energy is force times distance or displacement and that it does not take into consideration time or "how long" the energy must be consumed to do work. Energy with time taken into consideration is power. One watt (unit) of electric power (concept) is consumed when 1 volt (unit of force) pushes 1 coulomb (unit of charge) through a circuit in 1 second. Since one coulomb per second is equal to an amp, the more common definition of electric power is: watts equals amps times volts. For example, a 60-watt lamp draws .5 amp at 120 volts. One half amp times 120 volts is 60 watts. No problem! However, phase shift usually caused by inductance or capacitance disturbs this "watts equals amps times volts" relationship. A simple analogy may reveal the nature of this disturbance.

Did you ever forget a friend's birthday? One of those "I'm sorry," cards helps, but not very much. The card and/or gift should have arrived on time, not days later. A good excuse is, "I really care, but other things disturbed me, and I forgot what day it was." You might even throw in some excuse about phase shift. But phase shift is no trumped-up excuse when it comes to electric power.

Three sine waves are presented in figure 6-18. The applied AC voltage is represented by the wave labeled Volts. This wave is the 120 AC voltage available at any electrical outlet. The voltage scale on the left of figure 6-18 specifies the levels of voltage as the wave varies. At any instant of time during the 360 degrees, the "instantaneous voltage" can be found by placing a straight edge at the height of the voltage wave and comparing it with the voltage scale on the left. Current is represented in this figure by the wave labeled Amps. Notice that the current wave begins its rise above zero at 90 degrees after (to the right of) the voltage wave. Inductance accounts for this 90-degree current lag. Again, the instantaneous value can be found by looking to the left and finding the instantaneous value on the amps scale. The wave labeled Watts for watts is a result of amps times volts at any selected instant.

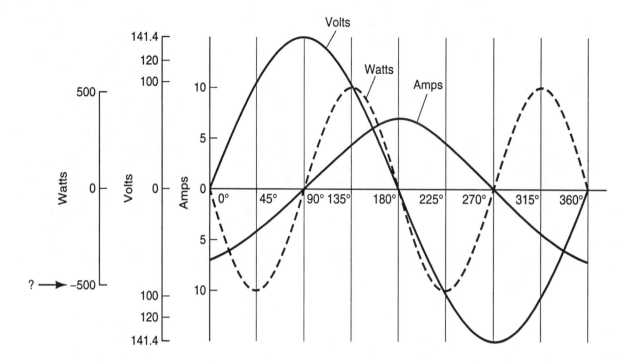

FIGURE 6-18 Power in an ideal inductor.

Let us reason through the wattage wave. Books contain all the ideas of humankind, and those ideas become yours only when you think them through.

Place a straightedge vertically through the 90-degree baseline. At 90 degrees, the voltage is 141.1 and the current is 0 amps. Now think about it—what is the power at 90 degrees? It is 141.1 volts × 0 amps = 0 watts. Notice that the wattage wave at 90 degrees is 0 watts. Check this same idea out at 180 degrees. At that instant, the current is 7 amps and the voltage is 0. Therefore, 7 amps × 0 volts = 0 watts. Wattage at both of the instants of time described (90 and 180 degrees) is zero because either the voltage or current is zero. Zero times anything is zero. Remember the forgotten birthday?

Move your straightedge to 135 degrees. At this instant, the voltage is 100 and the current is 5 amps. At 135 degrees, the power is 500 watts. This makes sense, since 5 amps times 100 volts is 500 Watts. Take a look at 45 degrees.

At 45 degrees, the voltage is 100 and the current is −5 amps. What could this mean? The voltage is pushing one way, yet the current is flowing the other. This is a practical result of the inductor's collapsing magnetic field. If the positive voltage is multiplied by the negative current, the result is a negative wattage. The wattage wave at 45 degrees is below zero and represents a negative wattage. Negative wattage is simply power being returned to the source or line. From a practical perspective, at 45 degrees the inductor's magnetic field is returning power that was stored in it during the previous cycle.

If the power curve is examined carefully, it will be seen that for half of the time, power is being consumed (stored in the inductor), and for half of the time, it is being returned to the source. The total continuous power consumed by this ideal inductor is 0 watts. From a theoretical point of view, this is directly due to the 90-degree phase lag of current. Another way to think of it is that the magnet balloon nature of the inductor is being blown up, and then deflated over and over again.

True and Apparent Power

The authors could have simplified the section on inductors to such a point that anyone would understand it on the first reading. Unfortunately, a true understanding of any idea, especially inductors, is not that easy. In short order, surface generalizations are revealed to be of little value. To learn what things truly are as compared with what they appear to be requires effort. The truth lies below the surface of an idea, particularly with power.

By now it should be clear that power (watts) is equal to amps times volts. It would appear to be simple matter to measure amps, with an ammeter and volts, with a voltmeter, and then multiply one by the other to establish the power in watts. Figure 6-19 shows a circuit connected to ordinary residential power. The load is an ideal inductor. All of the other components are meters. The ammeter indicates that 5 amps of current flow through the inductor. Notice that it is connected in such a way that current must flow through it. Amps can be measured anywhere around the loop of this circuit as long as it was open, and then the measurement completed with the ammeter. The amps must be made to flow through the meter. The ammeter is connected in series. Connected across (in parallel with) the load is a voltmeter, which is indicating 120 VAC of electrical force or push. This voltage could be measured across the source, across the load (inductor) or anywhere across the wires that connect the two together. Since this is a simple circuit, the voltage will be

the same as long as it is measured from the top wire to the bottom.

Apparent Archie takes these two meter readings and calculates that the power is 600 watts. Archie is the same person who got that great deal last week on the $400 used car, until he realized that the cracked windshield would cost $700 to replace. But this time, Archie is confident that the power is 600 watts. Are you? Look at the calculation on the right of figure 6-19. It appears to be correct. However, there is a little crack in the reasoning of this idea. Remember Archie.

Independent readings of amps and volts taken by separate meters do not take phase shift into account. The crack in this case is 90 degrees wide. The astute HVAC technician should expect phase shift whenever the load is inductive or capacitive. This ideal inductor is entirely reactive and causes a 90-degree phase shift. Two separate meters will not reveal this phase shift. Just as Archie should have looked into the cost of the windshield, you need to look into the effect of the 90-degree phase shift.

The wattmeter in figure 6-19 is actually an ammeter and a voltmeter interconnected. Look closely and notice that part of the wattmeter is connected like an ammeter and part of it is connected like a voltmeter. This wattmeter will only read upscale when volts and amps are present at the same time (in phase) and in the same direction (either both positive or both negative). It might be said that the wattmeter is crack sensitive, as it takes into account phase shift.

In this instance, the inductor's 90-degree phase shift results in a wattage that, like

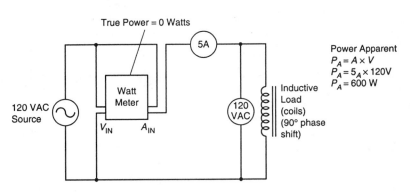

FIGURE 6-19 True vs. apparent power.

Archie's bank account, is zero—0 watts because of a 90-degree phase shift. If the load in this circuit were ordinary resistance like a heating coil, then there would be no phase shift and the apparent power of 600 watts would be indicated by the wattmeter as the true power. Therefore, a phase shift of 90 degrees always results in a true power of 0 watts, while no phase shift results in agreement between apparent and true power—and in between, it is in between.

Look back at figure 6-14, a real inductor, and determine the phase shift. Recall that both inductive reactance and resistance were present, resulting in a phase shift of 45 degrees. The load described by figure 6-14 would have a true power that is less than the apparent power. But how much less?

Trigonometry is an area of mathematics that deals with angles like those in figure 6-14. For now, all that needs to be known is that the cosine can be looked up on a chart of trignometric functions. Figure 6-20 represents a process that can be followed to find the true power once the apparent power and phase angle are known. First, the cosine of the phase angle is looked up on the trignometric function chart, a very brief version of which appears at the bottom of figure 6-20. Second, the cosine, in this instance .707, is multiplied by the apparent power to find the true power. Another, more HVAC-related name for the cosine of the phase angle is power factor.

P_A (apparent power) = amps × volts

P_T (true power) = amps x volts × cos θ
$$P_T = P_A \times \cos θ$$

The cosine of the phase angle may be looked up in a trigonometric table. Below are the cosines of interest.

$$\text{Cos } 0° = 1$$
$$\text{Cos } 45° = .707$$
$$\text{Cos } 90° = 0$$

FIGURE 6-20 Connecting phase shift to power.

Power Factor

Electrical devices such as motors, transformers, and capacitors frequently are not labeled as to the wattage they consume. Rather, the identification plate reads volt-amperes or VA. They are used to distinguish apparent power from watts of true power. Volt-amperes are simply volts times amps, ignoring the phase shift. However, the effect of phase shift is revealed on the identification plate by PF, or power factor. True or real power may be established by multiplying the VAs by the PF.

Generally, motors are made up of coils of wire that produce inductance. In actuality, all that is desired of the motor is rotary power. The inductance is a side effect. Expensive motors have less inductance, less phase shift, and a power factor closer to 1.

POWER TRANSMISSION

Some fundamental ideas about electricity were presented in Chapter 3. Foremost was the idea that electricity is a convenient transporter of power; which, through the application of technology, easily may be converted to heat, light, and motion. With the concepts described in the last few chapters, it is now possible to form a much sharper picture of "how" electric power is transported. All of the hardware with which the HVAC technician works is connected directly or indirectly to electric power.

The focus in this section is on a few clever ideas that provide great savings in the distribution of power—savings that are primarily related to the size and number of power distribution cables and the equipment at both ends.

High Voltage, Low Amperage, and Cost-Effective Wires

Transformers provide no gain in power; however, they do permit conversion, or the exchange of low voltage and high current for high voltage and low current. They can also perform this conversion in the opposite direction. For example, 1 megawatt of power at the input of a transformer could be composed of 500 VAC and 2000 amps. At the transformer's output, the 1 megawatt of power may comprise 500,000 VAC and 2 amps. This step-up transformer has raised the voltage, but dropped the current.

Wire size depends on current. This idea was mentioned in Chapter 3, when the copper part of a wire was described as the hole in an electrical pipe. Recall also that the insulation is the wall of the electrical pipe (see figure 3-9, wire size and wire gauge). The idea in figure 3-9 is that thicker copper wire can carry more current than can thinner wire.

With a technical understanding of the transformer and wire size, it is now possible to focus on the first cost-effective technique of power transmission. A clear restating of the problem is: "What is the least costly way to move electric power?" Cost in this case is related to stringing wires, and power means watts. Don't be fooled into thinking of just volts or amps. The work electricity performs is dependent on watts. It is watts of power that need to be moved. Any combination of volts and amps, within reason, that results, in this case, in 1 megawatt may be considered. Don't forget that wire size depends on current. Think about it.

High voltage and low current enable the use of thin wire. Thin wire is less expensive, is easier to work with, and requires less costly supports. High voltage requires better insulation, however, that problem is most easily overcome by stringing the wires on towers and poles. Edison's DC electricity could not be used because the transformer, which is used to raise or lower the voltage, requires AC to function.

The actual process goes like this: Power plant dynamos (alternators) produce 20 kilovolts, which transformers boost to 138 kilovolts for local area distribution and 345 kilovolts for long-distance distribution. Local area substations use transformers to reduce 138 kilovolts to 15 kilovolts, which is carried by wires on power poles. More transformers are used, either on the power pole for residential service or surface mounted for industrial and business

supply, which drop the voltage down to 240 VAC and 120 VAC. Each time the voltage is dropped, the current is raised. High-voltage/low-current transmission of power yields big savings.

Three-Phase Alternators

Efficient work in any field requires initiative and planning. For example, it would seem wise to do something else while the paint is drying. When this idea is applied to the simple alternator with one magnetic rotor and one-current producing stator coil, it becomes obvious that a large part of the rotor's 360 degrees of rotation does not produce current. There is space around the stator between the single coil windings that could be occupied by more coils. Figure 6-21 illustrates a simplified three-phase alternator that has three stator coils.

As the permanent magnet representing the rotor spins, it produces (induces) current, first in coil one, then in coil two, then in coil three, and then returns again to coil one. In this way, the three-phase alternator provides more electricity in one 360-degree revolution than does the single-coil or single-phase alternator.

Relationship of the Three Phases

Each of the three phases could be used independently of one another as single-phase AC. Indirectly, residential power, which is single phase is produced in this way. An accurate frequency of 60 Hz is provided by spinning the alternator at 3600 RPM: 60 Hz times 60 seconds in a minute. To see how other economies might be gained by three phase, it is important to understand how the phases relate to one another.

Look at how the coils are located within the alternator in figure 6-21. They are spaced at equal distances around the stator frame. Since the frame is a full circle of 360 degrees, the coils are separated by 120 degrees. This spacing results in phase separations of 120 degrees. First, phase one, then 120 degrees, later phase two, then 120 degrees later, phase three, to return after 120 degrees to phase one.

Be very careful not to confuse this phase separation with the 90-degree phase shift related to the reactance of inductors or capacitors. This 120-degree phase separation is due exclusively to the physical 120-degree spacing of the stator coils. To the right of the alternator in figure 6-21 are three sine waves representing the three phases following one another by 120 degrees.

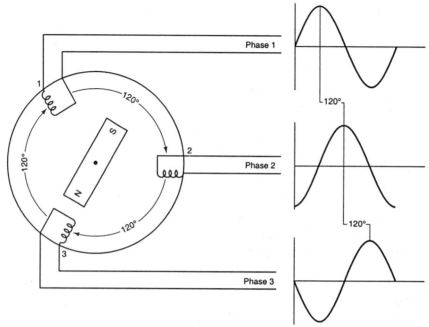

FIGURE 6-21 Three-phase alternator output.

exactly canceled by peak negative voltage. If they were added together, the sum would be zero. Move a straight edge across the graph in figure 6-22 to see that this three-phase sum is true at any position or instant. The three phases separated by 120 degrees always add up to zero.

The point of the above discussion is to recognize that three phases produced by alterna-

More Power Over Fewer Wires

Major technical innovations are often the result of a single insight. For example, cogeneration is based on the idea that heat produced by power generation could be used to heat buildings. In the case of three-phase AC, the innovation depends on looking at all three phases together. Figure 6-22 is an illustration of the sine waves of the three separate phases placed on the same graph. Each phase rises above zero at a point 120 degrees from the other phases. Now for the insight! Look at the vertical line placed over the graph at 75 degrees. The 75 degrees are, for practical purposes, an instant in time. At that instant, what is the total or sum voltage of all three phases? Numbers might confuse the situation, so let us just take a general look. At 75 degrees, phase one is producing a moderate positive voltage, and so is phase two. However, phase three is at its negative peak. If numbers were used, it would be found that two moderate positive voltages are

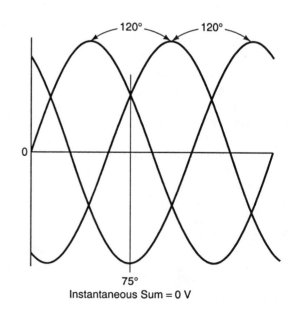

FIGURE 6-22 Phase sum.

tors balance out to zero. This insight can be used to provide a big savings in power distribution. Count the number of wires in figure 6-21 that come from the alternator. There are two wires for each of the three phases, for a total of six wires. It would be very cost-effective if the number of wires could be reduced.

Figure 6-23 shows a way in which the three coils that produce three phases might be connected together. The center connection common to all three coils replaces 3 of the 6 wires in figure 6-21. This can be done because in that wire all three phases cancel one another. This reduces the number of wires that are required to carry three-phase power to four. In instances where the loads on the three phases are balanced, it may not be necessary to carry the fourth common wire. In that case, the power that, without this technical trick, would require six wires, now requires only three. This wye-connected three-phase system provides great savings.

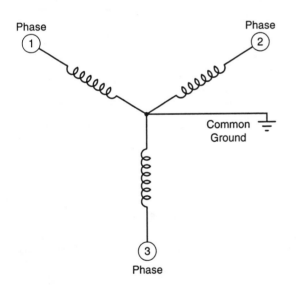

FIGURE 6-23 Wye-connected alternator.

The HVAC technician should also be familiar with a second method of three-phase power distribution called delta connected. Both wye- and delta-connected three-phase systems are widely used, and conversions can be made between them. Figure 6-24 illustrates how the

alternator coils are interconnected with the delta system.

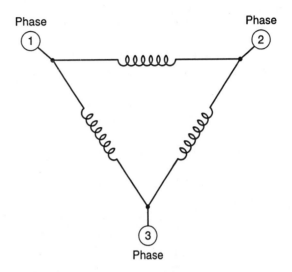

FIGURE 6-24 Delta-connected alternator.

Source to Load with Three Phases

For practical purposes, all electric power is produced as three phase. This may be either wye or delta connected. Transformers, often one for each phase, may readily convert between the two systems, as well as raise or lower the voltage. All sorts of combinations are possible. In Chapter 7, we will examine single- and three-phase loads such as lights and motors; however, at this point, it is necessary to get a general idea of both single- and three-phase loads.

The resistance heater shown in figure 6-25 is delta connected to the three-phase line. Indi-

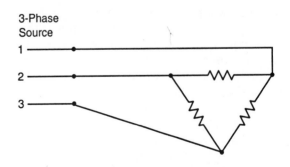

FIGURE 6-25 Three-phase resistance heater.

rectly, each leg of the heater is powered by one phase of the three phases produced by the alternator. A large quantity of power can be carried by only three wires to provide the high BTU's necessary in HVAC applications, as well as in many other industrial processes. Also, the heat produced is smoother and the electromagnetic fields around the heater are reduced by phase cancellation.

With very few exceptions all the electricity used for residential power is single phase. Few problems are encountered in converting one leg of three-phase to single-phase residential electric power. Figure 6-26 is a schematic of a pole transformer. The transformer's primary is connected to one of the three phases. The secondary is designed to produce 240 VAC from end to end. Notice that the secondary is

center tapped and grounded.

Appliances requiring 120 VAC are connected between the center tap, called the neutral and one of the two hot wires. Appliances requiring 240 VAC are connected between the two hot wires. In this way, two voltages are available for appliances that require low or high power. Also, additional safety is incorporated into the design. To get a shock (which often can be lethal) from the 240-VAC supply, both hot wires need to be touched. Ordinarily these wires are not easily accessible without tampering with closed electrical boxes. Shocks of 120 VAC between ground and one hot wire are more likely, and generally are not as severe. However, many people are electrocuted on 120 VAC. More details about how these connections are made will be presented in Chapter 8.

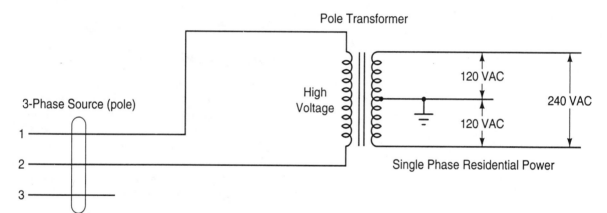

FIGURE 6-26 Single phase from three phase.

SUMMARY

The first topic discussed in this chapter was rotating generators. Spinning a magnet within a set of coils provides a practical way to produce large amounts of electric power. The spinning magnet could be permanent for low-power applications, or an electromagnet may be used if higher power is required. These devices, called alternators, produce AC electricity, which constantly reverses direction.

Sine waves of electricity are a result of the rotary action of the alternator, which is like a pendulum swinging between two opposite peaks.

Electrical sine waves are describe by the number of waves that occur in 1 second (frequency), the time it takes for one cycle (period), the maximum value in either direction (peak), and the effective or RMS value. All meters measure effective or RMS volts and amps, so AC electricity can be compared with DC on the basis of the work that it will perform.

Alternating Current permits the use of transformers, which are like transmissions or gearboxes. A ratio of the turns of wire in the input coil, called the primary, to the output coil, called the secondary, is directly proportional to the voltage and inversely proportional to the current. Transformers are used to provide high voltage for such HVAC applications as oil furnace ignition and air filtration devices. Step-down transformers are used to provide safe control level voltages with which it is easier to work.

Transformers, motors, relays, and many other components used in HVAC are made up of coils. The magnetic field of a coil stores energy, which produces, upon current termination, an inductive kick. Coils have resistance, but when powered by AC, also exhibit an effect due to magnetism that is called inductance. The unit of inductance, the henry, can be used to calculate ohms of inductive reactance. Inductive reactance ohms exist in addition to ordinary resistance, however, they cannot simply be added to ohms of resistance. Pure inductance causes a phase shift or current lag of 90 degrees, which also causes the ohms of inductive reactance to take effect 90 degrees after the ohms of resistance. The addition of these two types of ohms requires the use of a graphic technique called vectors. The total effect of a real inductor that has equal amounts of resistance and inductive reactance is a phase shift of 45 degrees (midway between 0 and 90 degrees). The total number of ohms, including resistive and reactive, is less than the sum of these two due to the phase shift. The reader is cautioned not to expect a full understanding of inductance without a great deal of study and experience.

Electricity comprises a magnetic field, as well as an electric field. The inductor depends on magnetic nature, while the capacitor depends on electric field. Two plates separated by an insulator can store an electric charge by one plate losing electrons and the other gaining electrons. A charged capacitor has a capacity to store electricity. The unit of capacitance, called the farad, is based on how much electricity the capacitor can store, as well as how the capacitor reacts to changes in the voltage source. Since an AC voltage source is constantly changing, a capacitor connected to it is constantly charging and discharging. In this way, the capacitor acts as though it is conducting AC electricity. The capacitor's voltage is always trailing behind the source voltage by 90 degrees. Capacitive reactance causes a voltage lag of 90 degrees. Ordinary capacitors are similar to ideal or pure capacitors in that no current is conducted through the capacitor by resistance. Rather, the primary effect is capacitive reactance.

Power is very much affected by phase shift. When the voltage and current sine waves are out of phase, the instantaneous time relationship of the two is disturbed. Therefore, the true power consumed by a reactive load is less than the simple product of amps times volts (apparent power). Volt-amperes, or VAs, represent apparent power and are usually stated when phase shift is present. Power factor (PF) is a conversion factor that may be used to convert apparent power to true power. Power factor is derived from the angle of phase shift.

Electric power transmission systems incorporate two ideas that result in reduced cost and increased flexibility. First, by increasing voltage and reducing current, smaller wire can be made to carry large amounts of electric power. Second, three-phase alternators provide more power over fewer wires, again reducing costs. Furthermore, three-phase distribution systems furnish both wye- and delta-connected loads with the high power necessary for large HVAC installations, as well as industrial applications. Single-phase power for residential services is easily derived from three-phase distributions systems.

PROBLEM-SOLVING ACTIVITIES

Review Questions and Assignments

1. Draw a simple diagram of the DC generator and the alternator. Identify and label the fixed and rotating parts of each. Describe the role that each of these parts plays in producing electricity.

2. Define the sine wave and make a list of things that produce this natural wave.

3. Why are AC volts and amps measured as effective or RMS values rather than peak values?

4. Construct a chart that compares a gearbox that has a driving gear with 100 teeth and a driven gear with 10 teeth with a transformer with a 10-to-1 turns ratio. On the chart, show RPM, torque, amps and volts.

5. Copy the formulas for henrys and farads, and then describe the cause for each and the resulting effect for each.

6. Why would a relay coil draw more current on DC than on AC?

7. What two effects are responsible for ohms in a coil of wire?

8. The phase shift in one coil of wire is 80 degrees while that in another is 20 degrees. Explain this.

9. Describe what happens when a capacitor is connected to a battery.

10. What conditions are necessary for a capacitor to be constantly charging and discharging?

11. Describe the connection between apparent power and phase shift. Use diagrams if necessary.

12. Why is there no difference between apparent and true power with DC circuits?

13. Where does power factor come from?

14. How can thin wire be made to carry more power?

15. Why are three-phase power generation and distribution used?

Calculation Problems

1. Calculate the inductive reactance (X_L) of a 30-henry coil connected to 60 Hz. Refer to figure 6-12 for the formula.

2. The coil in problem 1 has 3000 ohms of resistance. Follow the vector process described in this chapter and illustrated in figure 6-14 to establish the phase angle and the impedance (Z).

3. In figure 6-17, the capacitive reactance (X_c) of a 10-µF capacitor connected to 60 Hz is calculated to be 265.4 ohms. What would it be if the capacitor were 20 µF?

4. What is the true power of a motor that is labeled 1200 VA with a PF of .7?

CHAPTER 7

Single- and Three-Phase
AC Loads

In this chapter, the mental picture of electricity will be sharpened as it relates to loads. Electrical loads convert watts of electric power into other desirable and useful effects. Some of these effects, like resistance heating and motors, apply directly to HVAC. From an electrical standpoint, lighting is only indirectly related, however, the small part of this chapter devoted to lighting presents some important electrical ideas. Furthermore, the nature of various illumination sources has an important effect on room heating and cooling. The largest portion of what follows is about motors. Practically speaking, all motors make use of the magnetic effects of electricity to provide rotating motion. This look at how motors work presents a large selection of devices to the technician that serve as examples of technology's dependence on the magnetic nature of electricity. A more concrete understanding of three-phase power will be developed through discussions of how motors are designed to function on this power distribution system.

SINGLE-PHASE LOADS

Residential dwellings, small businesses, and industries depend on single-phase systems to provide their necessary power. Three-phase electrical service is furnished only when large amounts of heat, light, or motion are required. Single-phase resistance heaters find wide application not only in HVAC, but also in industrial processes. Electric heating is both clean and convenient.

Heaters

Electricity flowing through a resistance causes heat. It is helpful to think of resistance as a restriction in an electrical circuit. Any time this flow is restricted, friction develops and heat is the result. Thus, restricted electrical flow produces intended heat, as well as unintended heat. Unintended heat destroys components, wastes power, and starts fires.

High-current devices require clean, tight connections. Connections, switch contacts, relay points, and motor brushes in time may wear out or become loose. This causes resistance, which produces heat that generally accelerates the problem. Furthermore, the use of thin or damaged wire introduces unintended resistance into a circuit, which results in heat. The HVAC technician may spend as much time troubleshooting unintended resistance heaters as intended ones. Also, the acceptable but undesired heat produced by many electric devices such as motors is most often a result of current flowing through resistance.

The diagram in figure 7-1 shows a 240-VAC single-phase electric heater. For clarity, the heater frame and electrical box ground circuit have not been included. Take note of the symbols used to represent the resistance heater and the thermostat. For reference, other commonly used electrical symbols are presented in Appendix E. Within this simple circuit, current flow is controlled by the thermostat, which remains closed until a predetermined setting is reached. Ordinarily, each room has its own baseboard heater and a thermostat that allows for individual room temperature control.

Resistance heating devices are commonly constructed from nickel-chromium (NiCr) wire

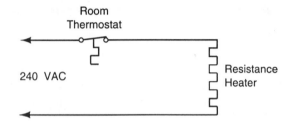

FIGURE 7-1 Electric heater.

that has much more resistance per foot than copper wire. Generally, these NiCr wire elements are insulated with a ceramic material and encased in finned tubes. Some baseboard heaters contain oil to increase the thermal mass.

Electric heat is 100% efficient. For practical purposes, all the electrical watts consumed are converted to heat. However, this does not mean that electric heat is cost-effective. Heating requires a great deal of electric power. For example, one 2400-watt heater uses the same electricity as six half-horsepower motors or 24 100-watt lamps. An electrically heated house may require many heaters. For this reason, such homes need entrance service panels with ampacities, or amp capacities, of 200 amps.

The necessity for additional electric power and individual heater circuits, as well as the required electrical licenses, leaves the installation of electric heating systems to electrical contractors. Furthermore, since very efficient building insulation is required, electric heat is usually installed during the initial construction of the dwelling. If the local electric rates are low and are likely to stay low, electric heat is a practical option.

Lights

There are a variety of devices that convert electricity to light. Each lighting system has particular qualities that define its appropriate use. However, common to all of these systems are a few basic circuits. We will examine these lighting circuits and then look briefly at the different sources of light.

Figure 7-2 shows two common methods used to wire overhead fixtures. The top diagram is a circuit in which power is fed to the switch box, where the hot black wire is controlled by the

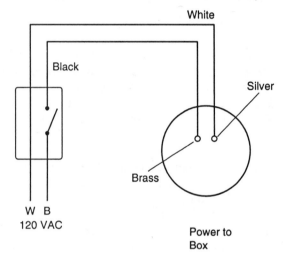

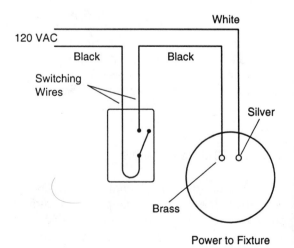

FIGURE 7-2 Overhead fixture.

switch and then carried to the fixture. Many circuits used in HVAC are similar to this one. When overhead lights are wired, power is often fed to the fixture as in the lower part of figure 7-2. In this case, wire is run from the fixture to the switch box in order to control the circuit. This type of circuit is used in situations where the switching device is located at a distance from the appliance. Though central heating thermostats are usually low voltage, they are connected in a way similar to that of the second type of overhead fixture.

Three-way switches are often used in situations where it is desirable to control a light from two separate locations, for example, at the top and bottom of a stairway. Figure 7-3 shows this type of circuit. It is completed only when both switches are up or down together. If either switch is in an opposite position to the other, the circuit is interrupted. Thus each switch has separate control over the lamp.

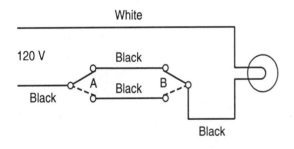

FIGURE 7-3 Three-way switch.

Overhead fixtures often use incandescent lamps. This type of lamp is actually a high-temperature resistance heater. Inside the lamp, a fine tungsten wire offers high resistance. When electricity is passed through the wire, it heats to a few thousand degrees. At this temperature, visible light is radiated. Incandescent light has a yellow-orange tint, which is referred to as warm. Pictures taken in this light have enhanced red colors. Many people find the warm color of this light pleasant. Generally, the longer-life lamps run with lower filament temperature and supply less light, while the high-output lamps provide whiter light, but have

a shorter operating life. Special quartz halogen incandescent lamps burn at such high temperatures that the bulb is made from quartz and filled with a gas that conducts heat away from the filament. This lamp provides brilliant white light. Unfortunately, all incandescent lamps produce heat as well as light.

One customer complained that a recently installed central air conditioner did not cool the dining room. The problem was due to a combination of kitchen heat and a 10-lamp candelabra fixture. Ten 60-watt incandescent lamps produce a great deal of heat, as well as light. Improved exhaustion of cooking heat and duct flow adjustments helped a great deal; however, when the fixture is off, the dining room is a bit too cool. Incandescent lighting converts a portion of the electric power into heat.

Fluorescent lighting is classified as cool light. It converts most of the power it consumes into light. Since the white light it provides is composed of less red-yellow color (the low-frequency light spectrum) and more green-blue (the high-frequency light spectrum), it may also be described as cool. The nature of this light makes it desirable when good vision is the primary concern. Offices, workshops, halls, and stairways seem brighter and larger with fluorescent lighting. However, where mood is a concern—for example, in dining rooms, living rooms, and bedrooms—the warm nature of incandescent light is often worth its inefficiencies. Fluorescent lighting may make a dining room look like an operating room. Its use, however, is a matter of taste.

The way in which fluorescent lights function depends on an electrical characteristic that's true of many gases and vapors. Neon, argon, krypton, and xenon, as well as vaporized mercury and sodium, can be made to conduct electricity if the voltage is high enough to knock outer-shell electrons free. Current is carried through these gases by ionization of the gas. Light is produced when the electrons return to their outer shells and release energy in the form of light. The voltage required to do this depends on the nature of the gas or vapor, the length of the tube containing the substance, and the temperature. Long neon signs require high-voltage transformers. Fluorescent lights in cold garages tend to blink until they warm up.

Ordinary fluorescent lamps use mixtures of substances in the glass tube, but the old standard is mercury vapor. Ionized mercury vapor produces invisible ultraviolet light, which, in ordinary fluorescent lights, is converted to visible light by a phosphor coating on the inside of the tube. Fluorescent lamps without this coating produce black light.

The conditions necessary for mercury to vaporize, as well as the voltage needed to ionize it, are provided by the circuit in figure

7-4. When the switch is first turned on, electricity flows through the switch, through the heater filament at one end of the lamp, through the closed contacts of the bimetal switch, and through the filament at the other end of the lamp, and returns to the power source by way of the bimetal heater and ballast coil. During this start-up time, the heaters in the lamp warm the mercury vapor and the ballast coil stores energy in its magnetic field. Also, the small heater located near the bimetal switch heats the switch, so that within a short time the switch contacts open.

When the contacts open, the circuit current stops, which collapses the magnetic field around

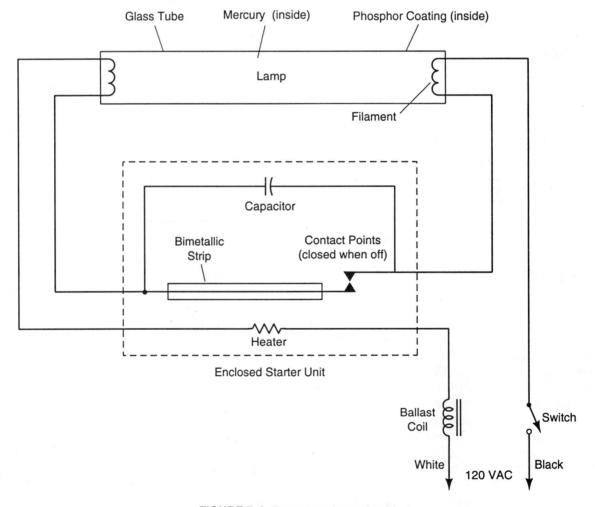

FIGURE 7-4 Flourescent lamp circuit.

the ballast, resulting in an inductive kick. High voltage produced by the inductive kick ionizes the heated mercury vapor and the fluorescent tube conducts electricity. Due to the opening of the contact points, the circuit acts very differently. Current now flows from the switch to the tube heater, not through the heater (as one end of each lamp heater has been disconnected owing to the open contacts), but through the mercury vapor to the other deenergized heater. The other heater now acts simply as an electrode to connect the ionized mercury vapor. Finally, current returns to the source through the bimetal heater and then the ballast coil. If you had trouble following this explanation, try redrawing the circuit in the start-up mode and then in the lighting mode. Quickly redrawing circuits often helps to clarify what happens when relays, switches, thermostats, and so on, open and close.

Where intense light is required, as for example, in industrial plants with high ceilings and in outdoor areas such as roads and parking lots, vapor lamps are used. Sodium vapor lamps provide three to four times the light of the same-wattage incandescent lamps. They produce the yellowish light that is characteristic of highway illuminating fixtures. These types of lamps employ a heater and ionization circuit that provide start-up characteristics similar to those of fluorescent lights. Aborted start-up in either of these lamps results in blinking. Mercury vapor lamps are also used where the addition of other gases can furnish a bluish-white light rather than sodium's yellow light. Some mercury vapor lamps use an electric arc to provide heat and ionization for start-up.

THREE-PHASE HEATERS AND LIGHTS

Most heaters and lights are inherently single phase. A single path of current flow heats a wire of high resistance to provide heat at lower temperatures and light at higher temperatures . A single current path through ionized gas furnishes light. When the power source is three phase, some arrangement is made to tap off single phase or to use three single-phase devices, each connected to

one of the three phases. Many motors are, by design, dependent on three phases in order to function. Three-phase motors will be discussed later is this chapter.

Balance is a key idea that needs to be considered when electrical loads are a mix of single- and three-phase devices. The proportion of that mix often defines whether wye or delta local distribution is used. Either system may be used efficiently as long as all three of the phases are loaded equally. Figure 7-5 illustrates this idea. The three boxes in each of the two circuits represent loads. As long as the loads are equal, all of the three feed wires carry the same current. Current imbalances would mean that some wires are carrying more current than they should, or that some wires are underutilized. Therefore, load balance is desirable.

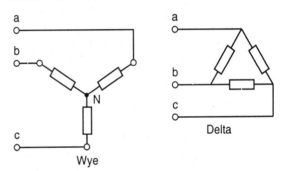

FIGURE 7-5 Three-phase load balancing.

In an institution such as hospital or an office building, a large portion of the power will be used to provide lighting and other single-phase devices that most often require 120 VAC, whereas large HVAC appliances and motors require three-phase power. Figure 7-6 shows a three-phase wye local supply arrangement that enables three individual single-phase circuits to be connected between the center of the wye and one of the phases. Equipment requiring three phases is connected to the outside three phases of the wye. Since each of the phases to the center neutral wire provides 120 VAC and the phases are 120 degrees apart, they will not add up to 240 VAC. The vectorial addition of two 120-volt legs of the wye at 120 degrees of phase separation provides 208 VAC single phase. Two of the legs of the wye may be used for 208-VAC single phase. Wye-connected three-phase system appliances should be selected that require 208-VAC three-phase power.

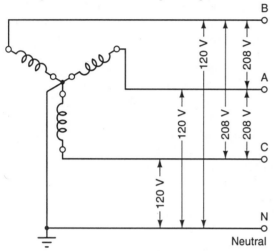

120 VAC Single Phase
208 VAC 3 – Phase
208 VAC Single Phase

FIGURE 7-6 Three-phase wye connections.

Industrial and commercial buildings that use power primarily for large motors and heaters often are connected to three-phase delta local systems. In this instance, a small portion of the power is used for single-phase 120-VAC loads. Single-phase 240-VAC devices may be connected across any two of the three legs shown

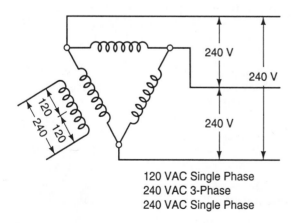

120 VAC Single Phase
240 VAC 3-Phase
240 VAC Single Phase

FIGURE 7-7 Three-phase delta connections.

in figure 7-7. Three-phase 240-VAC equipment is connected to all three legs. Single-phase 120-VAC equipment requires that the source transformers be center tapped to provide a neutral wire. As with residential pole transformers, 120 VAC is then available between the center tap and either end of that transformer. Many important elements are considered when the choice is made between wye and delta—both are common.

MOTOR CONCEPTS

Electric motors are confusing. To the uninformed, it is a mystery why they rotate at all. Furthermore, there are many different types of motors that look very different from one another and require different power sources. Some 40 years ago, one of the authors asked his father, "How does a motor work?" The answer, which involved a project that went on for some weeks, illustrated the most general idea about motors. They all spin as a result of the interaction of a stationary magnetic field with one that is free to turn.

Let us begin the process of understanding motors, as that boy did, with a handful of big nails and a few feet of enamel-covered thin copper wire (22–26 gauge). The reader can, by trial and error, actually build this motor, or simply build a mental picture of it while the writer builds it from a 40-year-old memory.

To begin with, two large nails need to be bent into an L shape, and then taped securely together to form a U shape. The nail heads should be the top of the U. This U may also be constructed from a piece of band iron. At this point, about 400 turns of copper wire are wound around the tape at the base of the U. Wind the coil around in one direction, going back and forth, forming layer on top of layer. Figure 7-8 shows the coil, just constructed, mounted on a small wooden base. It can be fastened to the base by any means that holds the U upright. Large wire staples were used on the original, and hot glue or silicon rubber would also work. Let us look at what we now have.

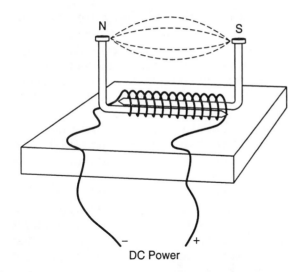

FIGURE 7-8 The motor field.

When connected to a DC power source, this part of the motor produces a stationary magnetic field, and hence it is referred to as the field. In this case, each nail head becomes a magnetic pole. Magnetic lines flow through the U-shaped nail frame and between the poles to complete the magnetic circuit. Almost all motors that have brushes would have a part that provides a fixed magnetic field. The heavy frame of an automobile starter motor is a much improved version of this nail frame. Many small motors, like those used in rechargeable screw drivers and toys, use permanent magnets for the field.

Returning to our construction project, select two nails that just fit between the field poles. Position them across a third nail, one on top and one below. Place the heads of the two nails at opposite ends. The third nail will act as a shaft, so be certain to position the two nails at right angles and centered across the shaft. Some fancy tape work was required 40 years ago, but you can use either hot or epoxy glue. Try to get the two nails parallel with the tape and/or glue. Next, wind a coil of about 300 turns of the enamel wire around the two nails. When winding the coil, always make sure that the turns are in the same direction, building layer upon layer, leaving a gap where the coil

crosses over the shaft. The loose wire ends of the coil should be insulated with tape and fastened to the shaft, again with tape. This part, the nail shaft with the nail arms, should look ssimilar to the one shown in figure 7-9.

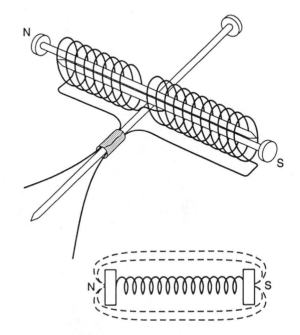

FIGURE 7-9 The motor armature.

This part of the motor, called the armature, is an electromagnet, just like the field. When it is connected to DC power, magnetic lines of force flow through the nails and through space from pole to pole. To see how this part works, it is necessary to locate it within the field in such a way that it is free to turn.

Bearings and supports may be constructed in any number of ways, however, one easy way is to hammer two nails next to each other into the wooden base. The nail heads act as the upper part of the bearing, while the lower part is simply a few turns of wire held in place by either glue or tape. Each end of the armature should be supported in this way. The tricky part of this step is to position the armature so that its ends turn as closely as possible to the field poles while clearing the field coil. Also, the bearings should spin freely. Tap the support nails around until the above conditions are met. Figure 7-10 shows this stage of construction. To prevent the nail shaft from sliding back and forth, put a small washer over the pointed end of the shaft and back it up with tape.

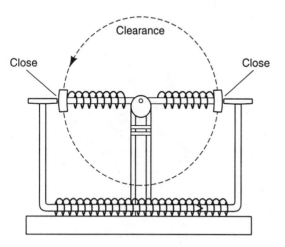

FIGURE 7-10 Armature support.

In order to see what is needed to complete this motor, let us try a few basic experiments. To do these experiments, a source of electricity is needed—either batteries or a low-voltage DC supply (6–12 volts). Use abrasive paper to clean the enamel from the last inch of the wire ends. Connect both the two wires coming from the field and the two wires coming from the armature to the source. Note that the coils may get hot, depending on battery voltage, wire size, and coil turns.

Interaction of Stationary and Moving Parts

With the coils powered up, slowly turn the armature and note what happens when it gets close to the field poles. When the armature is positioned near the field, it will attract, and when turned one half turn, it will repel. These magnetic forces of attraction and repulsion are what cause a motor to turn. When the poles of the armature are opposite the poles of the field, they provide a force that attracts the arms to the field. When the poles are the same, they

repel and force the arms away from the field. Disconnect the armature from power and reconnect it in the opposite direction. Take note that the effect is exactly the same, except that the armature is physically reversed. Reversing the current reverses the magnetic polarity of the armature.

Stop and think for just a minute. This motor could now be positioned in such a way that the magnetic interaction of the field and armature could force them away from each other, to make a quarter of a turn, and then to be attracted to each other through the next quarter turn. At this point, the armature now stops, since it is attracted to the field. How could that be changed? Recall that reversing the current reverses the magnetic poles.

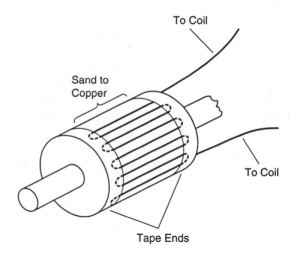

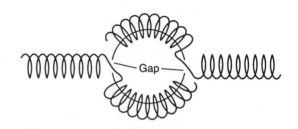

FIGURE 7-11 The motor commutator.

A rotary switch is required. It needs to change the electrical polarity of the armature every half turn. Let us build one. Increase the size of the shaft on the free wire end of the armature by carefully wrapping tape around it until it is about one half inch in diameter. Then shorten the wires from the coil until they are about 3 inches longer than the distance from the coil to the taped shaft. Clean the enamel from the last 3 inches. Refer to figure 7-11 for the next step.

Fold the last 3 inches of the wire back and forth to form a close zigzag that is as long as the tape is wide on the shaft. You should now position the folded wire ends on opposite sides of the tape-thickened shaft. It is important that the wire ends align with, or are on the same sides as, the arms of the armature. Then put narrow strips of tape around the ends of the zigzag wire to hold them in place. This completes the rotating part of the rotary polarity reversal switch called the commutator.

Getting Electric Power to the Moving Part

Finally, the last construction step! Get two pieces of thicker copper wire (14–18 gauge), each about 8 inches long. About 3 inches from the end of each wire, wrap each one around a small round head wood screw. Drive the screws into the base along side the commutator and

bend the wires straight up so that they rest against the commutator sides. For details, refer to figure 7-12. Adjust the copper wires so that they "spring" against the commutator. Practical motors use graphite brushes held against a segmented copper commutator by springs. In this way, electricity is connected to the armature so that it reverses polarity every half turn.

Testing the motor and adjusting it may require quite a number of trial runs. It is, therefore, necessary to connect the motor circuits in such a way as to permit easy connection to power. Figure 7-13 is a schematic of the motor field and armature connected to the power source through a switch. Follow this circuit as if it were a road map for wires. When it is completed, both the armature and field are connected to power through the switch. Make the connections clean and tight.

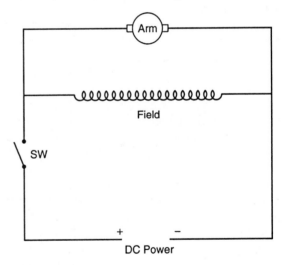

FIGURE 7-13 The motor circuit.

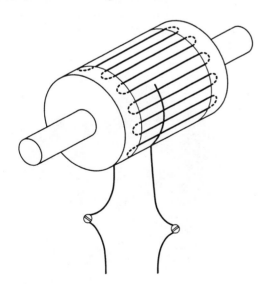

FIGURE 7-12 Brush details.

Turn the power on and give the motor a spin; try both directions. At this point, the motor should show some interaction between the field and armature, some magnetic attraction and repulsion. If it does not, use a small nail to see if the field is magnetic when energized. Also, look to make sure that the commutator and brushes are functioning, which will be indicated by sparking as the armature is spun. Lights may have to be dimmed to see sparking, especially if the source of power is low. If the motor "just about" spins, try a little more voltage, but be careful not to overheat the coils or go above 15 VDC. A very small amount of grease on the bearings will also help. If the motor still does not run, the problem may be brush timing.

The point within the rotation of the motor at which the armature's electrical polarity is reversed is important. Torque (rotating force) is produced only when the arms are near the field, first attracted to the field poles, and then repelled as the arm passes the field poles. At this place in the motor's rotation, force is produced; then for the rest of its rotation, it coasts due to momentum. The switching point in this sequence is important so that enough force provides the momentum required to continue rotation until the armature again approaches the field.

With this motor, brush position may be changed by bending one of the brush wires over the commutator top so that it contacts the commutator at a higher point. The other brush wire is bent so that it contacts the commutator at a lower point. A long process of adjustment, first in one direction, and then in the other, may be necessary. Many windshield-wiper motors have two brush positions to provide low and high speed. Each time the brush positions are changed, try the motor by applying power and giving it a spin. If you built rather than imagined this motor and it still does not work, seek assistance from your instructor.

Properly adjusted, this motor will run in one direction. Try to reverse its direction by reversing the power source. It still spins in the same way because the magnetic polarities of both the armature and field are reversed. These two changes cancel each other. Reversing the motor requires that the field wires be disconnected and swapped. Either the field or the armature may be reversed, but not both.

Improving the Simple Motor

The motor just constructed produces little torque (twisting force), uneven power, short life,

and a poor weight-to-power ratio. Changes may be made based on three general ideas. First, all that has been done could be done better. Better, more permeable steel coil cores, better bearings, closer tolerances, and a machined

FIGURE 7-14 Multiarmature coil motor.

commutator with low-friction carbon brushes and coils wound to provide greater magnetism all will afford major improvements in the motor. Second, the magnetic interaction may be enhanced by elongating the field and armature coils. This is why most motors have long cases. The third and last idea is that more coils may be added to this motor. For example, electric drills have a single pair of field coils and multiple armature coils. One set of brushes supplies power to the armature coil that is adjacent to the field at any instant. Figure 7-14 shows a motor of this type. In this way, torque is provided throughout the entire rotation of the motor.

FIGURE 7-15 Automobile starter motor.

Two field coils, two armature coils, and two sets of brushes may also be used to double the motor torque. This is like two simple motors wound on the same shaft and rotated 90 degrees from each other. Most automobile starter motors have two sets of brushes to energize two armature coils interacting with two field coils. Figure 7-15 shows an auto starter motor that in this way produces large torque in a small package.

A Basic DC Permanent Magnet Motor

Many small motors, generally those of less than 1 horsepower, do not use electromagnets to provide the stationary magnetic field. Rather, they depend on, in order of increasing strength and cost, Alnico, ceramic, or rare earth cobalt permanent magnets for this part of the motor. Several advantages result in this change of the motor's field. Motor direction reversal is simple. As the field polarity is fixed, due to its permanent nature, reversing the direction is achieved by switching the power connections, which reverses only the armature magnetic field. Since one less coil is used, heat due to resistance is also reduced. Also, efficiency is increased since less electric power is consumed. Many battery-powered toys and hand tools use this type of motor. It also finds application in the 12-volt automotive area where it may be used to power wipers, windows, seats, and antennas. A specialized version of this permanent magnet field motor is often used in computer-controlled robots and manufacturing equipment. This special motor has an armature that is designed to be low in weight so that it produces little inertia, enabling rapid acceleration and deceleration of the motor.

Generally, these motors are more expensive, larger, and heavier than their electromagnetic field counterparts. Another, more subtle disadvantage is that the permanent magnet field cannot be controlled by varying field current flow. The fixed magnetic polarity nature of this motor's field requires that it be powered exclusively by DC. If AC is accidentally connected to this motor, it may demagnetize the field and destroy it.

Series and Shunt Motors

Electric motors with armature coils and field coils may be connected internally in a number of different ways. How these coils are connected affects the characteristics of the motor. Speed (RPM) and torque (rotating force) are the most important of these characteristics.

Series-wound brush-type motors have the armature and field connected in a string so that current flows first through one, then the other. This arrangement is shown in figure 7-16. Note how the power is connected to this circuit. In this and all series circuits, there is only one path in which current can flow. For practical purposes, that means that the current that flows through the armature is the same current that flows through the field. To understand how this affects the motor, it is necessary to understand the connection between motors and generators.

The generator and alternator were explained in Chapter 6. Figure 6-1 presented the idea that to produce electricity, a coil can be rotated within a magnetic field. The motors we have discussed so far all have a coil spinning within a magnetic field. Even the names of the DC generator and motor parts are the same—an armature and a field. Under the right circumstances, the DC generator can be powered to spin like a motor, and the motor can be made to produce electricity. Considering these connections, it is not surprising that spinning motors also act like generators and produce a back voltage. A motor with no load spins very quickly, producing a great deal of back voltage, which opposes the applied voltage. As a result of the back voltage, less current flows. When the motor is doing such work as drilling a hole, it slows down and produces less back voltage. As a result, the current flow increases. Thus, the current draw of a motor depends on how heavy its load is.

Series motors are especially sensitive to load changes. Refer to figure 7-16 and imagine this motor starting up under a heavy load. At this point, the motor is not yet running, and hence it produces no back voltage. Heavy current flows through the armature, and, since it is in

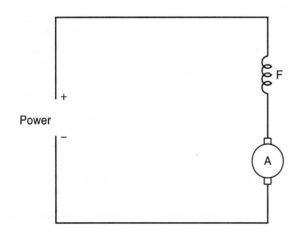

FIGURE 7-16 Series-wound motor.

series with the field, the same heavy current flows in the field. As a result, very strong magnetic fields are produced and tremendous torque is developed. Train engines, cranes, and car starter motors use this series-wound arrangement. Unfortunately, if this type of motor is run without a load, it may spin so fast that it destroys itself. Therefore, starter motors should not be tested without a load. Series-wound motors attempt to furnish enough torque to match the load.

Shunt-wound motors are connected in parallel. The motor constructed earlier was of this type. As with all parallel circuits, there are a number of independent paths through which

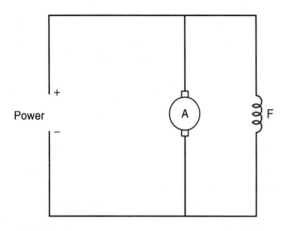

FIGURE 7-17 Shunt-wound motor.

current can flow. Look at figure 7-17 to see the two paths for current flow. The first path is from power through the armature and back to power. This path is completely independent of the second path, which also originates at the power source, and then flows through the field and back to power. This parallel arrangement isolates the field current from the armature current.

Back voltage is produced in the armature of the shunt wound motor, just as it is in the series motor; however, it does not affect the current flow in the field. As a result, this type of motor is less responsive to load changes and has a more consistent speed. Car heater fan motors and electric hand tools are often shunt-wound motors. Shunt-wound motors tend to provide more constant speeds and fewer load matching capabilities than series-wound motors.

It is possible to design a motor that employs some of the characteristics of both the series- and shunt-wound types. These motors are a bit more complicated to manufacture since they use two or more field coils. For this reason, they are referred to as compound wound. One of the field coils is usually connected in series and the other in parallel (shunt). Various designs are possible, so that a wide range of both starting and running characteristics is available.

Classifying Electric Motors

It is time to develop a mental file cabinet in which to store the motors already discussed and those that will follow. Two common schemes are used. The first is to classify motors on the basis of how they are used, for example, fan motors, compressor motors, and so on. The second method is to classify motors on the basis of their internal structure. Generally, the second method is based on how they work. Figure 7-18 classifies common motor types based on their internal structure. It is helpful to refer to this chart to bring together many ideas as you learn about motors.

Those motors discussed up to this point are classified as DC motors, since they all require electrical flow in only one direction. They

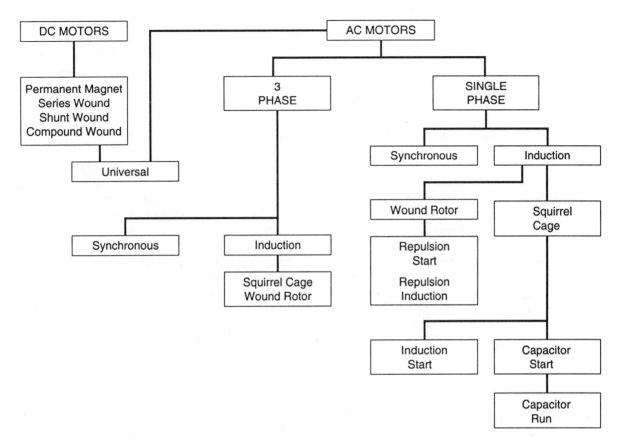

FIGURE 7-18 Motor classification.

appear in figure 7-18 under the "DC motor" heading. The permanent magnet motor will not run on AC since the field is permanently polarized in one magnetic direction. If AC is applied to this motor, it will hum while it attempts to turn first one way, and then the other, at 60 Hz. The rapidly changing AC magnetic field of the armature will demagnetize the permanent field. Series-, shunt-, and compound-wound motors designed for DC generally have too much inductance to permit AC current to flow. If they do run, this same coil inductance produces a great deal of brush sparking, greatly shortening motor life.

MOTORS FOR SINGLE-PHASE AC

Most of the motors with which the HVAC technician works are single-phase AC devices.

From a physical standpoint, the large majority of these motors may be separated into brush types and induction types. In figure 7-18, the AC brush-type motors are referred to as universal motors as they may be powered by AC or DC.

Universal Motors

Recently, one of the authors purchased a workshop vacuum. This vacuum is on casters and has the inlet and outlet on opposite ends. When first connected to power, it produced a high-pitched whine (very high motor RPM) and propelled itself across the floor. Since that time, it has swallowed up tools, whole boxes of screws, and just about anything that will fit through its 3-inch hose. It is a great portable tool for that all-important cleanup part of the job that leaves the customer satisfied. What is the source of the remarkable power of this vacuum?

Universal motors are actually specially designed series-wound DC motors. The design considerations include holding inductance to a minimum to reduce brush sparking and permit high current flow. Figure 7-19 illustrates the split-field series-wound universal motor. It can function on AC because the reversal of magnetic field caused by current reversal affects both the armature and the field. For example, if the north magnetic pole of the field coil is attracting the south magnetic pole of the armature and the AC current reverses, the poles simply reverse. All that is required for the motor's operation is that the poles be opposite and attracting, and then alike and repelling.

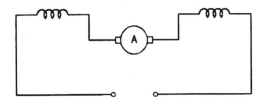

FIGURE 7-19 Universal motor.

Vacuums, mixers, drills, routers (20,000 RPM), saws, and just about all intermittent-duty portable devices use this popular motor. Its chief advantage is high horsepower in a small, light package. Many of these devices produce 2 to 3 horsepower. Unfortunately, the brush commutator part of the motor that provides power to the field limits the useful life of these devices. This is especially true in an environment containing abrasive and resinous air-borne dust. When, for example, sawdust or plaster dust gets on the brushes, it causes resistance, which then heats the commutator and accelerates wear. This can be overcome with careful design and prudent use; however, brush-type motors are generally not used in continuous-duty applications.

Induction Motor Basics

In this section, the motor that many of us may think of as a washing-machine motor is the

center of attention. On the motor classification chart, figure 7-18, induction motors are one of the two major subdivisions of both single- and three-phase motors. They furnish mechanical power in a way that, like the other motors discussed, depends on the interaction of two magnetic fields. However, the induction motor is unique in a very important way. To introduce this difference, let us do a little mind experiment.

Imagine a coil of wire connected to AC and the resulting magnetic field around it. As the current rises and falls, and then changes direction, only to rise and fall again, the magnetic field follows. While this is going on, you place your hand near the coil. In a short time, a ring on your finger gets warm. At first, you don't believe this, but soon it is hot enough to be annoying, so you pull your hands away from the coil. Is this magic?

Yes, but it's only transformer magic. The AC flowing through the coil causes it to act like the primary of a transformer and the ring on your finger is a single-turn secondary coil. A moving magnetic field produced by the primary coil has induced a current flow through the ring, which causes it to heat up. The ring doesn't look like a transformer secondary, but then neither does the induction motor rotor shown in figure 7-20.

FIGURE 7-20 Induction motor rotor.

The spinning part inside the induction motor, is called the rotor (not armature). Most of its heavy weight is due to the steel that is there to enhance the magnetic nature of this device (magnetic circuit). If the steel were removed, an assembly that resembles a small rodent exercise wheel would remain. Today, this might be called a gerbil cage rotor, however, it is commonly referred to as a squirrel cage rotor. Each of the bars that connect the end plates together acts like part of a ring circuit. Current flows through the rings, as illustrated in the lower right of figure 7-21. Also notice that figure 7-18 identifies the "squirrel cage" as a subcategory of induction motors.

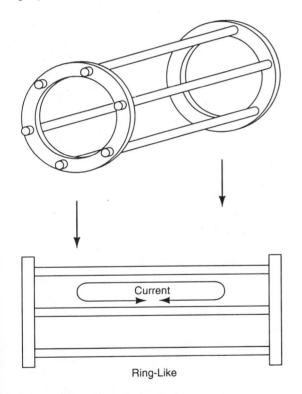

Ring-Like

FIGURE 7-21 Squirrel cage rotor.

The reason that your ring got hot is the same reason that induction motors are brushless. Current flow in the rotating part (called the rotor) of all induction motors is induced by a transformer-like action. Nikola Tesla realized that the brushes of the DC generator and motor would prove unreliable. In a single insight, he removed them from the generator, causing it to produce AC, and from the motor by perfecting

the induction motor. To understand how this motor runs, it is necessary to understand the stationary part.

The stationary part of the induction motor is called the stator (not field). Notice that in figure 7-22 a four-coil stator is connected in a series configuration. This is also referred to as a distributed coil in that it is actually one coil distributed to four locations. When this coil is connected to power, the current causes an expanding and contracting magnetic field to appear around the coil. Imagine what would happen if a single-turn ringlike coil were placed within the center of this stator. Just like the jewelry ring mentioned earlier, this ring would heat due to the induction of current. However, as well as producing heat, current flow also produces a magnetic field. To be certain that this process is clear, let us follow it through step by step. First, the expanding and contracting magnetic field around the stator induces a current flow in the single-turn coil (the ring), and then that "ring" current produces a magnetic field around the ring. This is very similar to what the transformer does, and, in fact, sometimes is referred to as transformer action.

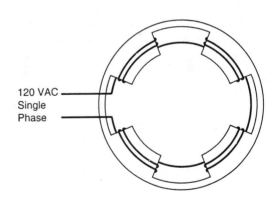

120 VAC
Single
Phase

FIGURE 7-22 Single-phase stator.

Remember Lenz's law, which may be thought of as the electrical version of "for every action there is an equal but opposite reaction." Fooling with Mother Nature in this case causes the ring and its induced current to provide a magnetic field that pushes away from the stator's magnetic field.

Now let us replace the ring with the squirrel cage rotor, which is actually a "bunch of rings." As a result of Lenz's law action, the squirrel cage rotor rings next to each stator coil push toward the center of the rotor. This action may be thought of as producing a magnetic repulsion between each of the stator coils and the rotor rings closest to them.

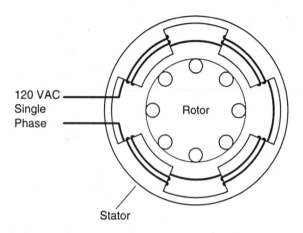

120 VAC
Single
Phase

Rotor

Stator

FIGURE 7-23 Stator and rotor.

If power is applied to the single-phase induction motor of figure 7-23, the rotor will not spin. Rather, it will periodically (60 Hz) be squeezed inward toward its center. In a general way, the rotor wants to escape the stator; however, it is trapped by its own inward magnetic forces on all four sides. As a result of certain malfunctions in starting circuits, ordinary split-phase induction motors exhibit this nonstart hum behavior.

If, at this point, the motor shaft is rotated manually, it will continue to turn. In other words, once the motor is started, it will continue to rotate. This surprising result is due to a complex change in the relationship of the stator and rotor's magnetic fields. The field interaction change that occurs between when the motor is still and when it is turning may be described in general terms.

When the motor is not moving, the stator acts like a transformer primary and the rotor acts like a transformer secondary. This results in a magnetic squeeze all around the rotor

(actually squeezed and stretched). Once the rotor is given a spin, it runs around in circles trying to escape the stator squeeze. This affects the stator's magnetic field, like four family members surrounding a pet house cat. Once the cat starts running in a circle, the four family members take off after it—that is, they react to its running by chasing it in a circle. In a like fashion, the stator's magnetic field is affected by (reacts to) the spinning rotor's moving magnetic field and takes off in pursuit. What was once a fixed stator field, as a result, becomes a rotating stator field. Once the cat starts running in a circle, everybody gets exercise, but the cat remains free. This is a bad situation for pet recovery, but exactly what is required in a motor. But how do we get the cat, in this case, the rotor, to start running in circles?

The trick has to do with phase shift, caused by either inductance or capacitance. Since this stator coil is made from relatively heavy wire and is assembled within a highly permeable steel frame (see figure 7-22), it is highly inductive. When it is connected to AC power, the current flows almost 90 degrees after the applied voltage. As a result, the expanding and contracting magnetic field around it is also delayed by almost 90 degrees. In this case, another magnetic field needs to be added to the stator to provide a simulated rotating magnetic field to start the motor, after which it will produce its own rotating magnetic field.

This second magnetic field is provided by a second set of coils, called the start windings. They are placed between the run windings. The coils are arranged start, run, start, run, and so on. Figure 7-24 show the location of the start windings. Start windings in this split-phase motor are constructed from thin wire that has high resistance. The ohms of resistance are so great that they diminish the effect of this coil's ohms of inductance. A vectorial addition of the two results in a phase shift close to zero. As a result of this difference in phase shift, the magnetic field of the start windings comes before the magnetic field of the run windings. And so we have a simulated rotating magnetic field.

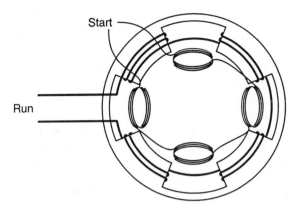

FIGURE 7-24 Start windings.

The high-resistance start windings provide a magnetic field first, then the high-inductance run windings provide a magnetic field, thereby providing a rotating magnetic field, causing the rotor to turn. Once the rotor is spinning through interaction with the stator run coils, a

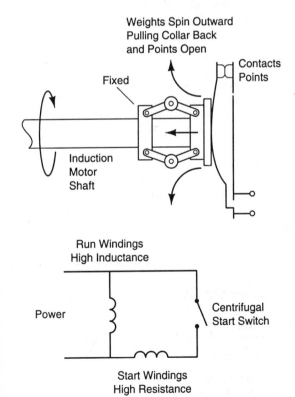

FIGURE 7-25 Centrifugal switch.

rotating magnetic field is set up and the start windings no longer are required. In many motors, they are disconnected after motor run-up by a centrifugal switch located on the motor shaft. This switch pushes against a set of fixed contacts in the motor end plate or bell, breaking the starter windings circuit. Figure 7-25 illustrates the circuit described, as well as the centrifugal switch.

The induction motor just described is called a split-phase induction start motor. Since the start windings are resistive in nature, this motor is sometimes referred to by the more technically correct name, split-phase resistance start induction motor.

Universal vs. Induction Motors

Comparing these two very different approaches to AC motors is rather straightforward. The universal motor connects power directly through to the spinning part, called the armature, while the induction motor uses induced transformer action to provide current in its spinning part, called the rotor. As a result, universal motors produce more horsepower in a smaller package, but are prone to brush and commutator failure. Induction motors are often in service for many years before they require attention. Their life depends primarily on bearing, starter switch, and coil quality. Unfortunately, since induction motors are big and heavy, they are not usable in such applications as portable power tools and vacuums. A drill press will probably be powered by an induction motor, whereas a hand drill will incorporate a universal motor.

SINGLE-PHASE INDUCTION MOTORS

All induction motors designed for use on single phase-power share certain design considerations. With all of these motors, current for the magnetic field of the rotor is induced rather than directly connected through brushes and a commutator. The motors discussed in this section have rotors that act like the secondary of a transformer. Also, these induction motors require some method to start the rotor spinning, which then results in a rotating magnetic field in the stator. A variety of single-phase induction motors are used in HVAC appliances that start and run by a number of different processes. As a result, each motor has different characteristics.

Squirrel-Cage Split-Phase Induction Start Motor

This is the motor that was discussed in the induction motor basics section. Take note of its location on the motor classification chart. Since it uses high-resistance start windings, its more precise name is split-phase resistance start induction motor. It is a reliable source of mechanical power, especially if it has ball rather than plain bearings. Ball bearings are generally lubricated for life, while plain or sleeve bearings require a drop of oil every year.

The most common version of this motor incorporates the centrifugal starting switch previously described. However, it is possible to connect the start windings through the contacts of a normally open relay. The relay coil is connected in series with the motor and is pulled in by heavy starting current. Once the motor has runup, the current draw is reduced and the relay disconnects the start windings. The direction of rotation may be changed by reversing the start winding connections. Ordinarily, the start and run winding leads are accessible near or below the power terminals on the motor.

Once this motor is started, it continues to run due to the rotating magnetic field in the stator. The speed of the motor depends on both the number of coils in the stator and the power frequency. Four- and eight-coil (two and four-pole) induction motors of this type are most common, with ideal or synchronous speeds of 1800 and 3600 RPM. The actual speed of the rotor is slower due to rotor slip. Rotor slip is the difference in speed between the rotating magnetic field of the stator and the rotor's speed. As a load is placed on the motor, the rotor slips through the rotating magnetic field.

This induces more current into the rotor and results in a stronger rotor magnetic field. In this way, the induction motor provides more power as the load increases, with actual speeds of 1725 and 3450 RPM.

The low starting torque of this motor relegates its use to devices that provide a light starting load, such as oil burners, drill presses, grinders, and small fans. Fractional-horsepower sizes are by far the most commonly used, but versions that produce a few horsepower are also available.

Split-Phase Capacitive Start Motor

Splitting phase is actually a result of phase shift, which, in the case of the inductance start motor, is caused by an inductive current lag. Unfortunately, though ideal inductors provide a current lag of 90 degrees, practical inductors produce current lags closer to 60 degrees. This results in a small time difference between the start and run winding magnetic fields. As a result, induction start motors provide low starting torques. Ideal and practical capacitors are more alike.

Capacitors cause a voltage lag or, viewed from the opposite perspective, a current lead. Split-phase induction motors that require higher starting torque use capacitors to cause a current lead. In this way, current flows through the start windings 90 degrees ahead of the run windings, providing higher starting torque. Capacitors split phase better than inductors do. Locate the capacitor start motor on the lower right of the motor classification chart of figure 7-18.

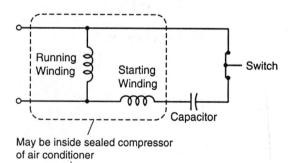

May be inside sealed compressor
of air conditioner

FIGURE 7-26 Capacitor start circuit.

The diagram in figure 7-26 illustrates how the capacitor is connected to the start windings and switch. Nonpolarized electrolytic capacitors are usually used. However, they may be connected in one direction to reduce short-circuiting caused by internal capacitor failure. Often the capacitor is mounted within a metal case on top of the motor, though it may be located near the motor within the appliance.

The capacitive start split-phase induction motor is used for large blowers, compressors, washing machines, pumps, and any devices that require high staring torque. Perhaps the most common application in the HVAC field is in refrigeration and air-conditioning compressors.

It is common practice to seal the rotor and stator of this motor in the same pod as the compressor. This provides a clean, protected environment for the motor, while eliminating the rotating compressor seals that would be necessary if the motor were external. Freon loss is far less likely on this type of system. However, the starting switch and capacitor must then be mounted externally. In this case, the starting circuit is connected by relay points that are normally closed. The relay coil is wired in parallel with the motor start coils.

Generally, all small sealed motor compressor units used in air conditioners and refrigerators follow this start-up sequence.

1. Thermostat-controlled power is applied to the compressor unit.

2. The normally closed starter relay contacts provide current to the start windings.

3. During start-up, the high current draw of the start windings results in low voltage across the start windings. Since the start relay coil is connected in parallel with the start windings, it remains deenergized.

4. When the motor spins up, voltage across the start windings rises and pulls open the start relay, disconnecting the start windings.

This process can often be heard when a refrigerator compressor turns on. The click is

the relay pulling in. If the relay or capacitor fails, the compressor will stall and trip a circuit breaker. Many compressors have been unnecessarily replaced when faulty relays and/or capacitors were the real culprit.

Split-Phase Run Motors

The split-phase induction motors discussed up to this point tend to be noisy. The constant reversal of the magnetic fields around the run windings of the stator results in a hum that is particularly noticeable when the motor is rigidly mounted. Rubber isolation mounts reduce, but do not entirely eliminate, this hum. Also, since these motors use high-inductance run coils, a relatively large amount of phase shift is introduced. Recall that phase shift causes a big difference between VAs (apparent power) and watts (true power). This results in a power factor of .6 to .8. The practical result is that both resistance start (commonly called inductance start) and capacitive start-split phase inductance motors are only about 60% efficient. Furthermore, the hum they produce is inherent to their structure.

At the lower right corner of the motor classification chart is an inductance motor referred to as capacitor run. It is also called a split capacitor or permanent capacitor motor. The capacitive run motor starts in the same way as the capacitive start split-phase motor. A conventional centrifugal switch or relay connects the start coils through a capacitor. As with the capacitive start motor, a current lead shifts the phase to initiate rotor motion. How-

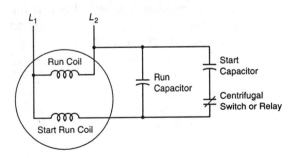

FIGURE 7-27 Capacitor run motor.

ever, after start-up, the capacitive run motor continues to provide power to the start windings through a capacitor that is usually smaller that the start capacitor. The capacitor run motor circuit is illustrated in figure 7-27.

A careful examination of figure 7-27 reveals that the capacitive run motor makes use of two sets of coils to provide continuous phase splitting. The run coils are inductive and cause a current lag, while the start–run coils are capacitively connected, which produces a current lead. Since the inductive and capacitive effects are opposite each other, the phase difference in currents flowing in the two coils is 90 to 100 degrees. With an ideal inductor (90-degree lag) and ideal capacitor (90-degree lead), a theoretical phase difference of 180 degrees would result. However, both the run and start–run coils have resistance that reduce this phase shift. Looked at from a practical point of view, the capacitive run motor is actually a two-phase motor. As a result, this motor is able to deal with load variations by producing more torque. In a very general way, the capacitive run motor has a greater phase split, which enables it to provide start-up–like torque when loads are intermittently increased.

In this type of motor the rotor is pulled around with 2 sequences of smaller magnetic tugs (the 2 phases producing magnetic fields) rather than with 1 single-phase series of larger magnetic tugs. As a result, the capacitive run motor produces less hum and the two opposite phase shifts cancel each other with respect to the source. From the perspective of the source, the current lag and lead sum to zero phase shift. Since total phase shift is zero with respect to the source, the apparent power (VAs) and true power (watts) are the same. Hence, the power factor is 1 and the motor is much more efficient.

The efficiency, quiet operation, and responsiveness to load variation make the capacitive run motor the "top of the line" of AC single-phase induction motors. It provides quiet, efficient, mechanical power for air conditioners, dish washers, refrigerators, and many other residential as well as industrial applications.

Wound-Rotor Induction Motors

With greater expense, the simple squirrel cage rotor may be replaced or supplemented with a wound coil rotor. An additional change is made in that the wound coil ends of the rotor are connected to a commutator. Resting against opposite sides of the commutator are brushes that are simply connected to each other (not to power).

During start-up, the wound rotor acts like two big coils that have current induced in them by the stator through transformer-like action. The stationary brushes are positioned at an angle with respect to the stator coils. The angle is selected so that it provides the greatest repulsion between the rotor and the stator. The repulsive forces are a result of Lenz's law action–reaction. Figure 7-28 shows a simplified schematic of the start-up and run circuits. It is important to realize that the two big rotor coils on the left are, in a schematic diagram sense, stationary. However, they are actually composed of constantly changing (spinning) wound rotor coils. Older versions of this motor moved the brushes a few degrees around the rotor's commutator to establish starting torque. In this way the necessary high starting torque for heavy machinery, including big refrigeration and air-conditioning compressors, can be developed.

Once the repulsion action gets the motor started, a centrifugal device short-circuits the commutator, as shown on the right in figure 7-28, converting it into a bunch of closed rings,

2 Big Coils   Many Small Coils

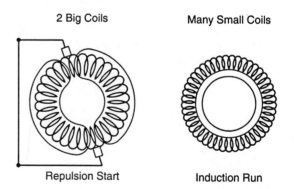

Repulsion Start   Induction Run

FIGURE 7-28 Wound rotor repulsion start-up.

which function like the squirrel cage rotor. This motor is most often referred to as a repulsion start motor.

Often the repulsion start motor incorporates a standard squirrel cage rotor. The addition of the squirrel cage improves speed regulation and eliminates the need for a commutator centrifugal short-circuiting ring. If a squirrel cage is included into the rotor, the device is referred to as a repulsion induction motor.

Shaded-Coil Induction Motors

Very small induction motors are used as a source of power for such items as bathroom and kitchen vent fans, as well as induced draft blowers. In these applications, low starting torque and low efficiencies can be tolerated. Often these motors start by using shading coils to simulate a rotating magnetic field. Generally, these motors are not encased in a housing, and thus they can be identified, as shown in figure 7-29, by a copper ring fastened to each stator field.

FIGURE 7-29 Shaded-coil induction motor.

The shaded-coil induction motor runs according to the same principles as the induction motor; however, it gets started in a rather unique way. To understand this requires a step-by-step look at what happens to a ring placed in an AC magnetic field.

Recall that inductors oppose a change in current by producing a back voltage (CEMF) during magnetic field expansion and a forward

voltage (EMF) during magnetic field collapse. This same effect occurs when a solid ring of wire is placed in a changing magnetic field. Let us take a closer look at one stator pole of the motor in figure 7-28.

Figure 7-30 illustrates one half of an AC sine wave that is the power source for the stator coil above it. The stator coil is pictured three times to represent what happens to its magnetic field as the power rises and then falls.

On the left, at 0 degrees, the stator current is rising and the stator's magnetic field is expanding across the shading coil. A back voltage is induced into the shading coil, which results in a current flow and a magnetic field that opposes the stator coil magnetic field. This reduces or cancels some of the magnetic field near the shading coil on the stator's right side. At the 90-degree peak of the sine wave, the current and magnetic field around the stator have stabilized, and hence no current is induced within the shading coil. As a result, the magnetic field is consistent across the face of the stator pole. Between 90 and 180 degrees, the current flow through the stator coil drops off and the magnetic field collapses. The collapsing magnetic field crosses the shading coil and induces in it a forward voltage that produces a current and magnetic field that aid the stator's magnetic field. As a result, figure 7-30 shows a stronger magnetic field on the right of stator pole. Notice that this is actually another example of an inductive opposition to current change.

These small shaded-coil induction motors are usually reliable, with the exception of the bearings' becoming gummed. This is especially true in kitchen vent fans, where a drop of oil will often freeup the motor.

Synchronous Motors

There are a great number of motors that fall within the synchronous category. All of these motors share the unique characteristic of very constant speed. Many of them are used for clocks and rotary mechanical timers. Since the devices of which they are a part are usually

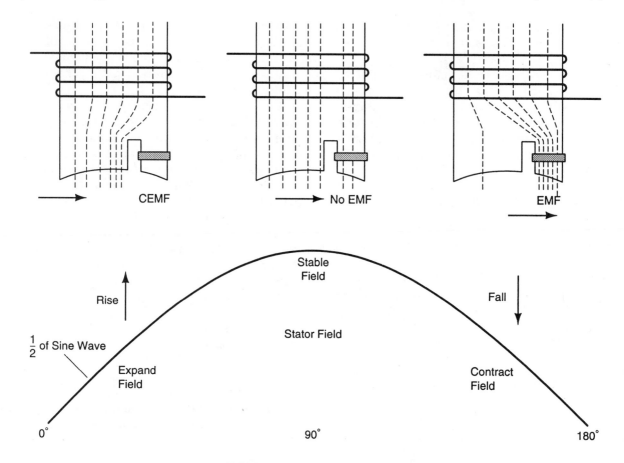

FIGURE 7-30 Shaded-coil action.

replaced as units, only a general idea of their functioning will be presented.

Synchronous means that two or more things are varying in some coordinated way. For example, all of the members of a band may be marching in synchronization, often to a synchronous drum beat. Induction motors that run in synchronization have a rotor that turns in step with the rotating magnetic field. As the speed of the rotating magnetic field depends on the power line frequency, these motors spin at precise RPMs. Unlike regular induction motors, synchronous rotors do not slip in the rotating magnetic field.

A simple synchronous motor can be constructed from an induction motor by replacing the rotor with a permanent magnet. If the permanent magnet rotor is manually spun up to synchronous speed, it will continue to turn in step with the stator's rotating magnetic field. To make the synchronous motor self-starting, a variety of techniques beyond the scope of this book are used.

THREE-PHASE MOTORS

Many electrical devices, like split-phase induction motors, are difficult to explain. Their principles of operation involve scientific concept on top of scientific concept. For example, induction supplies current to the rotor, but something must be added to get it started (phase splitting). Fortunately, some devices, like the three-phase motor, seem almost natural, in the sense that they apply scientific ideas in a direct way.

From the previous discussion of three-phase electricity, it should be clear that the peaks of the sine wave for the three phases follow one another in sequence, separated by 120 degrees. The phases rise to peak 1-2-3, then back to 1, and so on. If an induction motor is constructed of three pairs of coils (the pairs are opposite and across from one another on the stator), it can take advantage of the 1-2-3-1-2-3-1-2-3 phase relationship.

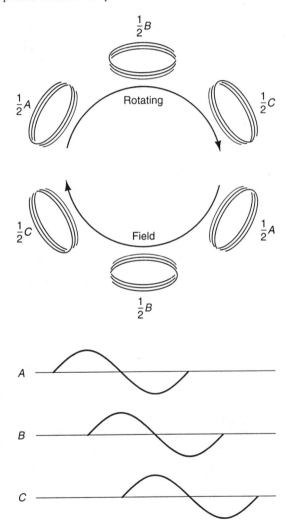

FIGURE 7-31 Three-phase induction motor stator.

From a practical perspective, the stators of all three-phase motors work in the same way. Figure

7-31 illustrates the three pairs of stator coils that are connected to the three phases of power. Since the phases, as shown at the bottom of figure 7-31 follow each other, so does the magnetic field around the stator coil—a 1-2-3 rotating magnetic field with no necessity for phase splitting tricks! A squirrel cage rotor placed within this stator will be towed around following the phase sequence. Motor rotation direction is a result of phase sequence, and thus it may be changed by reversing any two of the three wires feeding power to the motor.

Ordinarily, three-phase motors are available in sizes from 1 horsepower to over 100 HP. Smaller motors are not common since three phase is usually used where large amounts of power are required. As a group, these motors provide good starting torque and greater power per size and/or weight than do single-phase induction motors.

Three-Phase Squirrel-Cage Induction Motors

The squirrel cage rotor of a three-phase motor is, practically speaking, essentially the same as a single-phase motor. Since these motors are often used to provide high HP, all related motor characteristics become more critical. For example, inefficiencies with a 3-horsepower intermittent-use single-phase motor do not waste large amounts of power; however, the same inefficiencies in a 10 HP motor that is in continuous use are unacceptable. Also, the loads of fractional HP motors are generally small, low-mass devices that startup quickly. A 100-HP three-phase motor supplying power to a steel rolling mill may take many seconds, or even minutes, to get up to operating speed. The idea is a bit like comparing a rowboat and a battleship.

A variety of squirrel cage rotor designs are used. The most obvious differences are related to the shape of the bars that go from endplate to endplate of the rotor. Bar shapes vary from round to oblong, and are indicated on the motor identification plate by code letters A through V. These types differ as to their ability to produce torque during start-up and running. Also, double-bar squirrel cage rotors are used to provide very high starting torque. It is important that failed motors be replaced only by motors of the same or equivalent type.

Three-Phase Wound-Rotor Induction Motors

Starting up big machinery is like pushing a heavy car. The mass of the machinery or car exhibits a great deal of inertia. It is, for practical purposes, impossible to get a heavy car moving fast, even if the football team takes a running lead and crashes into the back of the car. In the same way, a large three-phase motor could exert a great deal of torque to get a big load started. However, in doing this, it would draw extremely high current and necessitate circuits well beyond those required for normal running of the motor. Control of both start-up torque and running speed of the three-phase motor is most commonly provided by wound rotors.

Rotors are usually wound in a wye configuration as shown in figure 7-32. The ends of the three coils making up the wye are connected to solid copper rings, called slip rings, on the motor shaft. Connected to the slip rings through brushes are large adjustable resistors. Induced current flow in the rotor is controlled by varying the amount of resistance. In this way the start-up, as well as running, characteristics of the motor can be controlled.

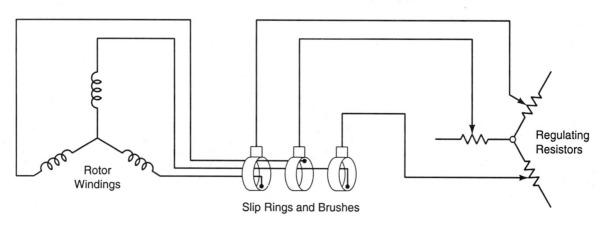

FIGURE 7-32 Thee-phase wound rotor regulation.

SUMMARY

This chapter expanded and further developed the concepts of single- and three-phase AC electricity as it relates to a variety of loads. Single-phase resistance heaters can be unintended loose connections or intended resistance elements that require a great deal of electricity. Incandescent as well as ionized gas and vapor lighting are in common use, however, the nature of the light they provide and the type of heat they produce are important considerations.

The idea that all motors produce power by the interaction of a fixed and movable magnetic field may be understood through the simple brush type motor. The fixed part of this type of motor is called the field, while the spinning part is called the armature. These motors are compact and provide high power for many portable applications, and are available as series-, shunt- or compound-wound types. The universal AC version of such motors is commonly used. However, like all brush motors, it suffers from brush and commutator failure.

Induction motors—which provide current to the spinning part, called the rotor, by AC transformer action—eliminate the necessity for brushes, but are large and heavy. Rotors may be of the squirrel cage or wound type used in the repulsion motor. Rotating magnetic fields in the stator are produced for starting by either inductive current lag or capacitive current lead. Once the split-phase motor gets started, it provides its own rotating magnetic field and a switch or relay disconnects the start windings. An exception is the capacitor run motor, which acts like a two-phase motor and furnishes quiet, efficient power.

Small induction motors may be started with the use of shading coils, which produce a moving magnetic field across the stator poles. Synchronous motors are also small induction motors that run with the rotor locked to the rotating magnetic field. A simple illustration is a permanent magnet spinning within a stator without any slip.

Three-phase motors are clear examples of the sequence of the three phases following one another and providing a rotating magnetic field in the stator, without the need for phase splitting. Since three-phase motors are generally used to supply larger loads, they are often designed with special squirrel cage rotors. Wound rotors may be used to enable reasonable motor start-up and speed control.

PROBLEM-SOLVING ACTIVITIES

Troubleshooting Case Studies

A. At a garage sale, you buy a nonfunctioning drywall screwdriver for $5. It is coated with plaster dust. What electrical parts are most likely in need of attention?

B. A refrigerator does not work. When the thermostat applies power to the compressor, a click is heard and the room lights dim, but the compressor does not start. What electrical circuit might be at fault? Describe the components of this circuit and their functions.

C. A friend bought a cassette player in Europe. With the appropriate plug adapter, it plays; however, all the tapes run fast. Considering the critical timing and speed nature of devices like this, what type of motor might it use and how could that explain the speed problem?

D. A small metal fabrication shop uses wye-connected three-phase power. When certain combinations of single-phase equipment are operating, the main breaker trips. Before a costly new service entrance is installed, what other solution should be investigated? (Note: the three-phase main breaker will trip when any one of the three feed wires draws more than the maximum allowable current.)

Review Questions

1. While wiping clean the fluorescent tubes of a fixture, you notice a flash of light. The tubes have been removed from the fixture and are not connected in any way to power. Explain this.

2. A thermostat is located above a table lamp. What problem might this cause? How might a different table lamp resolve this problem?

3. Some electric car door locks employ a permanent magnet as the movable core of a single coil. Compare the lock/unlock function of this device with the armature and field of a brush-type motor.

4. Why does a car starter have many armature coils, two pairs of brushes, and two pairs of field coils?

5. While testing a small toy motor, you accidentally connect it to low-voltage AC. As a result, it now lacks power. What has happened?

6. Induction motors contain much more steel than do equal-horsepower brush motors. Why?

7. Make a chart that compares the brush-type motor's and induction motor's moving and stationary parts. Correctly label these parts and describe the source of current or magnetism for each.

8. Describe the start-up process for split-phase motors.

9. In general, how inductance and capacitance affect the phase relationship of source current and voltage?

10. What are the chief advantages of capacitor run motors?

11. From what you know about the power transfer to the rotor of an induction motor, explain how the batteries of a pacemaker may be charged.

12. How do the brush (start-up) and short-circuiting ring (run) change the electrical nature of the wound rotor repulsion start single-phase motor?

13. An engagement ring is often given while the relationship is on the upswing, then returned if it goes downhill. How is this like the shading coil?

14. Why do three-phase motors have no start-up circuits?

15. A variety of squirrel cage rotor designs (A–V) as well as wound rotors connected to resistors, are used in large three-phase motors. Discuss these structures in connection with the typical load of this motor.

CHAPTER **8**

Circuits
and
Instruments

Previous chapters have presented the fundamental concepts of electricity, followed by the nature of specific DC and AC loads. This chapter connects various loads to circuits. Parallel circuits are discussed in relationship to residential electrical service. This affords a necessary general understanding of the common practices of electrical wiring, as well as of parallel circuit theory. The schematic of a basic air conditioner is used to develop series circuit theory. Much of the troubleshooting a technician performs involves determining which of a series of devices is malfunctioning. Determination of electrical values (volts, amps, and ohms) requires the use of meters.

Meters and instruments that provide the technician with "electrical ears and eyes" are presented in relationship to various devices and circuits. The measures they provide are useful only when they can be interpreted by the technician. Simplifying circuits and developing a common-sense approach to problem solving provide the important conclusion of this chapter.

TYPICAL RESIDENTIAL SERVICE

Electrical service entrances may be installed or modified by licensed electricians. Many electrical jobs require inspections to verify that the system, or modifications to it, meets the local and national electrical codes. Most new service entrances must be inspected before the utility company will connect new feed wires. As a result, it is often advantageous that the technician develop a professional relationship with a reliable and skilled licensed electrician.

The following discussion provides some general theory that enables the technician to "make sense" of residential electrical distribution. This section is not intended to provide enough detail to permit installation. Rather, the general ideas presented enable the technician to see the relationship of HVAC equipment to the building electrical supply system. For example, the installation of central air conditioning may require the addition of just a new circuit or of an entirely new electrical service in the building. The customer needs to understand the reasons for the difference in cost between these two systems.

In a more theoretical way, residential electrical service is used to illustrate the nature of a common parallel circuit. What happens to volts, amps, and ohms in the parallel circuit will add further detail to the mental picture of electricity.

Configuration of Feed Wires

Up-to-date electrical services have three feed wires that are run from the pole, or distribution transformer, to the house. If the service entrance is rated at 200 amps, each of these three wires is sized to carry 200 amps. Ordinarily, the primary of this transformer is connected to one of the high-voltage three phases that are used to distribute electricity throughout the local geographical area. This transformer's secondary is center tapped as shown in figure 8-1.

All the electric power consumed in the house is supplied by these three wires. Notice that the center tap of the distribution transformer connects to a wire called the neutral. This feed wire is neutral in two ways. First, there is no electrical potential or voltage between the neutral and grounded objects, such as water pipes or stakes driven into the earth (called grounding rods). Eventually, this wire connects with all the white wires run throughout the house. Therefore, they are all considered neutral and, ordinarily, there is no voltage between these wires and ground. Second, the neutral is

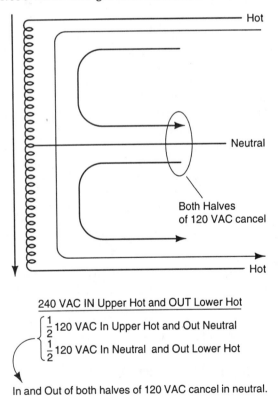

240 VAC IN Upper Hot and OUT Lower Hot

$\begin{cases} \frac{1}{2} \text{ 120 VAC In Upper Hot and Out Neutral} \\ \frac{1}{2} \text{ 120 VAC In Neutral and Out Lower Hot} \end{cases}$

In and Out of both halves of 120 VAC cancel in neutral.

FIGURE 8-1 Distribution transformer action.

a common wire to all 120-VAC circuits that are run throughout the house.

Assume that one half of the 120-VAC circuits are connected between the upper hot wire in figure 8-1 and the neutral, while the other half are connected between the lower hot wire and the neutral. Consider the nature of this well-balanced load (half connected to each hot wire) during one half of the AC sine wave. For half of the sine wave, current is flowing through the distribution transformer secondary in only one direction. The result of this is that one half of the 120-VAC circuits connected to the upper hot wire have current flowing out of the neutral wire, while the other half have current flowing into the neutral wire. In this way, properly balanced, 120-VAC load currents cancel each other in the neutral wire. However, each half of the 120-VAC load would draw 200 amps from its respective hot wire. Properly balanced 120-VAC loads neutralize each other in the neutral wire. For example, in a 200-amp service entrance with 120-VAC circuits (half on each hot wire), a total of 400 amps can be supplied. Each hot wire provides 200 amps, which cancel each other in the neutral. Currents above 200 amps on either hot wire will trip out the main breaker. The maximum power available from this service is 48,000 watts (400 amps × 120 VAC).

Appliances requiring large amounts of power are connected directly to 240-VAC circuits. Given the same current, a doubling of the voltage from 120 to 240 volts yields twice as much power. Remember it's power that does work. All 240-VAC circuits are connected between the two hot wires and have no direct effect on current flow in the neutral. If this were a 200-amp service, the maximum current draw at 240 VAC would be 200 amps. Currents above this level will trip the main breaker. The maximum power available from this service is also 48,000 watts (200 amps × 240 VAC).

At this point in our discussion, it is important to connect the ideas of 120-VAC circuits and 240-VAC circuits with respect to the total capacity of the 200-amp service entrance. Any combination or mix of these circuits that draws more than 200 amps from either hot wire will trip out the main breaker. For example, 110 amps of 120-VAC circuits all connected to one hot wire plus 100 amps of 240 VAC (carried by both hot wires) will overload that hot wire and trip the main breaker. A more careful balance of the 120-VAC load (55 amps on each hot) plus 100 amps of 240 VAC will load each hot leg to 155 amps and will not trip out the main breaker. Even with perfect 120-VAC load balance, the maximum capacity of a 200-amp service entrance is limited to 48,000 watts.

Service Panel

Overhead power wires are anchored to the house with insulators that support the weight of

these wires. Figure 8-2 shows how these wires are connected to an entrance head, which is mounted above the insulators. Service entrance cable of the appropriate ampacity is run through the entrance head and fastened down the side of the building to the meter box. The upper connector on the meter box is watertight. The meter completes the circuit of both hot wires, while the neutral wire runs directly through the meter box. If the meter is removed, all electrical service to the house is interrupted. From the bottom of the meter box, the service entrance cable is run a short distance through the wall into a service panel/circuit breaker box.

Service panels vary greatly in their basic design, as well as in their ampacities. Older

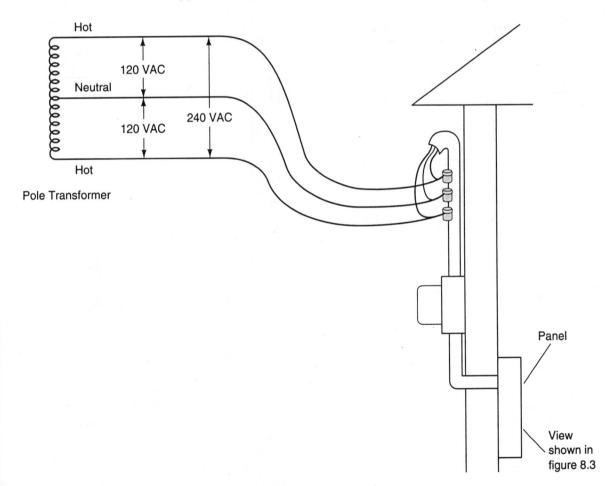

FIGURE 8-2 Residential power source.

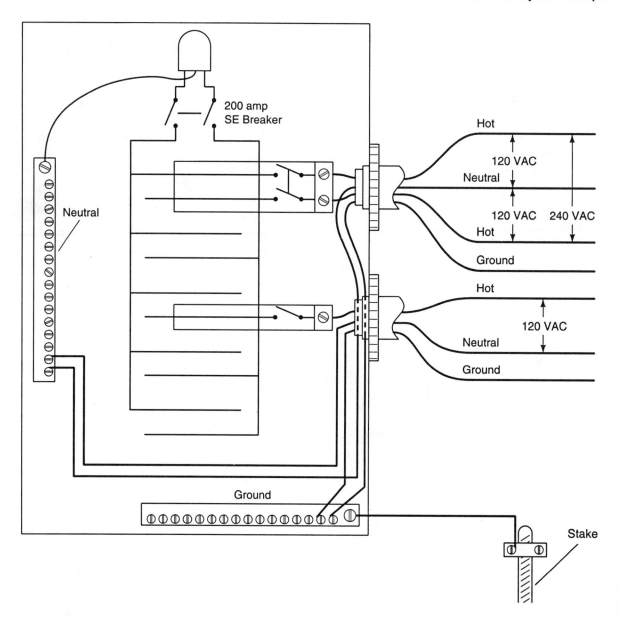

FIGURE 8-3 A 200-amp service panel.

panels may use fuses as overcurrent protection and provide only 120 VAC. Most service panels that use resettable circuit breakers and ground fault interrupters have ampacities between 100 and 200 amps. In a variety of ways, current service panels provide a main breaker that trips out both hot feed wires if either one draws more than its rated capacity in amps.

The service entrance shown in figure 8-3 is an example of a generic 200-amp panel. Notice that both hot feed wires are connected to the main breaker and the neutral wire is fastened to a neutral terminal strip. The neutral terminal strip is insulated from the chassis of the panel; however, older systems may tie together or bond the neutral to ground. Neutral wires from all branch circuits will be connected to this strip. Panels are designed so that the load side of the main breaker is connected to individual bus bars. Each of these two hot bars is designed to accept circuit breakers. A single-width breaker connects to one hot bar and is used to supply 120-VAC branch circuits, while a double-width breaker connects to both hot bars and supplies power to 240-VAC circuits. Figure 8-3 illustrates a common arrangement used with the hot bus bars that is designed to help provide load balance. As 120-VAC breakers are installed in the panel, they are alternately powered by each of the two hot feed lines. At the bottom of the panel, an second terminal strip connects the panel to a entrance water pipe or ground rod. All electrical boxes, appliance frames, and conduits or armored cables are connected through a ground circuit to the ground terminal.

Parallel Branch Circuits

The common practice in residential wiring is to feed all of the branch circuits from the main service entrance. As was the case with the examination of the service entrance, our look at branch circuits is intended to provide a general idea of 120-VAC and 240-VAC branch circuits. Many details in practice and theory have been omitted to permit an understanding of the key points. Actual work with these circuits requires adherence to national and local electrical codes and appropriate licenses.

120-VAC Circuits

Ordinarily, receptacles are connected with no. 12 wire (nonmetallic 2—no. 12 with ground), which can carry a maximum of 20 amps. (This cable is generally referred to as nonmetallic sheathed cable since the conducting wire is encased within a sheath of thermoplastic insulation for cable protection. The nonmetalic designation also differentiates this type of cable from metalic sheathed cable, or BX cable as it is usually called. BX uses a spiral wrapping of galvanized metal to protect the electrical wire within.) A single-width breaker rated at 20 amps

"taps" off one of the hot bus bars in the service entrance and feeds the black hot wire in the branch circuit shown in figure 8-4. Within the service panel, the branch circuit white wire is connected to the neutral bus and the green or uncovered wire is connected to the ground bus. From the service panel, the wire is run through and beside the house framing and held securely in place by insulated staples. At the receptacle box in figure 8-4, the grounded metal or plastic box has been omitted for clarity. Connections to the duplex receptacle are coded to provide easy installation.

[*Author's note*: One major cause of building fires can be traced to short circuits caused by

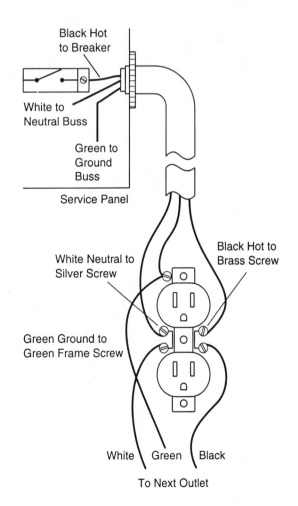

FIGURE 8-4 120-VAC receptacle circuit connections.

improperly installed cable staples. During installation, the staples are hammered too tightly against the cable. Over a period of time, as the plastic cable sheathing becomes brittle, the staple can work through cracks in the sheathing and short-circuit the electrical wires, causing a fire. Staples are not ordinarily a problem when BX cable is used to run branch circuits.]

Electrical devices, like receptacles, fixtures, switches, and plugs, are coded to indicate proper wire connections. The general installation rule is that the white, neutral wire is connected to the receptacle or fixture's silver screw. The black, hot wire is connected to the brass screw (see figure 8-4). Grounding of the circuit is achieved by connecting the green, or bare, wire from the nonmetallic cable to the green screw. Switches have two brass screws since they are intended to interrupt or switch the hot circuit. In this instance, the black feed hot wire is connected to one brass screw, while the black wire supplying power to the fixture or appliance is connected to the other brass screw. Devices that do not have screw terminals, such as ceiling fans and prewired fixtures, are usually supplied with short wires called pigtail leads. Connections to these devices are achieved by matching like color wires and securing the connection with a wire nut of the appropriate size.

240-VAC Circuits

Many stationary devices that require large amounts of power such as water heaters, clothes dryers, and air conditioners, are designed for 240-VAC operation—for example, a 30-amp clothes dryer circuit, which uses no. 10 wire (nonmetallic 3−no. 10 with ground). Overcurrent protection is provided with a double-width breaker that "taps" off both hot bus bars in the service entrance panel. This double-width 30-amp breaker is actually two breakers toggled together and is designed in such a way that currents over 30 amps from either of the hot legs will trip out both hot leads. The wire used for 240-VAC circuits ordinarily has two hot wires—one black and one red—that are con-

nected to the 240-VAC breaker. Figure 8-5 also shows that the white neutral and green ground wire from the branch feeder cable are connected at the service panel in the same way as 120-VAC circuits.

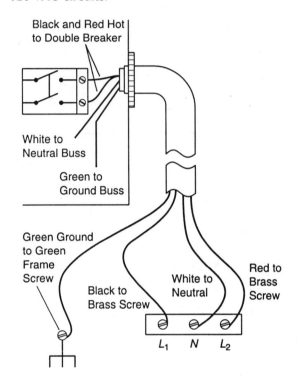

FIGURE 8-5 240-VAC circuit connections.

The cable for the 240-VAC circuit is routed through the house framing and held in place by insulated wire clamps or staples in the same manner as the 120-VAC circuit cable. At the clothes dryer, the wires are connected as shown in figure 8-5. The connections to the dryer are often made to a special 30-amp 240-VAC receptacle rather than directly to the appliance. Both of the hot wires (black and red) are connected to brass screws, often labeled line 1 and line 2 (L1/L2), while the white neutral is fastened to a silver terminal labeled neutral (N). The green or uncovered ground wire is connected to the green ground screw of the terminal strip. In this way, heaters within the appliance can be supplied with 240 VAC from the two hots (L1/L2) while 120-VAC devices, such as lamps and timers, are connected across one of the hots and the neutral. Outbuildings generally receive power either from branch circuits or from subpanels that are wired in a manner similar to that of the clothes dryer. Some devices, like water heaters, require only 240 VAC, therefore, the neutral shown in figure 8-5 would be eliminated from these circuits.

PARALLEL CIRCUIT THEORY

Parallel means alongside each other. For example, wall studs are parallel to each other. When parallel electrical circuits are diagrammed as in figure 8-6, the electrical loads are alongside each other. Well-constructed circuit diagrams draw the attention of the viewer to the

nature of the circuit, which may be hidden by the physical arrangement of all of the circuit components. Simply looking at the lamp, television set, and drill plugged into receptacles would not reveal the nature of their arrangement.

Parallel Circuit Voltage

Residential electric wiring and automobile electrical systems are primarily connected in a parallel configuration. An important clue in automotive electrical systems is that most of the electrical devices used in the automobile require 12 VDC. All of these devices are connected, either directly or indirectly, across the 12-VDC electrical supply. All the portable electrical devices in the home are "plugged in" to 120 VAC.

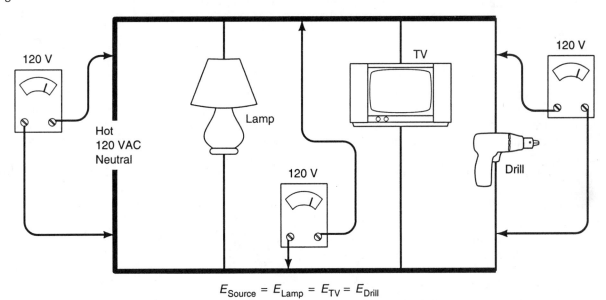

$$E_{Source} = E_{Lamp} = E_{TV} = E_{Drill}$$

FIGURE 8-6 Parallel voltages all the same.

It is helpful to think of the loads in parallel circuits as all being "hung" across the power source. One way or another, all 120-VAC appliances are "hung" across the hot and neutral wires. Directly or indirectly, all these devices are connected across the 120-VAC power source. Notice that in figure 8-6, 120 VAC could be measured anywhere between the top

and bottom feed wires. It is almost as if every device were connected independently to the power source. Mathematically, the formula in figure 8-6 indicates this idea in that the source voltage is equal to the voltage across the lamp, which is equal to the voltage across the TV set, and so on. In parallel circuits, the voltages are all the same.

Parallel Circuit Current

The flow of current for the lamp, TV set, and drill have been either read from the specification plates or calculated, and are noted to the left of each load listed in figure 8-7. Recall that the voltage (electromotive-force) provided by the source (on the left in figure 8-7), pushes the current through each of the loads.

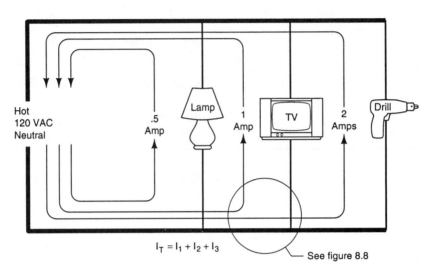

$$I_T = I_1 + I_2 + I_3$$

See figure 8.8

FIGURE 8-7 Parallel currents sum.

Follow the current flow from the source through each load and back to the source. Don't be confused by the arrow's direction. This is AC, and the real current flow is constantly reversing. Since all the current is being supplied by the source, all the lines representing current come from and return to the source. What is the total current flow supplied by the source?

If all the water in your house were supplied by the main feed pipe, it would be simple to calculate the total amount of flow by adding the gallons per hour of the shower, faucet, and lawn sprinkler. In the same way, figure 8-7 shows the formula for calculating the current in parallel circuits. To arrive at the total current the source is providing, the currents of all the loads are added (summed). All the individual currents .5 amps + 1 amp + 2 amps = 3.5 amps of total source current.

Kirchhoff's Current Law as a Diagnostic Tool

George Simon ohm described the linear (straight-line) relationship of electrical force

(volts) and flow (amps) in the early 1800's. In the middle 1800s, another German physicist, Gustav Kirchhoff, built upon Ohm's work by describing the nature of current as it relates to junctions of wires. From a practical perspective, junctions are terminals that connect a number of wires together. The service entrance panel has three primary current-carrying terminals that are junctions. At each of the hot bus bars, as well as at the neutral bus, a single main feed wire is connected to branch load wires. Any place at which a number of wires are connected together is a junction. To understand Kirchhoff's idea about junctions and current, it is necessary to focus on a small part of figure 8-7. The circled area on that figure is enlarged in figure 8-8. Notice that the current coming into the

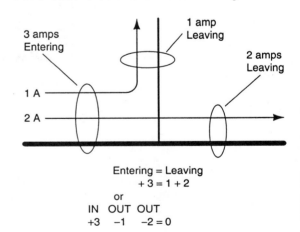

Entering = Leaving
+ 3 = 1 + 2
or
IN   OUT OUT
+3   −1   −2 = 0

FIGURE 8-8 Current entering and leaving a junction are equal.

junction from the left (3 amps) is exactly equal to the current exiting to the right (2 amps) plus the current traveling upward (1 amp). This idea can be described in a more general way. All the current going in is equal to all the current going out. Think for a moment, is there any place in this junction that can store current? There are no capacitors, batteries, or buckets for electrons! Since that is the nature of this junction, then what comes in must be equal to what goes out.

A more formal way to describe Kirchhoff's current idea is: "All the currents at a junction sum to zero". If the incoming currents are positive numbers and the outgoing currents are negative numbers then the junction in figure 8-8 looks like this: +3 amps −2 amps −1 amp = 0 amps. No matter how this idea is described, it means that at any junction, the total of the current arriving must equal the current leaving. This is true if the junction has only two wires or 1000 wires. This idea can also be used to describe the nature of water and gas pipe junctions.

Imagine it is a hot summer day and the phone rings. A customer complains that the central air conditioner does not work. Arriving at the site, you see that the air-conditioner branch circuit breaker is tripped off (midposition indicating overcurrent). Kirchhoff's idea immediately comes to mind—too much current is being drawn by the air conditioner. Since you know that the air conditioner has two main devices that draw current, you suspect that one of them is at fault. The current draws of the compressor and air circulation blower equal the current flowing through the breaker. You turn the breaker off and, at the appliance, disconnect the compressor. On power up, the breaker again trips. The problem must be the blower or associated wiring. A careful power-off inspection reveals a loose connector on the blower motor feed wire, which has caused it to wear through the insulation (abrade) to ground and cause a short circuit. You correct the condition with a little extra effort to be certain that it does not reoccur.

The billing on this job is primarily labor, since minimal material was required to correct the problem. Labor, in this instance, is thinking,

and the subcontractor was Kirchhoff. The accomplished HVAC technician is a thinker as well as a doer.

Parallel Circuit Resistance

New ideas are often confused by starting out with the wrong question, such as: "How do you find the total resistance of a parallel circuit?" Resistance is the quality of electrical components that resists the flow of electricity. As resistance goes up, the current flow drops. In other words, a good (high number) resistor is a bad conductor. Look back at figure 8-7 and notice that each time another resistive device is connected, an additional path for current flow is established. Adding a drill to the lamp and TV set increases the total current flow, even though the drill itself has resistance.

The core of this idea/problem lies in the direction of the number scale representing how well or poorly a device passes electricity. Adding resistors in parallel is not really adding resistance to the circuit, because, as resistance increases, the current flow goes down, while in the parallel circuit, when resistors (lamp, TV, and drill) are added, the total circuit current actually goes up. The solution to the problem makes use of a mathematical technique called the reciprocal.

As an example we may ask the question, "How do you feel about square dancing music?" As shown in figure 8-9, you may respond that on a love scale of 1 to 100, you would rate it 50, where 100 is high on the scale and 1 is low. A reciprocal of the love scale might be the hate scale. Bigger numbers on the hate scale would indicate more hate. The love scale may be modified so that its number order (low to high) goes in the opposite direction. If a 1 is placed over all the numbers on the love scale, it becomes a hate scale. Notice in figure 8-9 that the 100 is divided into 1 and becomes .01 on the hate scale. In this way, the hate scale is the reciprocal of the love scale. What is the biggest number on the hate scale? The numbers on the hate scale are backward as compared with the numbers on the love scale, so the highest hate is 1. Taking the reciprocal by putting 1 over a number changes the direction of a scale—high becomes low and low becomes high. A big lover is a small hater, and visa-versa.

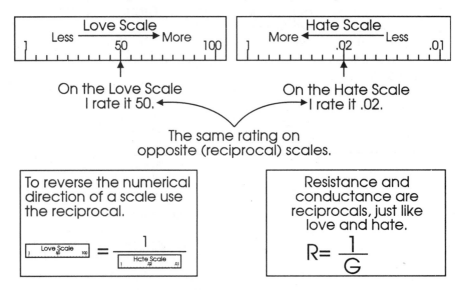

FIGURE 8-9 Reciprocals.

Since resistance is "how much something resists," its reciprocal (a scale with numbers in the opposite direction) describes how much something conducts. Note on the left of figure 8-10 that the reciprocal of resistance is G, conductance. Realize that the quality being discussed is the effect of a load (lamp, TV, and drill) on current flow. This effect could be described in numbers that represent how much it resists, or its reciprocal, how much it conducts. Which one we choose (resistance or conductance) depends on what we want to do with it.

Let us suppose that the iced tea in your glass is at midlevel. You are asked, "Do you want some more iced tea?" If you do not want any more, you might say, "No, my glass is still half full." On the other hand, if you want more you might say "Yes, my glass is half empty." In this case, fullness and emptiness are reciprocals, and which one you use depends on thirst.

Every time another device is added to a parallel circuit, the total current of the circuit increases. Even though each additional load has

resistance, it adds another path (in parallel) for current. Each additional device causes the circuit to conduct more current. Each additional device adds conductance. As a result, the question "How do you find the total resistance of a parallel circuit?" is confusing both conceptually and mathematically. The question to ask is, "How do you find the total conductance of a parallel circuit?" Keep in mind that R (resistance) and G (conductance) are reciprocals of each other.

To find the total conductance of a parallel circuit, the individual load conductances are summed or added. Figure 8-10 indicates the resistance, not conductance, of each device (lamp, TV, and drill). It is simple to convert these resistance measures to conductance by dividing each into 1. If a calculator is available, enter 1, and then divide it by 240 ohms (lamp resistance). Hit the = key and the answer is .0042 siemens. Some years ago, the unit of conductance was the mho (ohm spelled backward); however, the siemens (named for German electrical engineer Werner Von Siemens) is

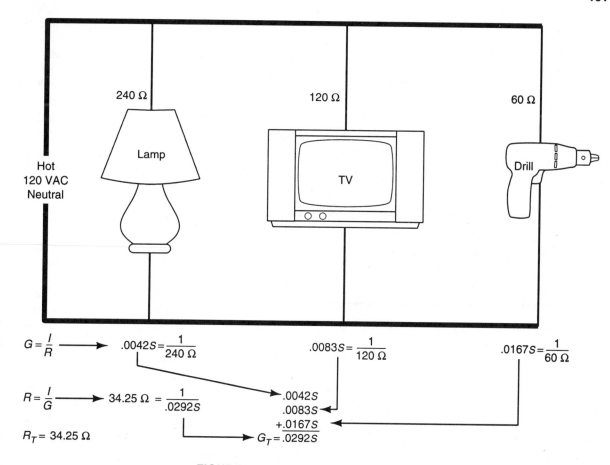

FIGURE 8-10 Parallel conductance sums.

the accepted term in current use. Look at figure 8-10 and find the calculation that converts the resistance of the lamp to conductance. The resistance of the TV set and drill, in this figure, are converted to conductance in the same way. Adding the three individual conductances together (the lamp, TV set and drill) results in the total conductance, in this example, .0292 siemens. The total conductance of a parallel circuit is the sum of the individual conductances.

Unfortunately the most common measure in electricity is ohms of resistance, not siemens of conductance. As a result, it is necessary to convert the .0292 siemens of conductance to ohms of resistance. Again, enter 1 into the calculator and divide it by .0292 siemens. Hit

the = key to see that the total resistance of this circuit is 34.25 ohms.

Recall that the initial question was, "How do you find the total conductance of a parallel circuit?" The answer is the caption to figure 8-10, "parallel conductance sums." But the process must be begun by converting all resistances to conductances and completed by converting the total conductance to resistance. To clarify this process, observe the following steps.

1. Convert each of the parallel resistances to conductances by using the reciprocal process.

2. Sum the individual conductances to find the total conductance.

3. Convert the total conductance to the more commonly used resistance by the reciprocal process.

The parallel resistance (and conductance) discussion just completed is what actually happens within a parallel circuit. The approach taken was as if the circuit were a machine and we removed its cover to examine the workings directly. What's inside is what really counts. As a review of this topic, let us look at it with the cover on.

$$R_T = \cfrac{1}{\cfrac{1}{R_{Lamp}} + \cfrac{1}{R_{TV}} + \cfrac{1}{R_{Drill}}}$$

FIGURE 8-11 Formula for parallel resistance.

The formula for parallel resistance is presented in figure 8-11. If the resistances of the lamp, TV, and drill are "plugged in" to the formula and a little calculator work is performed, the answer is easily arrived at. However, all you get from this approach is an answer. Applying a formula in this way does not increase your understanding of the circuit, nor does it sharpen your mental picture of electricity. Look closely at this formula to identify the reciprocals for the individual conductances, as well as for the total conductance. You could memorize the formula, or realize that in a parallel circuit the individual conductances sum to the total conductance, while resistance and conductance are reciprocals.

SIMPLE AIR-CONDITIONER CIRCUIT

Diagrams of electrical circuits are confusing. A brief survey of manufacturers' electrical diagrams reveals a wide mix of approaches. Some diagrams use the standard symbols as they are presented in Appendix E, while others use electronic circuit symbols. Many diagrams include a legend, or a table that identifies symbols, wire codes, and switch positions.

Appendix E shows some common electric and electronic symbols. Other manufacturers provide wiring diagrams that do not use symbols at all; rather, they represent components with simplified pictorial drawings or icons of them. The pictorial type of wiring diagram is very helpful in tracing actual wire connections, however it often hides the internal electrical nature of components. For example, relays and switches pictured as closed boxes with terminals do not reveal the internal configuration of these components. Wire tracing may be easy, but circuit understanding is nearly impossible. Interpreting the symbols on diagrams is often difficult.

Wiring diagrams also differ in their portrayal of circuit configuration. Some diagrams are drawn as if they were started at one corner and items were added until they were complete. No attention is given to what parts of the circuit are in parallel or series. Occasionally, electrical drawings are done in such a way that "how the circuit works" almost pops right out of the drawing. Companies sometimes provide a number of drawings of the same circuit. A schematic usually provides circuit configuration and internal component details, while a wiring diagram shows how parts are connected together.

Since the general aim of the first section of this book is to develop a clear mental picture of electricity, at this point, it is necessary to connect circuit ideas with circuit drawings. Figure 8-12 is a schematic drawing of a window air-conditioner unit that has been redrawn to emphasize circuit function and configuration.

The main feed wires on the window air-conditioner schematic have been emphasized to add clarity. Notice that the circulation fan and cooling compressor comprise two load circuits that are parallel, electrically, to each other. These circuits might also be described as being "hung" across the main power feed wires. The actions of the fan and compressor are electrically independent of each other, but are mechanically linked in that their switches share a single shaft. On low or high switch settings, both the fan and the compressor function. Thus in the event of compressor failure, the fan would still function. Most HVAC appliance

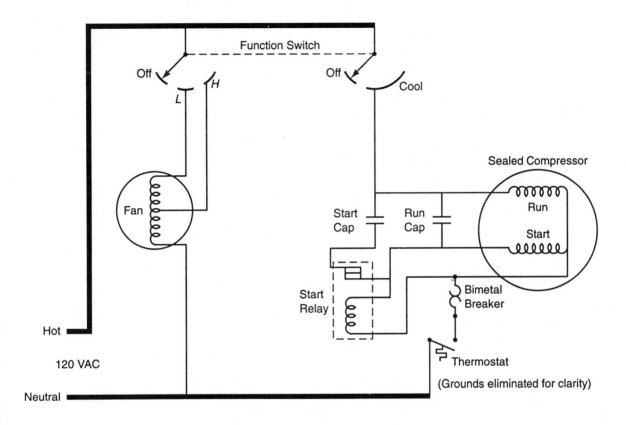

FIGURE 8-12 Simplified window AC unit schematic.

circuits consist of a number of load devices. Burners, fans, compressors, and heaters are all connected in parallel across the appliance main power feed wires. Marking the main feed wires of an appliance schematic with a colored pen often helps to reveal the parallel configuration of these load devices.

A String of Switches: Series Circuits

The World Series is played one game after another, a TV series is one episode after another. In series electrical circuits, components are connected in a string, one after another. Though HVAC appliances are fundamental parallel circuits, the individual legs or branches of these circuits generally consist of a string of components connected end to end—in series.

The word series has both a Greek and Latin origin, which means to join and string together. For example, a chain is a series of links.

The window air-conditioner schematic in figure 8-12 contains two series circuits, each of which is connected across the main power feed wires. To understand this circuit, it is necessary to concentrate on certain parts and ignore others.

Look carefully at the circuit and locate the two "legs" of the circuit. Notice that the fan and OFF-L-H switch are connected in series. These components constitute one leg of the circuit. To the right in figure 8-12 is the second leg of this circuit. For the compressor to run, current must flow from the lower feed wire through the thermostat, and then through the bimetal breaker and to the compressor. For the moment, the compressor motor, start and run capacitors, and starting relay all are considered as a single unit. From the compressor, current then runs through the OFF-COOL switch to the upper feed wire that completes the circuit. In this case, the series circuit string is: lower feed wire–thermostat–bimetal breaker–compressor–OFF-COOL switch–upper feed wire. This air conditioner is actually a parallel circuit composed of two series circuits.

The series circuit that supplies power to the compressor in figure 8-12 has only one path for current flow. Any interruption in that path will stop current. If either the thermostat is "open"

or the bimetal breaker has been "tripped," or if the switch is in the OFF position, no current will flow, and as a result, the compressor will not run. Since a series circuit is like a chain, a break or opening in any link will interrupt the chain of components and stop current flow. Series circuits have only one path for current flow. Therefore, an opening or break at any point in that path stops current. If any of the switchlike components (thermostat, bimetal breaker, and OFF-COOL switch) is open, the compressor cannot function. For the compressor to run, all switchlike devices need to be closed to provide a complete circuit.

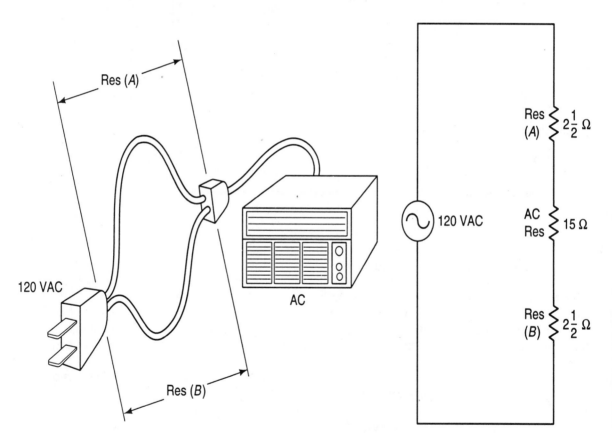

FIGURE 8-13 The thin extension cord.

The Thin Extension Cord

Often it seems that receptacles are never where you need them. Install a window air

conditioner and the wire is just a bit too short to reach the receptacle. That extension cord you just picked up for $1.89 should do it. Plug it in, switch it on, and cool down. Funny, the air conditioner seems to require a great deal of effort getting started, and, "What's that burning plastic smell?" Something is wrong.

The receptacle provides 120 VAC of electrical force, force that pushes current through the air-conditioner circuit. However, in this instance, a thin extension cord has become part of the air-conditioner circuit. A pictorial schematic of the extension cord and air conditioner is shown on the left in figure 8-13. To help clarify the nature of this circuit, the extension cord wire has been split lengthwise. The path for current in this circuit is from the one blade of the extension cord plug, through one wire of the extension cord, through the air conditioner, and back through the remaining wire of the extension cord to the other blade of the plug. Since the parts of this circuit are connected in a string (extension cord–air conditioner–extension cord), it must be a series circuit.

The schematic diagram on the right of figure 8-13 represents the extension cord–air-conditioner circuit in such a way that its series nature is clear. Above and below the resistor representing the air conditioner are resistors that represent the two wires of the extension cord. The resistance of the air conditioner was calculated with Ohm's law on the basis of 8 amps of current at 120 VAC, while the resistance of each of the wires of the thin extension cord was assumed.

Electric current flowing through resistance produces heat. As a result of too much current, the thin extension cord has become an unintended heater, as well as a fire hazard. Furthermore, the extension cord's resistance has affected the other important circuit characteristics of total resistance; current and voltage drop. A more detailed examination of series circuits is necessary.

SERIES CIRCUIT THEORY

If a circuit has only one path for current flow, then every resistive device (resistance) added to that circuit reduces current flow. A garden hose, since it has only one path for water, can be used to illustrate this idea. Stepping on the

hose and restricting the flow of water is like a resistor reducing current flow. If we step on the hose in two places, the water flow is further reduced. As more resistors are added to a series circuit, the flow of electricity is further reduced. Recall that as the resistance of a device goes up, it becomes a poorer conductor. In the same way, adding resistances in series circuits makes the circuit a poorer conductor. A good question to ask is, "How do you find the total resistance of a series circuit?"

Series Circuit Resistance

Resistance in a series circuit is simple to calculate. A series circuit is a single loop for current flow, so every resistance added to the circuit "gets in the way" of current flow. Each addition of resistance (resistive devices) adds resistance. The total resistance of a series circuit is the sum of the individual resistors. To arrive at the total resistance, simply add the individual resistances.

The air conditioner with the thin extension cord described earlier is a good example of a series resistance circuit. Figure 8-14 is a schematic of the circuit that includes the original air conditioner (15 ohms) and the extension cord going to (2.5 ohms) and returning from (2.5 ohms) the appliance. As illustrated in figure

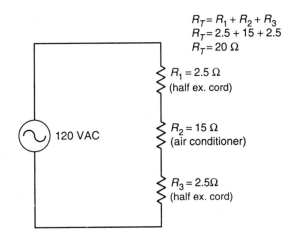

$$R_T = R_1 + R_2 + R_3$$
$$R_T = 2.5 + 15 + 2.5$$
$$R_T = 20\ \Omega$$

$R_1 = 2.5\ \Omega$
(half ex. cord)

$R_2 = 15\ \Omega$
(air conditioner)

$R_3 = 2.5\Omega$
(half ex. cord)

120 VAC

FIGURE 8-14 Series resistance sums.

8-14, the total resistance of this series circuit is 20 ohms. Without the additional resistance of the "too thin" extension cord, the resistance of the air conditioner by itself is only 15 ohms. Since the cord is in series with the air conditioner, it adds resistance to the series circuit. Series resistance sums.

Series Circuit Current

The current draw of the air conditioner (15 ohms) without the extension cord can be easily calculated with Ohm's law. Since $I = E/R$, 120 VAC is divided by 15 ohms, which results in 8 amps. By itself, the 15-ohm air conditioner draws 8 amps.

To calculate the current draw of the air conditioner and extension cord, the total resistance of 20 ohms must be used. As the box in figure 8-15 shows, 120 VAC divided by 20 ohms results in an air-conditioner and extension-cord current draw of 6 amps. The resistance of the extension cord has reduced the current of the circuit from 8 amps to 6 amps. It is important to realize that in both cases the current was calculated on the basis of the source voltage and total resistance.

Since series circuits have only one path in which current can flow, the current has to be

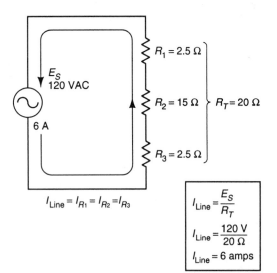

$I_{Line} = I_{R1} = I_{R2} = I_{R3}$

$$I_{Line} = \frac{E_S}{R_T}$$

$$I_{Line} = \frac{120\ V}{20\ \Omega}$$

$$I_{Line} = 6\ amps$$

FIGURE 8-15 Series currents are the same.

the same throughout the circuit. The current calculation for the air conditioner and extension cord indicated a flow of 6 amps, which is shown in the circuit diagram in figure 8-15. Follow the 6-amp current flow and notice that it passes through all three resistors. The current through R1, R2, and R3, as well as the line current, are all the same—series currents are all the same.

For a moment, let us look at the practical effect of the "too thin" extension cord on the air conditioner. First, additional resistance was introduced into the circuit via the thin extension cord. Second, the additional resistance reduced the current flow to the air conditioner. Reduced current flow through the air conditioner will "starve" it. The compressor will have difficulty starting, and, if it starts, will provide reduced and inefficient cooling. Furthermore, the extension cord's resistance will affect the voltage or electrical force applied to the air conditioner.

Series Circuit Voltage

A key to understanding series circuit voltage is to realize that whatever voltage (force) the source applies at one of its terminals is "used up" by pushing current through the circuit to the other terminal. The arrow in figure 8-15 represents current flow during the half of the sine wave in which the lower source terminal is negative (–) and the upper positive (+). At that point, 120 volts (–) pushes 6 amps (a large number of electrons) up through the circuit. As the electrons move upward, the force pushing them (voltage) becomes less and less as it overcomes resistances in the circuit. Finally, the electrons have just enough force (voltage) to arrive at the upper source terminal. One way to think of this is that the voltage (force) squeezes electrons through resistance (current flow), and in doing so is consumed.

Often, days in our lives are like voltage in a series circuit. The force you awaken with pushes you through the resistances of the day until, exhausted of force, you fall asleep. Sleep, like the battery, "pumps up" your force to enable you to begin another day.

One must think about this series circuit voltage idea to understand it. Imagine yourself as the electron being pushed through the circuit by voltage and gradually getting tired until your day is over and you fall asleep. A common question that occurs is related to what happens in the series circuit if a resistor, say R1 in figure 8-15, is removed. It would seem that force (voltage) would be left over (not used up), since the circuit contained fewer resistors. However, this does not happen, because the removal of a resistor is immediately compensated for by an increase in current. In other words, the nature of a loop circuit (series circuit) is such that fewer resistors result in greater current, and again the force (voltage) is "used up."

The largest portion of the force with which you start out the day is used up, overcoming the largest resistance during that day. For example, on weekdays, you may use up two thirds of the force at school and one third at a part time job. In the same way, the largest portion of series circuit voltage is used up pushing current through the largest resistances. Smaller portions of resistance use up smaller portions of the voltage (force). Don't forget that series circuits plan their day so that all the voltage (force) is used upon return of the electrons to the source.

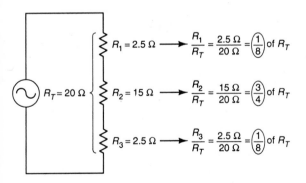

FIGURE 8-16 Each resistor is a portion of *Rt*.

With a series circuit, calculating what portion of the total resistance each resistor represents is necessary before voltage is even considered. Figure 8-16 illustrates the same air-conditioner–

extension-cord schematic discussed earlier. The total resistance (20 ohms) of this circuit is arrived at by adding the individual resistors. What portion R1 is of the total can be calculated by placing R1 over Rt, which, when reduced, equals one eighth. By the same process, R2 is three fourths of the total resistance and R3 is one eighth of the total resistance. If this calculation is correct, all three portions of resistance will add to the whole (1/8 + 1/8 + 3/4 = 1). You may be in for a surprise if you use the same type of calculation to find out to what things you devote most of your day.

Electrons in figure 8-17 awake with 120 volts of force. One eighth of that is used up pushing current through R3 (1/8 × 120 VAC = 15 VAC). Again, reasoning in the same way, three fourths of the 120 VAC is consumed pushing current through R2 (3/4 × 120 VAC = 90 VAC). The last portion of 120 VAC is consumed by pushing current through the remaining portion of resistance, R1 (1/8 × 120 VAC = 15 VAC). Notice that the voltage (120) was divided up proportionally

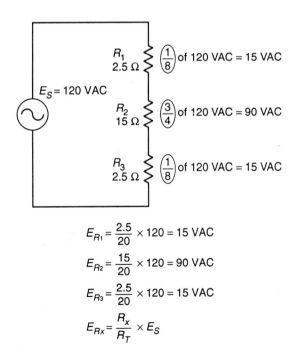

FIGURE 8-17 Portion of resistance equals portion of voltage.

the basis of how much of the total resistance lies in each resistor—a bit like dividing up a pie on the basis of each person's size.

All of the series circuit proportional division of voltage idea is mathematically described by the formula at the bottom of figure 8-17. It says, in numbers: the voltage drop across any resistor in a series circuit equals that resistor over the total resistance (portion of resistance) times (portion of) the source voltage. In series circuits, voltage is divided on the basis of each resistor's portion of the total resistance.

Kirchhoff's Voltage Law as a Diagnostic Tool

Look back at figure 8-17 and take note of the voltage drop across each of the three resistors (15 volts, 90 volts, and 15 volts). If these three voltages are added together, what will they equal? Both you and Kirchhoff may have come to the same conclusion. If you realized that the sum of the voltage drops across the resistors is equal to the source voltage, you shall henceforth be referred to as Kirchhoff Jr. With any series circuit, the voltage drops across the resistors (loads), when added, equal the source.

Figure 8-18 is the same air-conditioner-extension-cord schematic, only this time the voltages are illustrated. The voltage across the air conditioner could be easily measured with a voltmeter. It would indicate 90 VAC, well below the 120 VAC that the receptacle provides. Kirchhoff's idea about voltage in a series circuit indicates that somewhere in the series circuit 30 VAC is being dropped across unintended resistance. By now it is clear that 15 VAC is lost in each of the feed wires of the extension cord.

Troubleshooting series circuits can be made much easier by keeping Kirchhoff's voltage idea in mind. If individual voltages do not sum to the source, then somewhere there is unintended resistance. For example, if the voltage of a car battery is found to be 12 volts and the head lamps measure only 10 volts, then voltage is being lost. Measuring a voltage drop across the head lamp relay contacts, terminal connections, or wire would help a great deal in

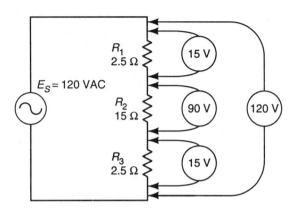

$$E_S = E_{R_1} + E_{R_2} + E_{R_3}$$
$$120 \text{ VAC} = 15 \text{ VAC} + 90 \text{ VAC} + 15 \text{ VAC}$$

FIGURE 8-18 Series voltage drops sum to the source.

locating and repairing the problem. Switches, wires, connections, and relay contacts are intended to be good conductors. In series circuits, voltages measured across these parts indicate unintended resistance. The absence of voltage drops across resistive parts may indicate a short circuit, no resistance, and hence no voltage drop.

One brief note before this section is finished: The air conditioner–extension cord used as an example throughout the discussion of series circuits is being seriously starved for power. It requires 120 VAC and 8 amps, which results in a power consumption of 960 watts. Due to the extension cord, it is receiving 90 VAC and 6 amps, or only 540 watts. It cannot run properly, and may well be damaged by its labored running. Furthermore, the extension cord has a drop of 30 VAC (15 + 15) and 6 amps flowing through it. As a result, the extension cord is a 180-watt heater (30 VAC × 6 amps). A serious fire or electrical short circuit is a likely result.

MEASURING VOLTS, AMPS, AND OHMS

While electrical components and the wires that connect them are real and can be visually examined, the electrical phenomena that occur within them are not. Meters enable the HVAC technician to "make visible" electrical circuit actions. Volts, amps, and ohms can only be seen through the skillful use of meters. However, before discussing meters, a brief but important point needs to be made. Common HVAC electrical failures are often revealed by thinking about the symptoms and utilizing the most important instrument of all, the human mind. Further, electrical malfunctions frequently are caused by, and result in, symptoms that can be detected without meters. A careful power-off inspection of the offending circuit can often reveal a loose connection, short circuits, burned components, tripped breakers, seized motors, broken belts, and many other problems that can be either seen or smelled. Don't immediately use "hi-tech" meter diagnosis because the problem is often visible. The most important instruments the technician has are the senses and the brain. Use them first.

In the less common situation where no visual evidence indicates a possible source of the problem, the use of meters will probably be necessary. Let us pause again at this point to consider a few important ideas. The purpose of a meter is to affirm or contradict a suspicion that the technician has about a malfunction. For example, if the proper functioning of the cad cell is in question, it may be removed from the circuit and tested with an ohmmeter. Test results bring the technician closer to solving the problem.

Nothing is more dangerous or pointless than probing around in a hot circuit with a meter to find the problem. A "bunch" of voltage and current meter readings provide no useful data. This approach brings the technician no closer to solving the problem than before. Rather, the results of this approach can lead to lethal electric shocks, damaged meters, and additional circuit problems. A common result of this type of approach goes as follows: a meter is incorrectly connected, it smokes, the technician panics and grabs for the meter, the technician gets a severe shock. This technique provides no useful information, and can result in a destroyed meter and a trip to the emergency room. Meters should only be used to affirm or contradict a suspected problem.

Meter Types

A wide variety of multipurpose meters are available. They permit the measurement of more than one electrical characteristic (volts, amps, ohms). Also, these meters differ in the nature of their displays: analog (moving needle) or digital (numerical). Furthermore, meters differ in less obvious ways, for example internal resistance. To use meters effectively, a more detailed discussion of their characteristics is necessary.

"Refer to manufacturer's instructions and specifications" is good advice when using a particular meter. However, to the beginner, the great variety of meters available and the particulars of these meters are bewildering. Let us limit our discussion of meters to the most common types, those that are useful in the HVAC field. In so doing, we will consider three general types. Each of the meter types will be described in general terms. Figure 8-19 is a summary of the characteristics of three meter types, as well as of a simple neon test lamp.

Perhaps the most common electrical tester is the multi-meter, which is sometimes referred to as a multi-tester. Ordinarily, the multi-meter contains no transistors or integrated circuits. Rather, it has a meter movement, range switch, function switch, and resistors. The meter requires no internal power or batteries to measure volts or amps. Instead, the needle is moved upscale by the volts or amps being measured. In other words, this meter uses a small amount of the electricity it is measuring to power itself. For this reason, multi-meters are described as having a low internal resistance. With electrical circuits, this small amount of electricity is of no consequence. However, electronic circuits, such as electronic thermostats, often run on very small voltages and currents. If the multi-meter is used to measure these circuits, large errors will be caused by the meter's low resistance. In practical terms, the multi-meter takes so much electricity from an electronic circuit that it acts like a short circuit. Multi-meters are intended for electrical, not electronic, work.

When the multi-meter is used to measure resistance, internal batteries supply the power. The measurement of resistance is accomplished by running a test current (supplied by meter batteries) through the external device being tested. The ohms reading on the meter scale is actually a result of the test current. Often, multi-meters have a continuity function that provides an audible "beep" when a circuit is completed between its probes. Locate the multi-meter column of figure 8-19 and study its characteristics.

Every HVAC technician should own a multi-meter to measure both AC and DC volts, as well as ohms. The DC amp function is not often used in HVAC work since most appliances use AC power. As figure 8-19 indicates, most multi-meters lack an AC amp function, which must then be provided by another meter. Some expensive multi-meters have special accessory probes that enable them to measure up to 300 amps AC. The biggest differences between the less and more expensive versions of the multi-meter relate to accuracy, durability, and maximum DC amps measured. Inexpensive meters are often accurate enough for ordinary work, however, they do not hold up well to long-term field use. High-quality meters, in addition to having higher DC amp scales, last longer and, in time, become extensions of the technician's mind. A good meter becomes a well-known and trusted friend.

Characteristics	METER TYPES			
	Multi-Meter	Electronic Vom	Clamp-On Amp	Neon Tester
DC Volts	0-1000 Vdc	0-1000 Vdc	No	Above 100 Y/N
AC Volts	0-1000 Vac	0-1000 Vac	0-600 Vac	Above 100 Y/N
Internal Resistance	Range Dependent 20K–2 meg ohm	High 10 meg-100 meg ohm	Range Dependent 20k–2 meg ohm	N/A
DC amps	0-12 Adc	No (see text)	No (see text)	No
AC amps	No (see text)	No (see text)	0-300 Aac	No
ohms	0-20 meg ohms	0-20 meg ohms	No (see text)	No
Continuity Buzz	Often	Often	No (see text)	No
Display	Analog (meter)	Analog/Digital	analog/Digital	NA
Auto Range	No	Digital Often	No (see test)	NA
Cost	$20-$200	$25-$200	$50-$200	$5

FIGURE 8-19 Basic meters for HVAC.

An electronic VOM (volt-ohm-meter) is very similar in appearance to a multi-meter. It differs internally in that it contains electronic components, usually integrated circuit (IC) chips. These components enable the electronic VOM to measure DC and AC volts without placing a load on the circuit being measured. The power required to move the needle of the meter or provide a digital readout comes from a battery inside the meter. This electronic type of meter is useful in measuring voltages in electronic circuits since it requires practically no power from the circuit being measured. As figure 8-19 shows, the internal resistance of the electronic VOM is very high. Most electronic VOMs can also accurately measure ohms. DVMs (digital voltmeters), TVMs (transistor voltmeters), and digital or FET (field effect transistor) voltmeters are all high-internal-resistance electronic VOMs.

Electronic VOMs are not necessary for HVAC electrical work. However, they are desirable in two ways. Electronic troubleshooting, which requires much advanced study, is impossible without this meter. Also, most digital (numeric) readout multi-meters are inherently electronic VOMs. They are often auto ranging. All the technician has to do is set the meter on the appropriate function, AC volts, for example, connect the probes, and the voltage is displayed. No other adjustments are required. While this is convenient, the digital readout is less desirable in certain situations. Intermittent loose connections often make the indicator needle of an analog meter bounce up and down while a digital display's blinking numbers are far less helpful.

Multi-meters and electronic VOMs are available from many manufacturers in a wide variety of configurations, and thus it is impossible to give specific instructions on their use. It is, therefore, important that the beginner study the specific instructions provided with each meter. However, the general ideas related to using either of these two meter types may be developed by discussing a simplified generic meter. Simplification in this case is achieved by removing extra functions. The meter shown in figure 8-20 is very much like an inexpensive multi-meter.

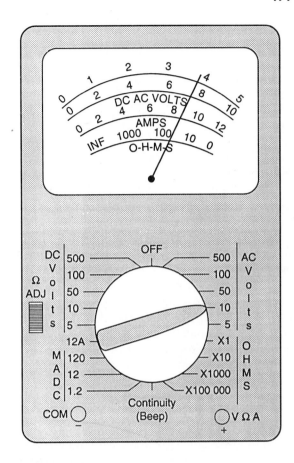

FIGURE 8-20 Generic multi-meter.

Notice that the meter has only one knob. The selection of function (volts, amps, or ohms) and range (0–5VAC, 0–100 VAC, etc.) is achieved by rotating this knob. For example, imagine that we want to measure a 9-VDC battery. The knob would be turned to a DC volt scale (upper left function) that is above 9 volts. On this meter, the 10-VDC scale would be appropriate. Lower scales would "peg" the meter, while higher scales would provide a reading so far to the lower left that it would be hard to read the meter accurately. The key idea is, if the voltage is known, select a scale just above it. This meter has two voltage scales on its face (0–5 and 0–10). Since the knob has been turned to the 10-VDC range, the reading would be taken from the 0–10-volt scale. At this point, the negative probe is connected to the negative battery terminal and the positive probe to the positive battery terminal. The voltage is read from the 0–10-VDC scale. Polarity (+ –) is important for DC volts, but not for AC. On this meter, all DC and AC voltages that are multiples of 10 (10 and 100) would be read from this scale, while scales that are multiples of 5 (5, 50 and 500) would be read from the 0–5 scale. In each instance, the decimal place would be adjusted to make the scale agree with the range setting. If the voltage being measured is unknown, start at a high scale and work down to one that permits accurate readings.

When the multi-meter or electronic VOM is used to measure resistance, three important ideas need to be understood. First, the device or component being tested must be disconnected from the circuit. For example, a relay coil would have to be disconnected from the rest of the appliance at its terminals. Circuit wires must be removed from the back of a switch being tested. Only when the component is disconnected can its resistance be accurately measured. Second, the meter needs to be adjusted so that when the probes touch each other, the needle goes to zero on the ohms scale (no resistance, 0 ohms). With multi-meters 0 ohms is to the right end, while electronic VOMs have the 0 ohms on the left. Most electronic VOMs also require that the meter be adjusted with the probes disconnected (infinite ohms). A second control, usually labeled "Ohms," is turned until the needle rests over the infinite or maximum far right on the ohms scale. Multi-meters do not require this second adjustment. Third, the range of ohm has to be selected, usually by trial and error, that causes the meter to read as close to midway as possible. As shown in figure 8-20, the ohms ranges are × (times) 1, 10, 1000, and 100,000. To read the ohms scale, simply multiply the number on the scale by the range setting. For example, if the scale indicates 4 and the range is ×10, then the resistance is 40 ohms.

While voltage is measured by placing the meter probes "across" the source of expected voltage, current is measured with the multi-meter in a very different way. It is important to understand that ordinary ammeters are connected in series with the current flow. Figure 8-21 shows a simple circuit with a battery as the source and a lamp as the load. The correct meter polarity is indicated with this DC circuit. Notice that the ammeter is in the path of current flow. Since the ammeter becomes part of the circuit, its internal resistance is extremely low (good conductor). If the ammeter is connected "across" the circuit, as if it were a voltmeter, the meter becomes a direct short circuit. This is an easy mistake to make, so extreme caution is necessary while using the amp function of the multi-meter. Since most multi-meters measure only DC amps, and HVAC equipment is powered by AC, there is seldom any good reason to use this function. Clamp-type ammeters are more durable and safer to use for these measurements since no direct circuit connections are necessary.

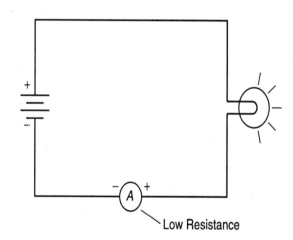

FIGURE 8-21 Ammeter in series.

Both meters discussed so far are useful for measuring volts and ohms, however, multi-meters and electronic VOMs generally cannot measure high AC amps. To provide AC amp measurements, the clamp-type ammeter is used. The meter chart in figure 8-19, displays the characteristics of clamp-on ammeters. This unique meter is not connected directly to a circuit, but is loosely clamped around one of the wires feeding current to the circuit being tested.

Figure 8-22 illustrates the lobsterlike claws of this meter and their position relative to the current-carrying wire. When the current carrying wire is centered within the clamp and the clamp is positioned at about 90 degrees, a reasonably accurate measurement of the current is displayed by the meter. In a general way, the AC flowing in the wire acts like the primary of a transformer and provides a constantly changing magnetic field. Inside the meter a coil wound on the clamp mechanism acts like a transformer secondary and provides power to the analog or digital readout of the meter. In this way, AC amp readings may be conveniently made with no direct circuit connections. Many clamp-on meters also incorporate AC voltmeters, which are used with plug-in accessory probes.

The clamp-on ammeter has many advantages. Appliances may be operated while the meter is monitoring current, allowing various circuit functions to be checked without internal probing. Adjustment of this meter is simple as it only requires the selection of an amp scale that is higher than the expected current flow in the wire. This meter tolerates errors in adjustment rather well, since current is induced rather than directly connected.

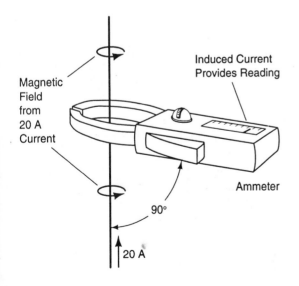

FIGURE 8-22 Clamp-on ammeter.

Magnetic Field from 20 A Current

Induced Current Provides Reading

Ammeter

90°

20 A

Our tool box contains a banged-up but trusted multi-meter ($125 20 years ago), a digital electronic VOM ($75), a name brand ($65 basic version) clamp-on ammeter, and a neon test light (see figure 8-19). The digital electronic VOM has an optional temperature probe. This assortment of meters enables us to measure volts (multi-meter) and amps (clamp-on ammeter) simultaneously or independently. Ohms and continuity are measured with the multi-meter. Temperature is measured with the digital electronic VOM and special probe. However, for most troubleshooting and critical thinking processes, the neon test lamp and continuity (multi-meter) provide the necessary information.

Meter Safety

Before the theory of connecting meters to circuits is discussed, attention needs to be devoted to the topic of safety. HVAC appliances and associated wiring are designed to be safe if properly installed and in good working order, and if the manufacturer's instructions are followed. Consumers who use common sense risk little chance of electrocution. HVAC technicians who install and repair equipment are in quite another situation.

During installation, the technician needs to adhere to the manufacturer's instructions, as well as to electrical, plumbing, fire, and building codes. This is difficult, and often requires on-site supervision by experienced and licensed experts who are qualified to checkout the job as it proceeds. In the same way, your instructor needs to show you the hands-on part of the profession in a step-by-step fashion, pointing out possible hazards on the way.

Repair work constitutes a large part of the technician's job. Safety becomes a primary concern, since the technician cannot be certain that the equipment was properly installed or is in good working order. The "do it yourself" installed or repaired system is especially dangerous to work on. Wires may be crossed, grounds missing, and breakers and relief valves either missing or inoperable. Often this type of customer wants you to "just make it work."

Don't! You are not a member of the bomb squad. Know when to refuse a job. Again, your instructor can offer advice on how to handle this sort of situation.

During the repair of equipment, it is necessary to remove covers from appliances. The second that the cover is removed, hazards multiply. Moving parts, hot components, sharp edges, and exposed electrical circuits await a careless hand. It is not only the technician who is in danger, but also anyone else who is present while the equipment is being serviced. Curious customers and children need to be kept clear of the area during repair. If it is necessary to leave a work site while the repair is in progress, be certain to replace covers and leave the site hazard-free.

Once the covers are off of the appliance, never try any test you are not absolutely certain is safe to operate. This means that you are familiar with the theory of the test, as well as the practice. In other words, you have seen it demonstrated by an HVAC professional and understand exactly what to do and what hazards are to be avoided.

In addition to the safety practices your instructor explains and demonstrates, there are some general and specific rules that should be followed.

General safety rules:

1. *Expect the unexpected.* Verify first hand what you are told or believe to be true. Consider that the breaker supplying power to the oil burner is off only when you yourself turn it off and see the burner shut down.

2. *Look first.* Carefully inspect the area you are about to work in and around. A hazard may be hidden near the fitting that needs tightening.

3. *Don't rush.* Take the necessary time to think through, step by step, what you are doing. Professionals get to know how long it takes to do various jobs and schedule accordingly.

4. *Work within your mental and physical capacities.* Do only those things that you are certain you know how to do. Surgeons perform only those operations that they trained and approved to perform. Furthermore, accidents happen much more often to people who are tired. Don't be tempted to take on too much work.

5. *Give your full attention to what you are doing.* "I did that a thousand times and never got a shock" indicates the danger that lies in the routine. The better you become at a task, the more likely it is that you will drop your guard.

6. *Consult experts.* Keep the telephone numbers of respected HVAC experts handy. Fear of asking a stupid question has killed many people.

7. *Whenever possible, don't work completely alone.* In the event of an accident, immediate attention is necessary.

Electrical safety rules:

1. *Don't work on hot circuits.* It is never necessary to work on circuits that have power connected. Test probes are available that permit easy connection to a circuit while it is turned off. The power may then be turned on and the meter readings taken. Live circuit testing is reserved for experts with many years of experience.

2. *Don't be grounded.* Leaning on the chassis of an appliance may be comfortable, but in the event of contact with an unexpected hot wire, you become a low-resistance circuit to ground. If you are grounded, an unpleasant shock can become lethal. Water, pipes, equipment frames, structural steel, and many other things are good grounds. Avoid them while working with electricity.

3. *Avoid using two hands.* Electricity flowing through the body from hand to hand

flows through the heart and is often lethal.

4. *Use safety equipment.* Approved insulating gloves and boots are useful in dangerous situations. Safety glasses should be worn at all times.

Connecting Meters to Circuits

No attempt should be made to connect a meter to a circuit until the technician understands basic meter use, the circuit being tested, and safety practices related to both. Selecting what to measure (volts, amps, or ohms) and where to measure them (test points) depends on suspected circuit or component failure. "Service Hints," the last section in this chapter, presents ideas related to the detective-like process of "solving the mystery by identifying the guilty person" (component). Preceding the

detective work, however, is the development of skills related to connecting meters and "making sense of their readings."

A good way to learn meter use is to imagine connecting meters to circuits at various points and discussing the resulting readings. From a practical perspective, each imaginary use of the meter can then be used to illustrate "how to," as well as what circuit clues are indicated by the meter readings. Since the simplified window air conditioner (figure 8-12) has been discussed, it will serve as the example for voltmeter, ampmeter, and ohmmeter use.

Measuring volts: Following are a series of brief comments on connecting the multi-meter to the air-conditioner circuit, adjusted to measure AC volts. Each meter connection is indicated in figure 8-23 with a letter identifying the associated comments that follow.

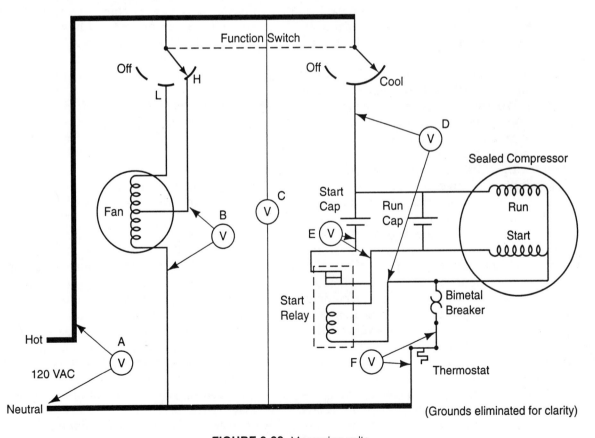

FIGURE 8-23 Measuring volts.

Voltmeter A: The meter should read 120 VAC (line voltage). A reading of zero indicates that no power is flowing to the appliance, which may be due to an open, either in the line cord or circuit. Also, check to be certain that the branch circuit breaker is on. Meter readings much below 110 VAC reveal low line voltage, a heavily loaded circuit, or perhaps a circuit that is a long distance from the entrance panel.

Voltmeter B: With the air conditioner function switch in the high position the reading should be 120 VAC (line voltage). A reading of zero indicates the possibility of an open fan switch or a loose connection in the fan circuit wires. With the fan switch in the low position, the meter will read about half the line voltage, since the other half of the line voltage will be dropped across the upper half of the fan motor stator coil. If, while the fan switch is in the low position, the meter reading is taken across the upper and lower fan motor connections, full line voltage should be indicated.

Voltmeter C: Readings taken across the main power wires throughout the appliance should indicate 120 VAC (line voltage). Breaks in the main power wires within the appliance can be located in this way. A break may also be indicated with a test lamp.

Voltmeter D: With the function switch of the appliance in the high-cool position, the meter should indicate 120 VAC (line voltage). A reading of zero indicates that the switch, bimetal breaker, or thermostat is open-circuited. Check the thermostat to be certain that it is set high enough to allow the compressor to function. Note that the meter is connected to the compressor, including the start circuit, and hence it indicates only power to the compressor. Power to the compressor may also be indicated with a test lamp.

Voltmeter E: This meter reading diagnoses compressor start-up function. When the compressor is first turned on, the start relay points are closed and the reading should be zero. As the compressor motor comes up to full operating speed, the voltage across the start winding rises, pulling the relay open and disconnecting the start capacitor. At this point, the meter should indicate voltage well below the line voltage, which results from an indirect circuit through the start capacitor. In this case, the specific voltage after start-up is not important, but rather is an indication of relay function.

Voltmeter F: Closed contacts on the thermostat should be indicated by a zero voltage reading on the meter. When the thermostat is warm and opens the circuit to the compressor, the meter should read close to the line voltage. Notice that the meter in this instance is acting like a high resistance in series with the low-resistance compressor motor and switches. Remember that with series circuits, voltage divides proportionally to resistance. Since the meter is a very-high-resistance device as compared with the air-conditioner components with which it is in series, all of the line voltage is dropped across the open thermostat. A test lamp would provide the same information.

Measuring amps: The use of the clamp-on ammeter is illustrated in figure 8-24. It shows how this meter may be clamped around wires in the circuit of the air conditioner to provide information useful in locating faults. Readings are based on the air conditioner looked at earlier, which ordinarily draws 8 amps at 120 VAC.

Ammeter A: The nature of the clamp-on Ammeter is such that it must be clamped around *only one of the two feed wires.* For example, lamp cord wire would have to be split. Properly clamped around the wire in position A, the meter should indicate 8 amps while the air conditioner is running. During start-up, the current may momentarily rise above 15 amps. If the current remains high after start-up, a short circuit or frozen motor may be the problem.

Ammeter B: Since meter position B measures current on a main feed wire, these

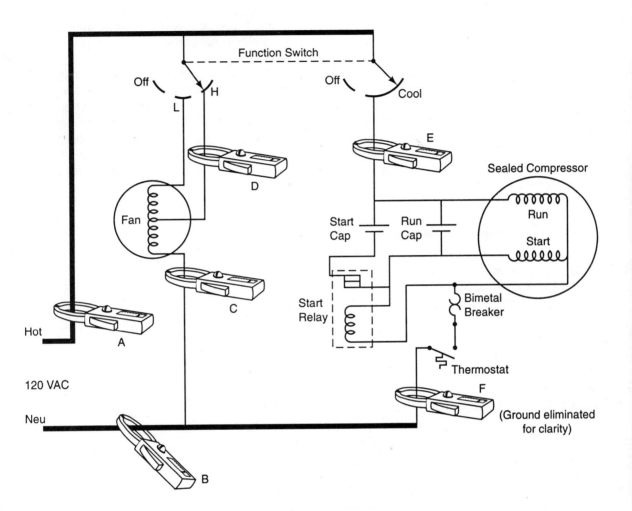

FIGURE 8-24 Measuring amps.

readings should be identical to position A. Differences in these readings would indicate an unintended current flow, most likely to ground. With this type of short circuit, the air conditioner may function, but it poses a safety hazard in that the case may be "hot." Ground fault interrupter (GFI) circuit breakers will trip in situations where the currents in both feed wires are not equal. They protect the consumer from shock between the hot and ground, but not from shocks between the hot and neutral. GFI trips indicate a "leakage current" short circuit that is dangerous.

Ammeter C: Positioned around the common fan feed wire, the meter should indicate about 1 amp with the fan on high and about ½ amp on low for this particular air conditioner. Ordinarily, the current draw of the fan is labeled on the motor. If the current is above normal, the fan may need bearing lubrication or have a short-circuited stator. Assuming that the fan switch is in good working condition, a current reading of zero indicates an open circuit in either the fan motor or wiring.

Ammeter D: High fan-speed current will be indicated with the meter in this position.

Low-speed fan current is not carried by this wire; hence, on the low fan setting, the meter should read zero.

Ammeter E: This meter position should read only compressor current. It should be high at start-up, and then drop to 7 amps. With this particular unit, the total current draw is 8 amps. Since the fan accounts for 1 amp, the remainder (7 amps) flows through the compressor. This is a good example of Kirchhoff's current law. High current usually indicates compressor problems. They may include short-circuited stator coils, overcharging, faulty start relay, or faulty capacitors.

Ammeter F: Position F, like position E, measures compressor current. Which position is used often depends on what wire is easiest to clamp around.

Many of the currents measured in figure 8-24 may be more easily and safely measured with a clamp-on ammeter accessory that is plugged into the receptacle and provides a single outlet into which the air conditioner is plugged. This device provides a ring that the clamp-on ammeter is clamped around. In this way, certain currents of the air conditioner may be measured.

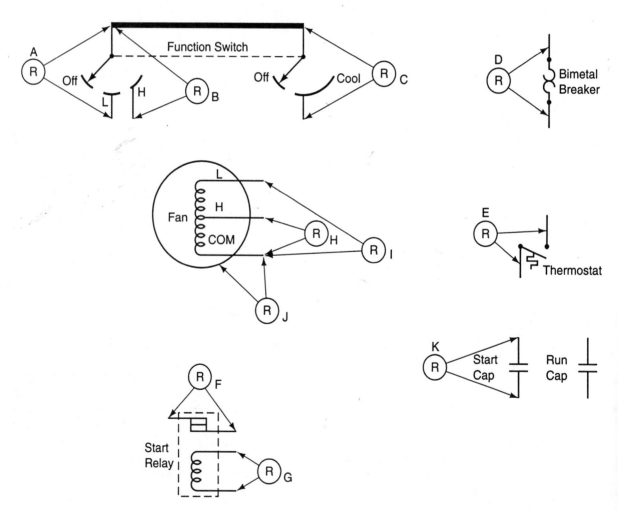

FIGURE 8-25 Measuring resistance.

Measuring resistance: Before you can measure resistance, components must be disconnected from their circuits. Figure 8-25 shows a number of the components removed from the air conditioner for resistance testing. As long as the air conditioner is unplugged, it may not be necessary to remove these components physically, however, they must be disconnected from the circuit. To prevent meter damage and shocks, the capacitors need to be momentarily short-circuited in order to remove any charge they may hold.

Ohmmeter A: With the fan switch in the low-speed position, the meter should read very close to zero ohms (good conductor). To achieve this reading, the meter needs to be carefully adjusted and the probes have to make good, firm connections. Infinitely high ohms (nonconductor) should be indicated when the fan switch is in the off or high position. Identifying switch contacts is often a problem that can be solved by reference to diagrams and wire tracing.

Ohmmeter B: The meter should read 0 ohms with the switch in the high speed position and infinitely high ohms in all other positions. Also, it is a good idea to check resistance between the low and high terminals to be certain that they are not short-circuited (0 ohms).

Ohmmeter C: Zero ohms should be indicated in the cool switch position and infinitely high ohms in all other positions. The continuity function of the multi-meter may also be used for this test.

Ohmmeter D: The bimetal circuit breaker opens the circuit in the event of too much current and/or a compressor overheat situation. It should measure zero ohms (good conductor) when cold. You may want to try heating a used breaker to see it function; however, it is ill advised to test this safety-related component in this fashion. If its function is in doubt, replace it.

Ohmmeter E: A gentle heating and cooling of the thermostat with a hair dryer will cause its resistance to go from zero to infinitely high.

As the thermostat ages, it often requires too much or too little temperature change to cycle. This results in an air conditioner that either cycles too often or permits too great a temperature change. If in doubt, replace it.

Ohmmeter F: This type of relay has normally closed contacts and should indicate this with a reading of zero ohms. If possible, operate the relay manually to see that the resistance rises to infinitely high ohms when the relay is opened. Also, check that the point terminals to coil terminals and relay frame are not short-circuited and indicate infinitely high ohms resistance.

Ohmmeter G: The relay coil should indicate moderate resistance in the high hundreds of ohms to the low thousands. Also check to see that the coil is not grounded to the frame of the relay.

Ohmmeter H: Connecting the ohmmeter to the common and high speed terminals of the fan motor should indicate a resistance in the low tens of ohms (15–30 ohms) for this small air conditioner. If the reading is much lower, suspect a short-circuited motor, especially if it smells burned. Since this is a coil intended to operate from AC, the current flow is limited by its resistance and its inductive reactance. The ohmmeter measures only resistance.

Ohmmeter I: This reading is the resistance of the high-speed motor coil (ohmmeter H) plus additional coil windings that provide low speed. The extra windings should increase the resistance measured in position H by 10–20 ohms. If the meter indicates infinitely high ohms (open circuit), the extra coil windings are open. This would result in no low-speed function of the fan.

Ohmmeter J: This test should result in infinitely high ohms between any of the fan motor terminals and ground. In some instances, very high resistances are acceptable; therefore, it is wise to refer to the manufacturer's specifications or to compare the unit with a new one of the same type. If the results indicate lower resistance, the fan motor coils are short-circuited to ground.

The unit may function, but the chassis presents a shock hazard.

Ohmmeter K: Before the capacitor is disconnected from the circuit, it must be discharged by short-circuiting its terminals. If this is not done, a painful shock and meter damage are likely to result. The instant that the ohmmeter is connected to the capacitor, it will indicate low resistance. In a few seconds, the meter needle will move to high resistance. This resistance change is due to the charging of the capacitor by the ohmmeter's battery. If the test is repeated, the charged capacitor will immediately indicate high resistance. Therefore, discharge the capacitor between repeated tests. In the event of a low ohms reading after a reasonable time for charging, the capacitor is short-circuited. The ohmmeter action indicates that the capacitor accepts a charge, but it does not determine the amount of capacity. Special meters are available that will conclusively test the capacitor, however, substitution is often easier.

When in doubt about any specific meter function or test to be performed, ask your instructor for a further explanation. Hands-on meter work should not be attempted until the appropriate background theory is understood, safety considerations are known, and the practice has been demonstrated.

SERVICE HINTS

An expert troubleshooter has experience—a lot of experience. Nothing can substitute for experience, however, some general ideas increase the value of experience. These general ideas provide a pattern into which each experience may be placed. In this way, the connections are developed and experiences have meaning. The following sections provide a pattern.

Symptoms

A cough, fever, and muscle aches are common symptoms of the flu. Each symptom by itself may be due to one of many other ailments, however, when they appear together, the possibilities are greatly narrowed. In the same way, the detective work of troubleshooting begins with a careful examination of the symptoms. Often the customer has to be questioned to reveal all the symptoms. Was there a smell before the furnace shut down? Did you hear any sounds? Be careful to inquire whether this happened gradually or all of a sudden. Failure that occurs "all of a sudden" is usually due to one component or circuit. Gradual system failure ("It got worse and worse over the past year") is often due to general system deterioration and/or lack of maintenance. Also, check to see if this problem has occurred in the past, and if so, what was done to correct it. Make a list of the symptoms, then imagine the system operating, start-up–run–shutdown. If possible, run the system and be alert to smell, noise, and any visual symptoms that appear. A variety of symptoms, visible stack smoke, for example, reveal problems (improper combustion), as well as properly working components (ignition transformer). Think about the symptoms first.

Causes

The actual cause of a malfunction is absolutely identified only when the HVAC system is back in good working order. Before this is achieved, all causes are tentative. Tentative causes of the problem that are likely to be the actual cause are generally obvious. Close visual examination and instrument testing of the suspected component should reveal evident symptoms of malfunction. In other words, when you find what is wrong, you will usually be able to see it with either your eyes or your test instruments.

Components in well-designed systems usually fail as a result of use and age. However, it is not uncommon to find the cause of system malfunction to be something outside of the system itself. Something as simple as cleaning the basement may result in disturbed switches or control settings, loosened wires, and bent or disconnected ducting. In one instance, the authors found a commercial air conditioner that had stopped operating as a result of wiring changes made during the winter months. One customer had heat distribution problems because duct dampers had been disturbed during the painting of the house. Consumers seldom relate cause and effect, whereas, to the technician, the relationship is obvious. Ask questions.

Probability

The master HVAC problem solvers have learned through years of work that certain problems and causes are more common than others. Through experience, the technician will also learn that particular systems, as well as particular brands, have weak points. Look at those components and systems that often fail first. The most common and likely causes should be examined first, and the least likely last. The nature of the customer's use of the HVAC system may directly affect what problems are most likely. For example, high utility costs for a small office building may be directly due to open access to thermostats. Both high utility costs and premature thermostat failure are likely results. What is most likely to go wrong, considering the peculiarities of the equipment and the nature of its use, is an important part of troubleshooting.

An Attack Plan

Once the technician has taken note of the symptoms, explored possible causes, and considered the probabilities, the next step is to develop an attack plan. This plan ensures that the diagnosis is achieved in an orderly, efficient manner. It may be written down, a good idea for the beginner, or just a clear mental list of the procedure that is to be followed to identify the malfunction.

Finding the faulty component is similar to the game of "guess the number." In this game, one person thinks of a number between 1 and 10. The other person makes a guess, which is either correct or incorrect. If it is incorrect, the first person tells the second whether the number is higher or lower than the guess. This process continues until the number is identified. The idea is to identify the number with the fewest guesses. What would be your first guess?

The best first guess is 5. If it's wrong, you will find out if the number is higher or lower and by this process have eliminated half of the possible ten numbers. In the same way, a problem with a heating, ventilating, and air-conditioning system is solved by eliminating or subdividing the system. Is it in the air conditioner? No. Is it in the ventilating and air movement system? No. Is it in the heating system? Yes.

Further subdivision of the heating system might go as follows: Check the thermostat—OK. Does the burner run? Yes. Is ignition spark provided? Yes. Does oil combustion occur? No. Is the nozzle spraying oil? No. Is oil being supplied to the burner? Yes. At this point, the technician has localized the problem to the oil supply function of the burner and nozzle assembly.

Every step of the attack plan should involve tests that subdivide the system and localize the problem. Divide and conquer!

SUMMARY

This chapter began with electric power entering the house. Three primary feed wires supply 240 VAC between the two hot wires and 120 VAC between the neutral and either hot wire. The service entrance panel interrupts both hot wires with a main breaker and branch breakers. In this panel, all neutral wires are fastened to a neutral bus and all circuit grounds are connected to a bus that is grounded to earth. The 120-VAC circuits provide power from one hot bus, tapped off by a single breaker (black wire), a neutral (white wire), and a ground (green wire). The 240-VAC circuits connect to both hots (black and red) through a double-width breaker, a ground (green wire), and, if 120 VAC is also desired, a neutral (white wire). Wiring should be done by licensed electricians and requires adherence to codes, as well as inspections.

Residential electrical service is fundamentally a parallel circuit since all the loads are connected across the feed wires. These loads, which are in parallel, are all subject to the same voltage. In parallel circuits, voltages are the same across each load while the current each load draws depends on its own resistance. In this way, all the load currents in parallel sum to the source current. With parallel circuits, individual load conductances (the reciprocal of resistance) sum to the total conductance (the reciprocal of the total resistance).

The compressor of a simple window air conditioner is connected in series with the function switch, circuit breaker, and thermostat. A series of components connected end-to-end like a string results in one path for current flow. In series circuits, the current is the same through all of the components connected in series. The resistance of a series circuit is the sum of all the individual resistors connected in series. Since the source voltage is applied across the series string, it divides up proportionally to the resistance of each component over the total resistance. All the individual voltage drops sum to the source voltage.

Voltmeters, ammeters, and ohmmeters are important diagnostic tools. They enable the technician to "see" into the circuit and analyze problems. The nature of meters varies, therefore, reference to instructions supplied with the meters is necessary. General servicing may be achieved with a multi-meter, which is used to measure VAC and ohms, and a clamp-on meter to measure amps. Care needs to be exercised in adjusting the function and range of each of these meters. Furthermore, all safety considerations and practices should be understood before actual troubleshooting is attempted.

Meters should be used to provide information relative to circuit function that helps the technician to evaluate suspected faulty components. Symptoms point to particular causes that localize the problem within the HVAC appliance. Tests should be preformed in accordance with an attack plan that both considers what malfunctions are most likely and focuses in on the problem.

PROBLEM-SOLVING ACTIVITIES

Case Studies

A. After a severe storm, a customer complains that the central air conditioner does not function. While examining the air conditioner you notice that a number of lighting circuits in the house do not function either. What is the most probable cause of the malfunction, and what help would be required to resolve the problem?

B. You have installed an attic circulation fan. The next day, the customer calls, complaining that the lights in two rooms dim momentarily when the fan is turned on. What might be done to minimize this effect, and how might it be explained to the customer?

C. Your stereo system has two sets of identical speakers. The remote speakers are lower in volume than the main speakers. Why is this so, and how might it be corrected?

Review Questions

1. Why is it desirable to evenly divide a 120-VAC branch circuit breaker between the two hot service panel buses?

2. How much current flows through the green ground wire in a branch circuit? Explain your answer.

3. Describe the similarities between a parallel circuit and a river with tributaries.

4. What is the relationship between resistance and conductance? Describe it in your own words and with a mathematical formula.

5. Draw a simple diagram of a parallel circuit and describe the currents at any junction.

6. What is wrong with the question, "How is the total resistance of a parallel circuit found?"

7. If three 10-ohm resistors are connected in series across a 9-VAC power source, what portion of the voltage is dropped across each resistor? What is the voltage across each resistor?

8. Why are switches connected in series with the devices they control?

9. What is the primary difference between a multi-meter and an electronic VOM?

10. Describe the purpose of the function switch on a multi-meter.

11. If an unknown voltage is to be measured, on what range should the meter be placed?

12. What are two advantages of a clamp-on ammeter?

13. In your own words, list the safety precautions that should be considered during meter use. Discuss this topic with other class members to make as complete a list as possible. Review this list with your instructor.

14. Describe an efficient attack plan to fix a nonfunctioning oil-fired furnace (the burner gets power, but no components are energized—the customer is present).

15. A parallel circuit is composed of a 5-ohm and a 10-ohm resistor connected across 50 VAC. Draw the circuit and label the parts. What is the voltage across each resistor? What is the current through each resistor? What is the total current draw on the 50-VAC source? What is the conductance of each resistor? What is the total conductance of the circuit? Convert the total conductance to resistance.

16. A circuit is composed of four 10-ohm resistors connected in series across 20 VAC. Draw the circuit and label the parts. What is the total resistance? What is the source current? What is the current through each resistor? What voltage would be measured across each individual resistor?

CHAPTER 9

Electronic Components and Circuits

Electronics is a field that involves small and complicated circuits intended to manipulate and present information. Just as the stereo and TV set entertain and inform us, the electronic controls used in HVAC inform the heating and cooling equipment of changes in temperature, relative humidity, building and room usage, time, and even window and door position. Furthermore, this information can be processed and evaluated by computer-like microchips, which make decisions related to heating and cooling requirements. Electrical circuits provide the power for contemporary technology, but electronic circuits provide the brains.

Because of the ever-increasing use of electronic controls in the HVAC field, the technician needs to have a basic understanding of their function. This is a difficult topic, involving many concepts and ideas. Electronic technicians may attend five or six courses just to develop general electronics knowledge, as well as special courses in the application, programming, and repair of controllers. Enthusiastic students are encouraged to continue their education in this area of technology.

SIMPLIFICATIONS AND LIMITATIONS

In the next two chapters, fundamental ideas are covered that relate to all electronic circuits, with an emphasis on process controllers. This helps to prepare readers for industry-sponsored workshops at which they can gain specific information on particular brands and types of controllers. The cost of these courses, which ordinarily run eight hours a day for several days, is often covered by the employer, especially if the technician shows initiative and skill. Also, many manufacturer-published manuals provide the necessary information. Rather than developing vague ideas about a generic controller, Chapters 9 and 10 provide a common-sense electronic foundation.

FURTHER STUDY

If in this and the next chapter, you find ideas that are interesting, don't stop here. Take more courses, read manuals, attend all the workshops you can, and continue to learn. Technology changes rapidly. When the authors were born, there were no TVs or computers. Today, we communicate with light (fiber optics), play music on CDs (digital), estimate HVAC jobs on spreadsheets (software), and use highly reliable, very complex electronic computers (microchips). All of these innovations have occurred within the past 15 years. Fifteen years from now, you probably will be a third of the way through your working life. Change and learning go hand in hand.

DISCRETE SEMICONDUCTOR DEVICES

Semiconductors are components manufactured from N and P "doped" silicon. They may be individually packaged (discrete) devices such as transistors or diodes, or integrated, when millions of them are etched photographically onto tiny slabs of silicon the size of a fingernail. For practical purposes, almost all electronic circuits are semiconductor in nature. About the only exception is the cathode-ray tube used as a display in TVs and computers. And gradually, even in those applications, flat semiconductor displays like those on portable TVs and computers are taking the place of the conventional picture tube. Circuits composed of semiconductors are commonly referred to as solid state.

Recall from earlier chapters that electrical wire was compared to pipe. In electronic circuits, printed circuit (PC) boards carry electricity on thin copper foil patterns called circuit traces. The detail of circuit trace patterns on current electronic boards is extremely complex. However, as complex as they are, circuit patterns on PC boards are still like pipes. Connecting these microscopic electronic pipes together are semiconductor devices that act like electron gas valves. What follows is a plumber's view of the most common discrete semiconductor devices.

Thermistors

Ordinary conductors, for example, NiChrome and copper wire, increase in electrical resistance as their temperature increases. In the case of NiChrome heating elements, their "cold" resistance is lower than their hot resistance. Incandescent lamps generally double their cold resistance when heated to normal operating brightness. These examples exhibit a *positive temperature coefficient*, that is, an increase in temperature causes an increase in resistance. Superconductors apply this positive temperature coefficient by dropping the temperature close to absolute zero. With no thermal activity, some materials lose all their resistance and become perfect conductors. Most material and electrical devices exhibit a positive temperature

coefficient. As the temperature increases, so does their electrical resistance.

The thermistor, a semiconductor device, is an exception. As the thermistor is heated, its resistance drops and its electrical conductivity increases. A cool thermistor may have a resistance of tens of thousands of ohms that, when heated, falls to tens of ohms. Since the thermistor's temperature and resistance vary in opposite directions, it is described as having a *negative temperature coefficient.*

The symbol for a thermistor (thermal resistor) is a circled resistor with a T, indicating its temperature dependence. Circuit A in figure 9-1 illustrates a series circuit composed of a battery source and a thermistor. A meter is included to measure the small current flow in the circuit, and an adjustable resistor called a potentiometer is used to calibrate the circuit. In this example, the thermistor's temperature is dependent on its environment. It is measuring ambient air temperature—the temperature of the air surrounding it. The meter indicates current flow, which is thermistor dependent. The ambient air temperature is sensed by the resistance changes of the thermistor and indicated by the meter. Temperature measurement with this circuit is rather imprecise for two reasons. First, the circuit does not contain any reference temperature standard. Second, the thermistor may be unintentionally self-heated.

The self-heating nature of a thermistor is often used to advantage. Circuit B in figure 9-1 shows the capacitor bypass method used on some capacitive run motors. During motor start-up, the thermistor is cool, and hence its high resistance allows most start winding current to flow through the capacitor and permit phase shift. By the time the motor runs up to speed, the thermistor's current flow has "self heated" it. The lower resistance of the hot thermistor permits current to bypass the capacitor. This reduces phase shift and provides more efficient and smoother capacitor run motor operation.

Time delay relays are often used in HVAC equipment. For example, a group of high-wattage heaters during cold start offers low resistance and draws a very high starting

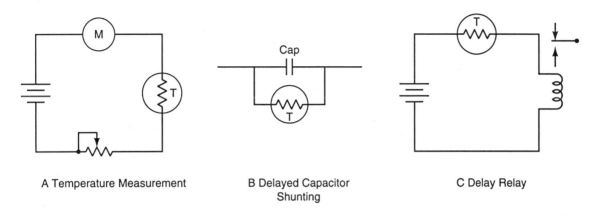

A Temperature Measurement   B Delayed Capacitor   C Delay Relay
                                    Shunting

FIGURE 9-1 Thermistor circuits.

current. Delay relays permit the sequential 1–2–3 start-up of multiple-coil heaters, easing the start-up current surge. One of the many ways in which relay closures may be delayed is illustrated in circuit C of figure 9-1. Notice that this circuit is powered by a battery. When the circuit is first powered up, the cool thermistor's resistance is too high to permit enough coil current to pull in the relay. However, after a short time, the thermistor's resistance drops due to self-heating. Relay coil current now increases and permits the relay to close. A drawback to this method of delay is that the thermistor needs to cool down before the delay relay can be cycled a second time.

Thermistors are like cheap garden hoses that swell in the hot sun and become bigger pipes. Unlike garden hoses, however, when properly used, thermistors do not burst. Furthermore, while thermistors may self-heat from current flow, garden hoses do not. Look at the three examples in figure 9-1 and notice that thermistors may be used in DC or AC circuits. They, like ordinary resistors, are not polarity dependent. Rather, the thermistor's main role in circuits is to decrease electrical resistance with increased temperature and thereby sense temperature or control current.

Diodes

One of the most useful tools the mechanic has is the rachet-and-socket set. It could be described as a one-way wrench; it turns a bolt in one direction, while the handle operates in both directions. One-way devices are used throughout HVAC technology. For example, the back-flow damper permits air circulation in only one direction. Mother Nature has made use of this one-way idea in the human circulation system to prevent the back flow of blood. The heart, as well as veins in the extremities, especially the legs, incorporate flapper-like valves ensuring one-way circulation. Water pumps, kitchen exhaust fans, and many narrow streets operate in a one-way direction.

The silicon diode is the one-way valve used in electrical circuits. It permits electrical flow (amps) in only one direction. Figure 9-2 shows that electrons flowing from the negative battery terminal are permitted through the diode only when its arrowlike symbol is opposite that of the arrow representing electron flow (amps). In other words, the diode conducts electron flow when it is against the arrow. Some actual diodes are labeled with the symbol shown in figure 9-2, while others have a band at one end. The banded end of the diode is called the cathode, and it has to be connected to negative (–) to permit current flow. The other terminal (nonbanded), called the anode, is connected in the positive (+) direction of the circuit. Connected in this way, the diode is forward biased.

The physical size of a diode is directly related to the amount of forward current it can safely carry. Bigger diodes can dissipate larger

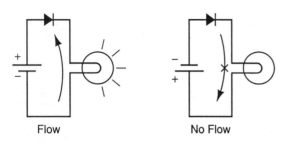

Flow   No Flow

Figure 9-2 Diode action.

quantities of heat, especially if they are fastened to metal heat sinks. Large stud-mounted diodes are available that can handle hundreds of amps, while small diodes are often limited to fractions of an amp. The most important forward (conducting direction) rating of a diode is its maximum continuous forward current capacity.

When reversed biased (connected in the no-current direction), diodes are not conductors. The reversed biased diode on the right in figure 9-2 has the battery source voltage across it. Connected in the reverse direction, the primary rating of a diode is peak inverse voltage (PIV), the maximum voltage level that the diode can block. A voltage above PIV will force (remember, voltage is a measure of electrical force) the diode to conduct, often destroying it in the process.

Ordinary silicon diodes are one-way electrical valves that have a maximum current rating in the forward direction and a PIV in the reverse direction. A good diode will indicate high resistance in one direction and low resistance in the other. Diodes have many applications: one of the most common is to convert AC to DC. Circuits that convert AC to DC are called rectifiers.

The half-wave rectifier diagrammed in figure 9-3 is used to supply DC power for simple low-current applications, which include NiCd battery charges and battery eliminators. It provides pulsating DC power by permitting current flow during one half of the AC sine wave. Notice that

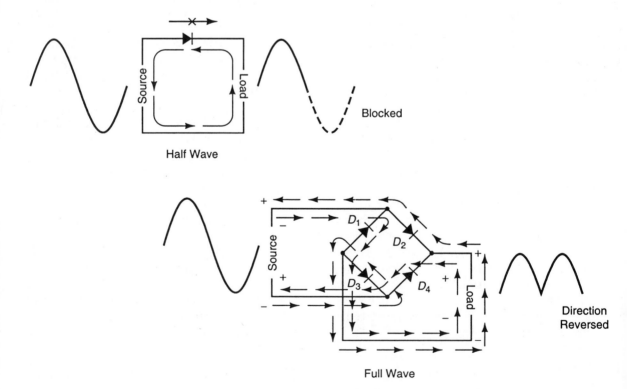

Half Wave

Full Wave

FIGURE 9-3 Half- and full-wave rectifiers.

only the positive top part of the AC wave is passed through the diode, while the reverse current, indicated by the lower half of the wave, is blocked. Thus, one half of the sine wave is blocked and the remaining half supplies spurts of DC. Frequently, this type of rectifier has a capacitor connected across its output to collect the spurts of DC and provide a smoother output. Another way of providing smoother output is to use a full-wave rectifier.

The four diodes that make up the full-wave rectifier in figure 9-3 may be separate, or they may be packaged as a single device called a bridge rectifier. The full-wave bridge rectifier takes the reverse-direction lower part of the AC sine wave and flips it over. To understand this process, trace the current when the top of the source is negative (–) and follow it through D1, up through the load, and back to the source through D4. During this half cycle, D2 and D3 block flow. When the bottom of the source is negative, current flows through D3, up through the load, and returns to the source through D2. During this half of the cycle, D1 and D4 block flow. In this way, the bridge full-wave rectifier supplies higher and smoother DC current flow than the half-wave unit.

A variety of special diodes are in common use, two of which are important enough to be mentioned. The light-emitting diode (LED) requires about 2 volts in the forward direction to conduct and produce light. Ordinary LEDs have a maximum forward current of 30 milliamps and a PIV of 7–30 volts, and are available in red, green, and amber colors. They are used primarily as indicator lights, although recent improvements in brightness make their use in such applications as automotive stop lights likely. LEDs are manufactured in many shapes and sizes, and are available at most electronic parts stores.

Zener diodes are special devices that act like pressure regulators. They are designed to go from a nonconducting to a conducting state at a specific voltage level. For example, a 12-volt zener diode may be connected across a power supply that varies from 12 volts to 14 volts. When the voltage rises above 12, the zener conducts and applies a load to the circuit, which results in voltage reduction. Zener diodes are often included on integrated-circuit voltage regulator chips.

Silicon-Controlled Rectifiers and Triacs

A part-time job that helped put one of the authors through school involved the operation of stage lighting equipment. Sitting in front of a very large lighting control console and raising and lowering lights on cue was an interesting task that paid well. However, the job was hot. Thirty years ago, the lighting panel was made up of giant adjustable resistors call rheostats. Large handles that were connected to these rheostats stuck out of the control panel. The panel itself was a very large black box, about the size of a big desk, with a metal chimney that vented the heat it produced through the theater roof. Indeed, it was a beast of a machine. This long-gone beast lighting control panel has been replaced by a control unit not much bigger than a desk-top computer. These units produce little heat and have small slider controls. How did this major change come about?

The circuit shown in figure 9-4 comprises a large wire-wound resistor connected in series with a bank of stage lights. Since the rheostat

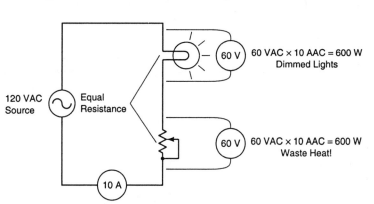

FIGURE 9-4 Rheostat function.

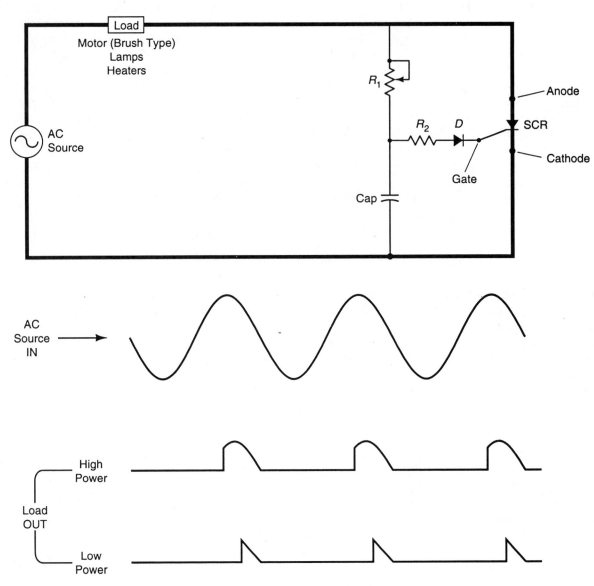

FIGURE 9-5 SCR action.

is adjusted to a resistance that is equal to the resistance of the lights, the 120-VAC source voltage is evenly divided across the two components in series. Also, the rheostat's resistance lowers current flow in the circuit. This was the way in which stage lights were controlled thirty years ago.

Unfortunately, the nature of a rheostat is such that although full current flows through it,

half of the voltage is dropped across it. In this example, the results are the production of 600 watts of unwanted heat. Stage lighting controllers had 10 or more rheostats, producing, in this instance, 6000 watts of heat—6000 watts that had to be discarded. This was not at all a good arrangement, especially for the operator. The rheostat is still used in low-current applications, such as auto dash lamp dimmers and sewing machine speed controls. In all other applications, it has been replaced by a solid-state power-controlling device commonly referred to as an SCR.

The SCR controls power by rapidly switching on and off. It is either fully conducting, like a closed switch, or nonconducting, like an open switch. The rheostat's resistive power-choking nature is replaced by the SCR's power-chopping nature. Since the SCR is either fully on (no resistance heating) or fully off (no current flow, no watts of heat), the device produces little waste heat. Furthermore, the SCR's power-chopping nature provides pulses of full power. As a result, variable-speed drills that use solid-state speed control, develop full torque at slow speeds. The chopping nature of solid-state controls can be observed when a variable-speed drill is operated at its lowest speed and rotates in pulses. Light dimmers may also exhibit this effect on the lowest settings by blinking the lights.

The circuit and basic action of the SCR are shown in figure 9-5. The heavy current flow in this circuit is identified by the thick wire that connects power to the load through the SCR's cathode and anode terminals. Ordinarily, these are the two bigger terminals of the SCR, and the gate, the third terminal, is the smallest. The SCR, like the silicon diode, is a one-way device. As a result, the bottom half of the sine-wave source voltage, shown below the circuit, is lost. The remaining half of the AC wave is permitted to pass through the SCR as it switches on at different points along the wave. Current flow stops when the sine wave reaches zero, which results in the SCR's being turned off. SCR turn-on is controlled by the gate terminal and its associated components (R1, R2, D, and the CAP). Let us try to get an idea of how the control circuit works.

As the sine wave rises, current flows through resistor R1 and starts to charge (fill) the capacitor. In this case, R1 is like a faucet filling the electrical bucket capacitor. When the voltage of the capacitor is high enough, it triggers the gate of the SCR through resistor R2 and the diode D. The gate is protected by R2, which limits excessive gate current, and the diode D,

which prevents reverse gate current. If resistor R1 is adjusted to low resistance, the capacitor charges quickly and triggers the gate early in the positive half of the sine wave. Early turn-on of the SCR provides the high-power output illustrated in figure 9-5. High-resistance settings of R1 take longer to charge the capacitor and result in late turn-on of the SCR. Late turn-on provides a small portion of the positive half of the sine wave to the load, thereby resulting in the low power shown in figure 9-5. In this way, the SCR controls power without wasting it or producing unnecessary heat. Things have cooled down backstage.

Since the SCR passes only one half of the sine wave, most power controllers that are advertised as SCR units do not actually use the SCR. Although they function in almost the same manner, the SCR is replaced by a two-way solid-state device called a triac.

Figure 9-6 illustrates the triac power controller, which is very similar to the SCR circuit in figure 9-5. Notice that the SCR has been replaced with a triac that controls current flow between its heavy terminals, labeled anode 1 and anode 2. The smaller gate terminal triggers the device into conduction on either the positive or negative half of the source sine wave. Like the previous SCR circuit, the charge time of the capacitor is controlled by the setting of R1. Also, like the SCR circuit, R2 limits current to the gate. The diac or a common alternative, the SBS (silicon bilateral switch), is used to improve circuit performance. Neither the diac nor the SBS will conduct in either direction until a predetermined voltage level has been reached. This provides a more definite signal to the gate and prevents intermittent low-power triggering.

Examine the high- and low-power output shown in figure 9-6 to understand how earlier or later turn-on of the triac provides more or less power to the load.

The HVAC field makes frequent use of the triac circuit in AC power control. Solid-state relay modules are often triac circuits. They eliminate unreliable relay points and are usually operated in full off or on modes. Computers

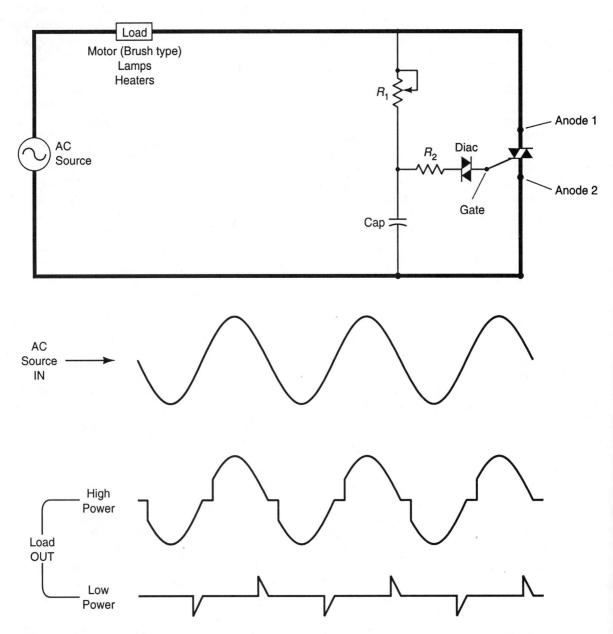

FIGURE 9-6 Triac action.

can provide the small electric current necessary to control triacs, thereby controlling HVAC devices that require up to 200 amps AC. The DC use of the SCR or triac is generally not practical, as they remain conducting as long as current is flowing through them. Since DC is constantly flowing, once triggered, the SCR or triac cannot be turned off. The current must be stopped to permit a second triggering. The transistor, another semiconductor device, is often used to control DC power, and, in some instances, AC as well.

Junction Transistors

Early radio receivers depended on a diode made of natural lead ore crystal called galena. The output of the crystal set was so weak that one had to use headphones to listen to it. The invention of the electron tube enabled, among many other things, the amplification of relatively weak radio signals and ushered in the age of electronics. Unfortunately, the radio tube was fragile, power hungry, and unreliable. Many people believed that a crystal-like device could be developed that would replace the tube. Beginning in 1900, a great deal of crystal research was done in scientific laboratories, especially in the United States. Finally, in 1948, three Bell Lab scientists—Bardeen, Brattain, and Shockley—perfected a device they called the transistor (transfer resistor). Though the importance of this device was not immediately recognized, over the next 20 years it entirely replaced the radio tube. The field of electronics, which had turned away from crystals and toward tubes in the 1920s, turned back to crystals with the development of the junction transistor. How does this semiconductor work?

The junction transistor has three terminals or connections that internally connect to 3 layers of semiconductor material. These three layers may be arranged as NPN or PNP. Both types work in the same way, however, all DC circuit polarities are reversed.

Output current of the transistor flows from the emitter terminal, which emits current, to the collector terminal, which collects current. The main output current flowing from the emitter to the collector is controlled by a third terminal called the base. Figure 9-7 depicts current flow in an NPN transistor. To understand the junction transistor, it is very important to realize that its function depends primarily on current rather than on voltage. It might be said that it is primarily a current-operated device.

As figure 9-7 shows, the junction transistor provides control of a large current flow from the emitter to the collector with a small current flow from the emitter to the base. In other words, a feeble emitter-to-base current has control of a large emitter-to-collector current. The transistor is like a valve in that a small action, turning the handle (base current), has a great effect on flow (emitter to collector flow). Through this action, the junction transistor can, much like a relay, use a small current to control a large current. The junction transistor can be thought of as a semiconductor electrical valve.

When used as a valve, the junction transistor can turn on and off loads that may be related to lights, motors, or heaters. As figure 9-8 shows, these loads have a separate source of power that is controlled by the resistance between the emitter and collector. If the base current is high enough, the resistance between the emitter and collector drops very low, thereby switching on the load. If the base current is very low, the resistance between the emitter and collector is very high, which results in no current flow to the load. In this way, small electric currents supplied by computer-like electronic process controllers can be used to switch on/off large electrical HVAC devices, especially those that operate from DC.

Junction transistors may also be used to amplify (make bigger) small electrical signals. For example, microphones convert sound, the vibration of air, into electricity. In figure 9-9, the weak electrical signal of the microphone is an electrical model (analog) of the sound vibrations, however, it is too small to be useful. It can be boosted, or amplified, by the transistor. For the purpose of simplification, some necessary resistors have not been included in figure 9-9. Small base current variations result in

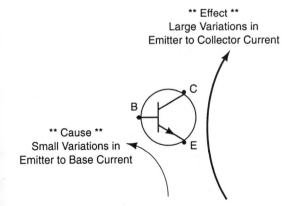

** Effect **
Large Variations in
Emitter to Collector Current

** Cause **
Small Variations in
Emitter to Base Current

FIGURE 9-7 Junction transistor action (NPN).

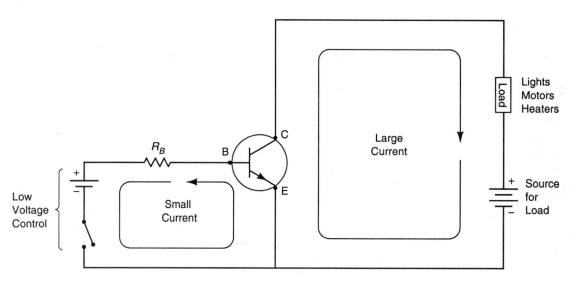

FIGURE 9-8 Junction transistor switch (NPN).

variations in emitter-to-collector resistance. Since the emitter-to-collector resistance is in series with the speaker and its source, a variation in this resistance produces variations in speaker current. The speaker converts the electrical variations into air vibrations (sound). In this way, weak signals from radios, CDs, and cassette players, as well as sensors and computer signals, can be amplified. Notice that amplification circuits use the transistor between the full-on and full-off (moderate-resistance variations) modes, while switching circuits use the transistor in the full-on (low-resistance) and full-off (high-resistance) modes.

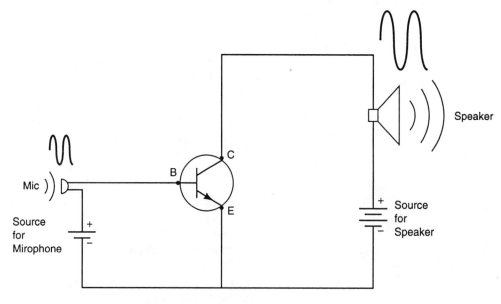

FIGURE 9-9 Junction transistor amp (NPN).

MOS Transistors

In the 1960s, a second type of transistor became popular—a semiconductor device consisting of a nonconductive channel connected at one end to a terminal called the source and at the other to the drain terminal. Current between the source and drain terminals is controlled by the gate terminal. Internally, the gate is connected to a microscopic metal plate that is separated from the source-to-drain channel by a very thin silicon dioxide (glass) insulator. It is called MOS, for *m*etal (gate) *o*xide (glass insulator) *s*emiconductor (source-to-drain channel). In other words, the gate sits on top of the source-to-drain channel, but is entirely insulated from it by the thin glass insulator. Because the gate is insulated, no current can flow from the gate to any other terminals of the MOS transistor.

The symbol for the MOS transistor is shown in figure 9-10. Notice how the structure of the insulated gate is indicated by its separation from the three short lines representing the nonconducting channel. The first time we saw it, we wondered how this little device could possibly work, since the gate connects to nothing. That's true. However, the gate operates

by a voltage charge, not (like the junction transistor) by current.

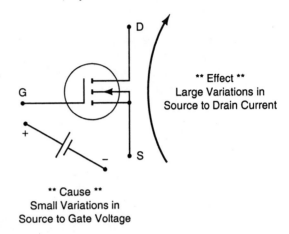

FIGURE 9-10 MOS transistor action (N Channel).

Imagine a positive charge placed on the gate. What would be attracted by it? The positive electric field on the gate attracts negative electrons in the channel. Since the gate is insulated, electrons cannot actually flow to it. Rather, they line up in the channel below the gate's glass insulator. Attracted in this way, the electrons serve as conductors from the source to the drain. The MOS transistor's gate–insula-

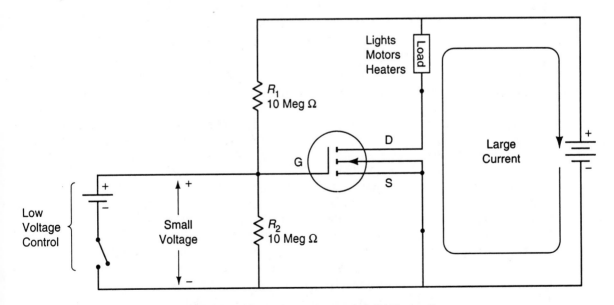

FIGURE 9-11 MOS transistor switch (N Channel).

tor–channel action is like a capacitor in that a positively charged gate results in a negatively charged channel—the glass insulator prevents conduction between the two. As a result of this action, electrons are provided in the channel that permit conduction from the source to the drain. In other words, a large source-to-drain current is controlled by voltage (*not current*) on the gate. In the N-channel MOS transistor, the gate is charged positive to make the N (negative) channel conduct.

The MOS transistor's gate is so sensitive that tiny stray electric charges may turn the device on and off. Some table lamps use the touch-sensitive nature of the gate as an on/off switch. To prevent the effect of stray charges on the gate, resistors may be added, as shown in figure 9-11. Resistors R1 and R2 have very high resistance and conduct practically no current. Rather, they ensure that any stray charges picked up by the gate are conducted away. Notice that these resistors are equal and

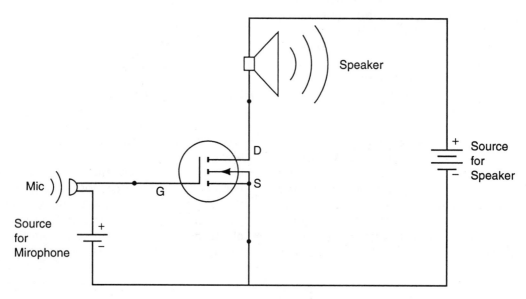

FIGURE 9-12 MOS transistor amp (N Channel).

thereby keep the gate fixed electrically between the source and drain. During the following explanation, ignore the presence of these resistors, since their only purpose is to keep the gate's voltage from wandering.

Any low voltage can provide the necessary electricity for the gate to turn on the source-to-drain channel. Microscopic sensors that measure pressure and temperature, for example, produce low voltages that can provide gate control. Also, computer controllers provide the feeble voltages necessary for the MOS transistor to operate. The large source-to-drain current indicated in figure 9-11 may commonly go to 10 amps. Furthermore, unlike junction transistors, MOS transistors can be connected in parallel,

enabling ten MOS transistors rated at 10 amps each to control 100 amps. Imagine controlling 100 amps with a voltage charge so feeble that it could come from the touch of a finger.

As illustrated in figure 9-12, the MOS transistor can also serve as an amplifier. Note that, in this case, the gate voltage depends on the output of the microphone. Think of the gate as the handle of a valve that is being turned in and out rapidly, thereby controlling source-to-drain current. The speaker converts the varying current to sound. Remember that, with the MOS transistor, a small voltage on the gate controls a large current from the source to the drain.

Over the past ten years, MOS transistors have become very popular, and are rapidly replacing junction transistors for a number of reasons. The most important reasons are related to integrated circuits. Since MOS transistors occupy less space on circuit chips, over two million of them can fit on a chip about a half inch square. Furthermore, MOS transistors require fewer production steps, providing chips at lower costs.

INTEGRATED CIRCUITS: INTRODUCTION

Tiny, reliable, very complex, and inexpensive all describe the integrated circuit (IC). It was invented in 1959 while scientists were researching ways to construct microelectronic components. At that time, transistors were being manufactured from thin slices of highly refined silicon. Many transistors were made on a disk of silicon about 2 inches in diameter. The individual transistors were then separated by scoring and breaking the disk into hundreds of transistors. Each transistor was packaged in a three-lead case, and eventually soldered to a circuit board. Robert Noyce and Jack Kilby, working independently, hit upon the idea of interconnecting transistors and other components right on the silicon disk. Every year since that time, the number of components on a single chip has doubled (to about two million in 1992) and the cost of the circuit has been cut in half. The IC has given birth to microcomputers, VCRs, robotics, guidance electronics, laser printers, process controllers, pacemakers, computer animation, and a miraculous new electronic world.

ANALOG INTEGRATED CIRCUITS

Integrated circuits may be divided into two major subcategories: analog and digital. Indeed, some ICs employ both circuit technologies on one chip. However, the majority of chips are either analog or digital. HVAC technicians will find the use of both analog and digital IC chips increasing at a rapid pace in the near future. Therefore, it is necessary to understand the general ideas of analog and digital circuit technologies.

Analog Circuit Concepts

Everyday life experiences usually fall between the extremes. Sound varies from the very soft whisper of a gentle breeze to the ground shaking roar of a jet engine, but most sounds fall somewhere in between these extremes. Colors vary along a continuous spectrum of red–orange–yellow–green–blue–indigo–violet, with most colors falling somewhere along this spectrum. For example, as a banana ripens, it changes color from a yellowish green to a greenish yellow, then becomes yellow. The banana's color is difficult to represent on an old computer because the choices are limited to 16 colors. The old computer displays a banana that is not at all lifelike. Life is experienced somewhere in between the extremes of love–hate, rich–poor, smart–dumb, and even yellow–green.

An analog circuit uses electrical models to represent actual objects. For example, the color green might be represented by 0 VDC, while yellow might be 10 VDC. As the banana ripens, the voltage representing its color could be located anywhere between 0 and 10 VDC. In this case, the voltage is a model (analog) that represents color. Common entertainment electronic devices, such as radios, TV sets, and record players, are made up of analog circuits. The radio signal rises and falls in a way that models the vibrations of sound. Television pictures are composed of lines produced horizontally across the picture tube. As the voltage of the TV signal changes, the lines vary between light and dark. In other words, the voltage level represents the brightness level. The grooves in a record zigzag in a pattern that mimics or models the sound waves of what is being recorded. Analog circuits manipulate electrical (usually voltage) models of actual things.

For the faithful reproduction of actual things (sound, pictures, movement, etc.), the electrical model has to be precise. High-fidelity stereo uses more accurate electrical models to provide lifelike sound. High-resolution TV provides a more accurate rendition of the "real" picture because it uses a more accurate electrical model. Starting with the earliest cave drawings, humankind has been on a never-ending quest to provide models that are more and more life-like models of real things. In the not-too-distant future, three-dimensional holographic TV may provide images so lifelike that we may have trouble telling them from real life. One of the current "hot topics" is virtual reality: "It was almost like being there."

Application-Specific IC Chips

Flipping through the pages of an electronics supply catalog recently revealed a section on linear IC chips. Linear means varying between two extremes, like the banana's color. Almost all the chips in this section of the catalog manipulate or work with varying voltage signals. While these chips were referred to as linear (varying between a low and a high level), many of them would be used in circuits that use electrical models (analog). The terms analog and linear are often used interchangeably to describe circuit chips that work with voltages that vary between two extremes. In this section of the catalog was a subsection labeled "Application-Specific Integrated Circuits."

Single chips that contain all the electronic components and circuits for TV sets, radios, and amplifiers are referred to as application specific. For example, a toy manufacturer may use application-specific RC receiver and trans-mitter chips in radio-controlled cars. These chips are designed for one general purpose, and ordinarily are not usable in other ways. Some application-specific chips provide a com-monly required electrical circuit and are used in many devices. For example, most electronic HVAC circuits contain voltage regulator chips. Used in one way, these chips are only useful for regulating voltage, and hence they are applica-tion specific. However, they can also be used in other, widely varying applications, such as programmable controllers and drum sound synthesizers. A great many analog or linear chips fall into this gray area between very specific and very nonspecific uses.

Operational Amplifiers

The most important group of very nonspe-cific analog ICs are referred to as operational amplifiers (op amps). They derive their name from their extensive use in analog computers. Though digital computers have replaced the analog computer, the op amp is still widely used. Ordinarily, op amp chips have input, output, and power terminals.

In the analog computer, these devices accept a small signal or signals (voltages) at the input terminal and perform a mathematical operation on the signals. The output of the op amp is then a result of this mathematic operation. For example, a summing op amp might be used to add (sum) the voltages provided by two tem-perature sensors. If one temperature sensor were providing 2 VDC and the other were providing 1.5 VDC, the results of the summing op amp would be 3.5 VDC. To get a better idea of the function and application of op amps, it is necessary to examine a few different types of these chips.

The basic op amp in figure 9-13 is in reality an analog amplifier on a small chip. For example, the feeble electrical brain signals available on the scalp may be picked up with contact electrodes and amplified by the op amp, to a level high enough to be used in a chart recorder. Brain wave and heart monitor circuits employ op amps to boost these small biological signals. In the same way, op amps are used to boost or amplify HVAC sensor signals.

An op amp can be understood through an examination of its basic electrical characteris-tics. A perfect or ideal op amp exhibits the following characteristics.

1. *Very high input resistance.* This means that the op amp does not load down the source signal device. It takes so little signal (very high input resistance) that feeble voltages, often from sensors, are not suppressed. Like the MOS transistor, there is almost no input current.

2. *Very low output resistance.* Since the output resistance of the op amp is low, it is an excellent source for circuits that follow it. All of the power of the op amp's output, almost like a battery, is available to provide input for following circuits.

3. *Very high gain.* The op amp can boost a weak signal thousands of times, enabling the use of sensors that provide tiny voltages. This characteristic is commonly modified by the addition of a feedback resistor, which reduces gain.

4. *Fast response time.* If the input signal to an op amp changes, its output reacts very rapidly.

5. *Null input and output.* If the input is 0 volts, the output is also zero. Many other circuits have output voltages, even though the input voltage is zero.

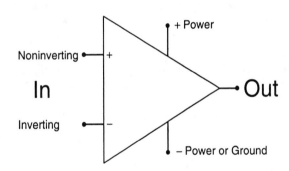

FIGURE 9-13 Basic op amp.

One of the unique characteristics of the op amp that is used to great advantage is the presence of two inputs, as shown in figure 9-13. When a rising signal is connected to the noninverting input (labeled +), the output also rises. Just the opposite happens if a rising signal is connected to the inverting input (labeled –); in that case, the output signal falls. In other words, the output's rise or fall can be selected by using a noninverting or an inverting input.

Power is connected to the op amp through two pins on the chip. Some op amps require a power supply that has a center ground with a positive supply voltage above zero and a negative supply voltage below zero. In certain applications, this puts the op amp at a mid-voltage position (zero) that enables it to rise to a positive voltage or drop to a negative voltage. Some op amps use an ordinary power supply, which is simpler, but less flexible.

A basic op amp amplifier schematic is shown in figure 9-14. Notice that the input is connected to the noninverting positive terminal, so that the output will rise as the input rises. The gain—that is, how much greater the output is than the input—is affected by Rf, the feedback resistor. This resistor takes some of the output and feeds it back to the inverting input, which works against the signal connected to the noninverting input. If Rf is a low-resistance resistor, a great deal of negative feedback results in much reduced gain. However, without the feedback resistor, a slight change in input voltage would result in the output's going to a full positive or a full negative supply voltage. Too much gain results in a switching rather than amplifying action. This switching action will be discussed in relation to the op amp comparator. Resistor Rin represents the source resistance and/or an additional resistor that affects the performance of the op amp, however, its function is beyond the scope of this basic discussion.

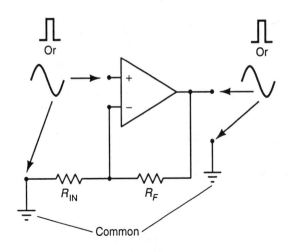

FIGURE 9-14 Op amp amplifier.

Removing the ground from the input resistor in figure 9-14 and connecting it to the noninverting input results in the signal rise's being connected to the inverting input. Changing the input from noninverting (+) to inverting (−) results in the amplifier shown in figure 9-15. The op amp inverting amplifier has gain, like the noninverting amplifier that is controlled by the feedback resistor. With the exception of the output's dropping when the input rises, this inverting amplifier is substantially the same as the noninverting amplifier.

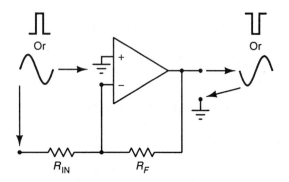

FIGURE 9-15 Op amp inverting amplifier.

The summing amplifier shown in figure 9-16 illustrates the idea that the op amp can be used to add two voltage sources. The analog computer uses such a circuit to perform addition. Say that 6 VDC represents the number 60 and 3 VDC represents the number 30. A voltmeter connected to the output would indicate 9 VDC, which would be interpreted to be the number 90. However, if the inputs represented 60.5 plus 30.02, the output would not permit an interpretation as precise as 90.52. The approximate nature of the analog computer limits its wide-ranging acceptance, however, it is precise enough for summing temperature sensors to provide more distributed temperature measurement. With changes in resistor values, this circuit also can be made to average multiple input voltages.

The last op amp circuit to be discussed is a bit different from the previous ones. It uses no feedback resistor. Rather, its gain is very high,

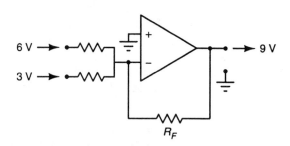

FIGURE 9-16 Op amp summing amplifier.

so that a slight change in the input causes the output to rise to full on (+ power supply voltage) or full off (− power supply voltage or ground). This circuit compares two voltages, one connected to the inverting input and the other connected to the noninverting input. The output is either on or off as a result of which of the two input voltages is higher. A drop in one input or a rise in the other input causes the comparator to change states. The general idea of comparing two voltages is shown in figure 9-17. With the appropriate sensors, this circuit could be used to switch fans on/off, depending on the temperatures of adjoining rooms, or to compare any of a great variety of phenomena.

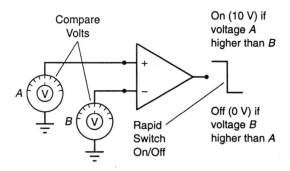

FIGURE 9-17 Op amp comparator.

The preceding examples of op amp circuits provide a very general idea of the nature of these devices and a few related simplified common circuits. However, it is important to recognize that all of these circuits, as they work with varying voltage, are analog in nature. Electronic technology is becoming increasingly digital. This means that phenomena like temperature are represented by numbers rather than by voltage variations. As a result of the popularity of digital circuits, the analog output of op amps is often converted to binary numbers.

Analog to Digital Converters

The nature of the world around us is analog. Colors have innumerable shades, rather than just the 16 colors provided by some computers. Temperature varies from low to high in a continuous way. Indeed, it would be strange if temperatures changed in steps of 10 degrees. Imagine the thermometer jumping from 70 degrees to 80, with no possible temperatures in between. As odd as this may seem, if electronic, digital-computer–like process controllers are to be used, all of the analog phenomena of nature must be converted into numbers (digital).

Before we examine this process of conversion, two general ideas need to be explained. If numbers are to be used to represent analog phenomena, the range has to be large enough to provide the accuracy necessary for that

application. For example, the telephone company converts analog speech into the numbers 0 to 255 (256 steps). This resolution is fine enough for voice, however, CDs are recorded with 0–65535 steps. This provides the high-quality sound reproduction required for this application. Resolution, or accuracy of representation, is enhanced by increasing the number range.

Some analog phenomena, such as temperature, change slowly, while others, such as video signals, change rapidly. The sample rate is simply how often (time) the analog phenomena are converted to digital (numbers). For example, temperature could be accurately represented by a sample rate of one conversion per second. During 1 second of time, the temperature does not change much. Both sample rate (how often conversion occurs) and resolution (how many numbers or steps are used) are important analog-to-digital conversion ideas.

Figure 9-18 presents the idea of converting temperature (analog) to numbers (digital). As indicated on the left of the figure, the temperature level is increasing. The temperature is converted electronically to a number between 0

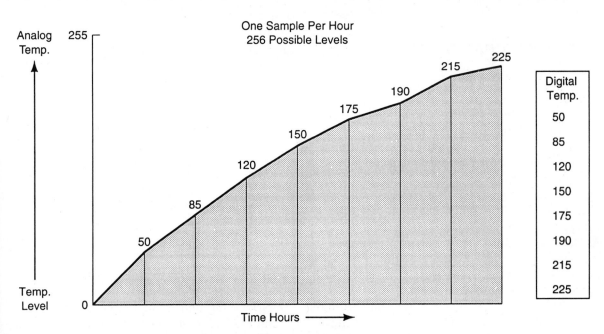

FIGURE 9-18 Temperature to numbers.

and 255. In this case the resolution is dependent on 256 steps. Furthermore, the sample rate (how often the conversion occurs) is one sample per hour. What was once analog temperature change has been converted to a sequence of numbers by a chip called an analog to digital converter (A/D).

The analog-to-digital converter chip is often a part of computerized controllers. As figure 9-19 shows A/D chips are commonly used in con-

junction with sensors and op amps. The block diagram of this circuit begins with analog temperature between 0 and 100 degrees celsius that provides a sensor output of 0 to .5 VDC. This small analog voltage change is amplified or boosted ten times (gain) by the op amp. The A/D chip then converts 0 to 5 VDC to a number between 0 and 255. It is important to recognize that with the A/D chip we have moved away from analog and toward digital circuits.

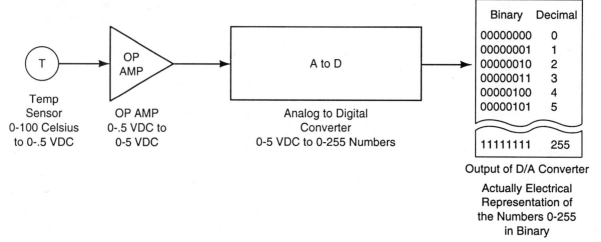

FIGURE 9-19 Analog to digital.

DIGITAL INTEGRATED CIRCUITS

Almost everything that is analog in nature can be converted to a number scale. The preceding example showed how temperature measurement is converted into numbers. A similar process can be used to convert pictures, sound, humidity, rotary motion, straight-line motion, weight, air velocity, fuel consumption, and just about anything that can be imagined into numbers. The technology of converting things into numbers is very highly developed. As a result, there has been a tremendous increase in the use of circuits that process numbers. Digital IC chips work with electrical voltages that represent numbers.

Digital circuits use a strange numbering system called binary. This is not an entirely new way to count. Rather, it is a different way to

represent quantities. Let us look at a simple example of binary representation.

Binary	Decimal
00000000	0
00000001	1
00000010	2
00000011	3
00000100	4
00000101	5
11111111	255

FIGURE 9-20 Binary and decimal numbers.

Notice in figure 9-20 how all the binary numbers are composed of 0s and 1s. While decimal numbers are made up of the ten symbols 0–9, binary numbers have only two symbols—0 and 1. The trick to understanding binary lies in the idea of weighted number positions. Weighted number position means that the value of a symbol, say 7, depends on how many places it is to the left of the decimal. For example, 7. = seven ones, 70. = seven tens, and 700. = seven hundreds. Each move to the left in the decimal system multiplies the 7 symbol's weight by ten times. Binary also has weighted positions, however, they start immediately to the left of the binary point (binary equivalent of the decimal point) and are multiples of two. For example, 1. = one 1, 10. = one 2, 100. = one 4, and so on. The weighted positions of an eight-position binary number are (128)–(64)–(32)–(16)–(8)–(4)–(2)–(1)–(. binary point). Since there are only two symbols in binary (0 and 1), you either have or don't have that weighted value. For example, the binary number 10000101 means 1×128 plus 0×64 plus 0×32 plus 0×16 plus 0×8 plus 1×4 plus 0×2 plus 1×1 equals 128 + 4 + 1 equals 133. Look at the comparison of binary and decimal numbering systems in figure 9-20 and see if you can understand this process. If you don't, don't worry too much, as the most important idea is that counting can be performed by using just zeros and ones.

Digital circuits represent a binary "0" with 0 VDC and a binary "1" with 5 VDC. Other voltages are used, but 0 and 5 volts are the most common. Look back to figure 9-19 and notice that the analog-to-digital (A/D) converter has an output consisting of eight places of binary. These binary places would comprise eight pins on this IC chip labeled D0–D7 (data 0–7, eight pins). Each of those eight pins would have either 0 VDC (binary 0) or 5 VDC (binary 1). In this way, an eight-place binary number is electrically represented as 0 or 5 VDC on eight pins, each of which represents one of eight weighted positions. These eight pins (D0–D7) are connected to eight circuit board traces called the data buss. Digital circuits manipulate these 0- and 5-VDC representations of binary through the use of digital logic.

Digital Circuit Ideas

Digital circuits became popular in the 1960s, at which time they encompassed many separate transistor switches on a circuit board. The development of IC chips enabled designers to condense these circuits into single chips. These chips were commonly provided in a large variety of compatible logic function versions that were called a family. All the chips in the same family had the same power requirements and could be directly connected to one another. The most popular family of chips called TTL (transistor–transistor logic), are used far less today than in the past. All members of the TTL family have numbers beginning with 74. Let us look at a few examples to get the idea of digital logic.

TTL IC Chip Family

The 74XX TTL chip family has a few hundred members. Each member chip requires 5-VDC power and provides logic levels of 0 and 5 VDC. Generally, the output of one chip may be directly connected to the input of another chip. In this way, complex logic can be performed by interconnected TTL chips. To clarify this idea, let us look at a few 74XX chips.

TTL chip 7404 is referred to as a hex inverter. It contains six (hex) inverter circuits. The inverter changes 0 VDC (binary 0) to 5 VDC (binary 1) or 5 VDC (binary 1) to 0 VDC (binary 0). The output of the inverter is always opposite the input. When working in digital logic, individual 0s and 1s are commonly referred to as false (0) and true (1). It also could be said that the inverter changes a false to a true or a true to a false. As a result, another name for the inverter is a NOT circuit. If it is true, the NOT circuit (inverter) makes it false (NOT true). In electronic circuits, the 7404 chip inverts 5 VDC. In digital numeric circuits, a 1 becomes a 0, and in logic circuits, a true becomes a false. Don't be confused—it is always the same circuit chip doing the same thing. The difference depends on how we describe what is going on. What is going on is electrical, numerical, or logical. It all depends on the application. No matter how we look at it, this chip flips or inverts the signal.

Figure 9-21 shows the pin-out of the 7404 chip and the logic symbol for inversion (NOT). Notice how the chip contains six (hex) of these inverters. The pins are numbered counterclockwise from the notched end of the chip, as viewed from the top. Power is connected to the chip through pins 7 and 14. If we connected 5 VDC between ground and pin 1, a voltmeter would read 0 VDC between ground and pin 2. If 0 VDC was connected to pin 1, then 5 VDC would be measured at pin 2.

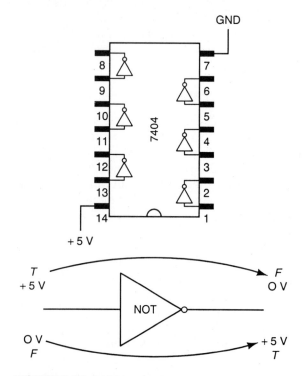

FIGURE 9-21 7404 hex inverter.

The logic of the AND gate requires that two or more inputs be at 5 VDC (true) for the output to be 5 VDC (true). Figure 9-22 reveals the symbol used for the AND gate. In this case, there are two inputs, however, AND gates are also available with five inputs. All five inputs must be at 5 VDC for the output to be 5 VDC. The 7408 chip presented in figure 9-22 is called a quad (4) AND gate. It contains four two-input

AND gates. If pin 1 AND 2 were both at 5 VDC, then pin 3 would be at 5 VDC. A true AND a

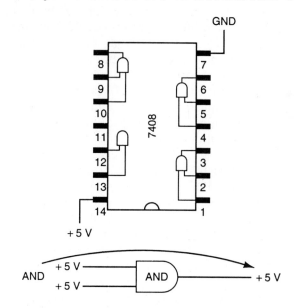

FIGURE 9-22 Quad AND gate.

true will result in a true. Any other combination of inputs will result in an output of 0 VDC (false). A simple home alarm system could use the inputs of the AND gate to indicate, through

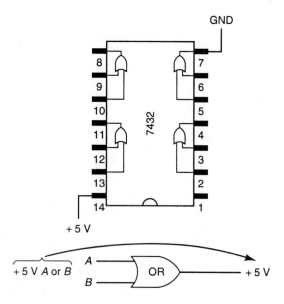

FIGURE 9-23 Quad OR gate.

switches, that the front and rear doors of the house are closed.

The last example of the TTL logic family is the 7432 chip. As indicated in figure 9-23, this chip contains four (quad) OR gates. This two-input OR gate produces a true (5-VDC) output if either input A OR B is true (5-VDC). Like the AND gate, OR gates are available with many inputs. As long as any or all inputs are true, the output is true. An OR gate could be used to indicate that one of two cars doors is open.

In the recent past, large numbers of TTL logic chips were interconnected to provide the hardwired logic necessary for machine controllers, calculators, and computers. Unfortunately, any change in the desired function of the circuit required redesigning the hardwired logic. Improvements in IC technology provided a solu-tion. Hundreds, or even thousands, of logic gates could be fabricated on single chips. However, these chips could not be application specific, since any change in function would require a complete redesigning of the chip. These new chips had to be designed in such a way that their function was universal. Further changing chip functions had to be reasonably easy and not require changes in the chip's circuit design.

Flexible Microprocessor IC Chips

An early example of universal-logic–based digital IC chips was the pocket calculator. It appeared in the early 1970s and was an instant success. As demand for these calculators increased, manufacturers devoted large portions

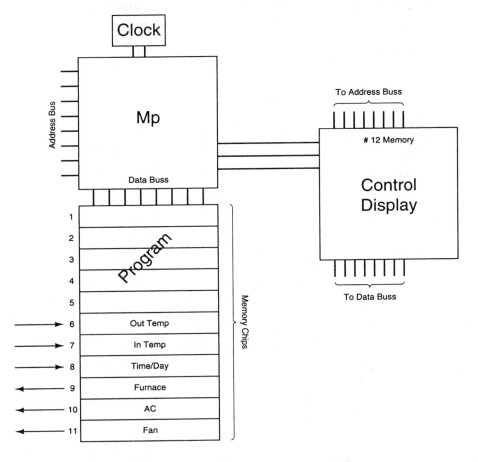

FIGURE 9-24 HVAC microprocessor controller.

of their research budgets to developing better calculator chips. As the price of these calculator chips fell, many electronic hobbyists experimented with them. Let us try to imagine what might have gone on in the mind of one of those experimenters.

"Let me think now . . . when I do calculations with this chip, I go through a process. First, I identify the formula required, then I obtain the numbers (data) that will go into the formula, and finally I step through a series of mathematical operations to eventually arrive at the answer." Ohm's law $I = E/R$ (he enters 20 volts), divided by (he hits the divide button), resistance (He enters 10 ohms). To get the answer (he hits =) answer (he reads 2 amps). Looking at his digital clock, our experimenter remarks, "It's late. I better get back to work."

With a little more time, our experimenter might have hit upon the idea of a true microprocessor chip. The experimenter was very close, but did not connect the process necessary to use the calculator with the clock. Microprocessors are very much like calculators that incorporate additional circuits to make use of a list of instructions (process). They are caused to step through the instructions by a clock.

The microprocessor controller: Figure 9-24 is a simplified block diagram of a microprocessor used as a digital HVAC controller. First let us look at the function of each part.

Clock: This circuit sends a tick-tock–like on/off signal to the microprocessor (MP). The clock controls the speed at which the process is executed. The faster the clock runs, the faster the process is stepped through. For example, a 33-megahertz clock runs a microprocessor chip twice as fast as a 16-megahertz clock.

Memory chips. Memory is composed of a series of sequential cubbyhole-like electrical storage locations. Each location can hold an eight-place number. Also, each location is identified by another number called an address. In figure 9-24, the first five locations in memory contain a list of the steps required to provide climate control. This is called the program.

Often the program has thousands of steps. Memory locations 6 and 7 contain numbers provided by outdoor and indoor temperature sensors. These sensors are constantly updating memory locations 6 and 7. Memory location 8 contains a real-time clock, which provides the time of day, as well as the day of the week. The last three locations in memory contain numbers that result from the program. These numbers serve as outputs that control the off/low/medium/high functions of the furnace, air conditioner, and ventilating fan.

Microprocessor. The microprocessor contains the necessary digital logic circuits to perform mathematical (+, −, =, >, <) and logical (NOT, AND, OR) operations. In addition, it can specify a location in memory by placing a number on the address bus that connects to, as well as enables, the single memory location (cubbyhole). Through the data bus the microprocessor can read or write (change) the contents of the memory address specified.

Control display. This unit often contains a key pan that permits data and programs to be entered, as well as a numerical display to provide information related to the ongoing process and memory contents. The wires connecting the control display unit to the microprocessor enable it to interrupt the program and take control. Notice also that the control display unit connects to the data bus. This permits the display to communicate with the microprocessor and memory. The control display unit is also connected to the address bus. In this instance, the control display is assigned memory location 12. This provides a communication address for the control display unit, which the microprocessor and the program can use.

With this brief look at the individual parts of the HVAC microprocessor controller, it is now possible to get the general idea of how it works.

The process goes like this. Upon system start-up, the MP reads the first location in memory and begins processing the program. The program instructs the MP to read memory location

8 and check to see the time and day. With that information, the microprocessor decides what the desired temperature is. The desired temperature has been previously entered into the program through the control display unit. Then the MP reads memory location 7 to see if the present temperature is above or below the desired temperature. Next, memory location 6 is checked to obtain the outside temperature. Now the program takes into account time/day, indoor temperature, outdoor temperature, and predetermined settings to calculate HVAC needs. At this point, the MP reaches program steps that load new numbers into memory locations 9, 10, and 11. These numbers cause the furnace, air conditioner, and fan to function. As a last program step, the microprocessor is instructed to go back to the beginning and run the program again. In this way, the program is constantly being repeated, always checking the inputs and calculating and adjusting the outputs. As you can imagine, program changes can radically change system function.

This brief introduction is intended to provide a general idea of HVAC process controllers. Many ideas were necessarily vague, and perhaps you have more questions than answers. That's good. Take more courses in electronics. Fill out the missing parts of this mental picture of analog and digital electronics. But don't be surprised to find there is always more to learn. Imagine how boring life would be if you knew everything.

SUMMARY

The reader was cautioned at the beginning of this chapter that it was intended to provide a general knowledge of electronics, a complex and complicated field. Additional courses and study are necessary to work with electronics.

Discrete semiconductor devices include the thermistor, which decreases resistance with a rise in temperature. The diode is a one-way electricity valve and can change AC to DC by its rectifying action. A special rectifier called a silicon-controlled rectifier (SCR) is a device that can be turned on at different places along the sine wave. In this way, the SCR and its bidirectional counterpart, the triac, can control power by chopping it up. Junction transistors are devices that permit the control of a large emitter-to-collector current by a small base current. Junction transistors amplify by increasing the power of a feeble signal. MOS transistors also amplify by controlling a large source-to-drain current with a very weak gate voltage. In 1960, scientists devised ways to construct entire circuits on tiny silicon wafers called integrated circuits, (ICs); these may be divided into two major groups called analog and digital.

Analog ICs are chips that use electrical models to represent pictures, sound or motion. While they are manipulating these signals, their voltage varies between a high and low level. They are also referred to as linear ICs. Application-specific analog ICs are available that include the entire circuit of a radio, TV, or other device on a single chip. Operational amplifiers (op amps) are general-application analog amplifier chips that have high input resistance, low output resistance, high gain, quick response time, and input–output voltages that remain at zero when no signal is present. Since an inverting input and a noninverting input are available, many unique circuits can be constructed. Amplifiers, inverting amplifiers, summing amplifiers, and comparators are a few examples. With the increased use of digital circuits, the conversion of analog to digital is accomplished with the A/D converter chip. This chip converts the analog variations to specific numbers. More numbers provide a finer resolution of the analog signal and more conversions provide a higher sample rate.

Digital circuits work with off/on electrical voltages that represent binary numbers 0/1 and the logical false/true. An entire family of compatible chips called TTL logic is still in wide use. All of these chips begin with the number 74. By connecting together NOT, OR, and AND gates, hardwired logic circuits may be constructed. These circuits are hard to modify if the required logic changes. Early calculators are a good example of many logic circuits on one chip. Furthermore, they provide flexibility in that they can perform a number of mathematical operations. Microprocessors are chips that provide the additional circuits necessary to make a calculator-like chip step through a program. This permits the reading of sensor inputs and their comparisons with predetermined settings. Additional program steps calculate the required HVAC functions to provide equipment with a power-controlling signal. The microprocessor HVAC controller cycles through the program over and over, constantly sensing the climate and adjusting the equipment functions.

As a result of this chapter, you should have more questions than answers. In a complicated field like electronics, that is natural. Take courses—study—learn.

PROBLEM-SOLVING ACTIVITIES

Review Questions

1. How can the thermistor be used to indicate fluid level? (Hint: Fluids can act as "heat sinks.")

2. How might a thermistor be used to provide slow warm-up for high-wattage incandescent lamps?

3. Diodes are sometimes connected in series with DC equipment to prevent accidental damage from reverse polarity hook-up. Explain how this would work.

4. Special light-emitting diodes are available that glow green with one polarity, red with the opposite, and amber when connected to AC. Explain how this could work. (Hint: Tricolor LEDs contain a green and a red LED connected in parallel.)

5. The output of a full-wave rectifier is only a half wave. What might be the cause of this malfunction?

6. Draw a simple diagram that compares the rheostat power controller with a triac power controller. Explain the advantages of the triac.

7. Why are SCRs and triacs not usually used to control DC power?

8. Make a simple chart that compares the terminal of a junction and MOS transistor with a valve. Label across the top VALVE – JUNCTION TRANSISTOR – MOS TRANSISTOR. On the left side, label INPUT CONTROL, OUTPUT FEED, and OUTPUT DRAIN. Fill in the terminal names for each of the devices.

9. What is the primary electrical difference between the input for a junction transistor and that for a MOS transistor?

10. What is the primary difference between discrete component circuits and IC chips?

11. What are the two major subdivisions for IC chips?

12. Comparing electricity to gas is an analogy. What does *analog* mean?

13. List the characteristics of a perfect op amp.

14. Why does an op amp usually require a feedback resistor?

15. Invent a game that makes use of the comparator.

16. Identify and explain the two important characteristics of A/D converters. (Hint: accuracy and time.)

17. Digital IC chips represent phenomena with _____.

18. List the characteristics that make the TTL chips a family.

19. Make up a chart to identify all the possible input and output conditions for a NOT, AND, and OR gate. The columns should be labeled IN A, IN B, and OUT. Use enough rows to show all possible combinations for a two-input AND and OR gate.

20. What is the major drawback of hardwired logic?

21. How is a microprocessor like a calculator?

22. What controls the speed at which the microprocessor executes steps?

23. What is the sequence of steps the microprocessor follows called, and where are these steps stored?

24. What are the names and the purposes of the two major buses in the microprocessor controller?

25. Whom would you contact to find a course about a particular brand of controller?

CHAPTER 10

Control Theory

Poor Mother Nature gets very little respect. For centuries, we humans have been borrowing her best scientific principles to build the technical world in which we live. It seems as though we have forgotten that these scientific principles were first used by nature. In a rather self-centered way, we refer to these principles as laws, usually named after the inquisitive people who first recognized them. It almost seems that we believe nature must obey our laws. For example, Isaac Newton's gravitational laws simply describe the nature of gravity. With or without Newton the planets will continue to stay in orbit. Gases act they way they do, with or without Boyle's law and Charles's law. Even in the field of electricity, it's easy to forget that Ohm's law is a description of the nature (Mother Nature) of electricity. All of the great scientists were people who observed, directly or with instruments, the workings of Mother Nature and "figured out" what was going on. Hans A. Bethe received a Nobel prize when he "figured out" how the sun works. By so doing, he described how Mother Nature warms her people. Mother Nature, it seems, got no prize. We think Mother Nature deserves more respect.

The point of this idea is that humans, in particular, technologists and scientists, are not so much law makers as they are "copy cats." Steel frames that support giant skyscrapers are remarkable structures, but not nearly as remarkable as the human skeleton. Further, buildings don't move. Computers, as amazing as they are, seem childlike when compared with the human brain. All the building plans, electrical schematics, maps, charts, and plans we have devised are elementary when compared with the genetic code. Imagine every detail of the entire structure of the human body stored in a chemical code millions of molecules long. We may, one day, modify these codes, but it's important to remember that we did not make these codes.

The things we do make are based on principles that have been copied from nature. In this chapter, which completes the theory section of this book, the focus will be on control theory. Control theory assembles the separate electric devices and theories into a complete whole, in the same way that a human being is made of parts—sensors to provide information, a brain to process it, and muscles to take action. Machines that are based on control theory mimic this "human nature". Anthropomorphic means to give humanlike characteristics to nonhuman things. HVAC systems are becoming more and more anthropomorphic in that they make extensive use of control theory. Let us examine this humanlike nature of machines.

BASIC PRINCIPLES

To provide a general introduction to control theory, it is helpful to understand how machines can be made more humanlike. For example, the ability to stand and balance depends on the coordination of muscles and mind. In the same way the coordination of microprocessor controllers and HVAC equipment can result in systems that balance temperature, humidity, and air quality.

Humans and Machines

The chart in figure 10-1 portrays the key control-theory similarities between humans and machines. During the following discussion, refer to it frequently, and give some thought to the comparisons presented there.

Both humans and machines require a power source to function. Machines in this discussion include furnaces, air conditioners, and all HVAC equipment. Humans primarily use food and oxygen, while machines may use a variety of

Concept	Human	Machine
Power-Energy	Food & Oxygen (All Purposes)	Electric (Action & Heat) Pneumatic-Hydraulic (Action) Gas, Oil & Coal (Heat)
Sensors	Sight, Sound, Taste, Smell, Touch, Tension, Chemical	Temperature, Humidity, Pressure, Flow, Light, Position, Gas Content, Level
Communication	Nerves	Wires, Fiber Optic, Remote (IR & RC)
Actuators (Motion)	Muscles	Motors, Solenoids Cylinders, Bellows
Actuators* (Heat)	Metabolism (Slow Burn)	Fuel Combustion, Resistance (Rapid Burn or Consumption)
Actuators* (Cooling)	Perspiration (Evaporation)	Air Conditioner Cycle (Heat Movement)
Thinking	Brain Mind	Microprocessor Program

* Note: More commonly referred to as converters

FIGURE 10-1 Humans and machines.

power sources. Machines in this discussion include furnaces, air conditioners, and all HVAC equipment. The machine power source for heat is usually relegated to the combustion of gas, oil, or coal, and resistance heating. By far the most common power source for machine motion is electricity. Pumps providing air pressure (pneumatics) or oil pressure (hydraulics) are considered power sources for machines, even though the pumps are generally driven by electric motors. In the cold of winter, people require more food (fuel) since metabolism is increased to keep warm. In the same way, as more heat is required from the furnace, fuel consumption rises sharply.

Humans are aware of their surroundings though sensors. Things look good, sound bad, taste funny, smell like gas, and are hot when touched. In addition to the five senses, humans are equipped with a variety of sensors that are unconscious. For example, these sensors provide information to the automatic part of our brain that controls the digestion of food, respiration (CO_2 sensors), and muscle tension. Though you may not realize it, tension sensors in many of your muscles are providing information to your brain, which uses this information to control the muscles that maintain your present position. Relax all your muscles and you will slump to the floor. Machines also need to be aware of important phenomena that surround and influence their function. Machine sensors are highly developed. They can be designed to measure all of the phenomena presented in figure 10-1, as well as just about anything else. Acceleration, viscosity, radio activity, velocity, and even smell sensors are widely available. Microelectronic IC fabrication techniques are used to produce tiny, highly accurate, and inexpensive sensors. For medical purposes, these sensors may be implanted in humans.

Nerves serve as the primary communication system in humans; however, chemical messages are also sent through the circulation system (hormones), as well as directly between cells (neurotransmitters). Humans are collections of millions of cells that Mother Nature has provided with a highly complex communication system. Machines are simpler. They ordinarily depend on wires and circuit traces to carry electrical messages. Digital messages are also carried by light flashes through fiber-optic cables. And one can expect that HVAC equipment will depend increasingly on fiber-optic communication. Though of less importance, some remote communication is carried by infrared light (TV remote) and radio signal (garage door opener). With special codes, telephones will also be used to control house climate functions.

As Grandfather would say, "Action is the magic word." People make events happen by thinking, and then taking action. Muscles provide us with motion—motion to act on our best judgment. Just as people have unconscious sensors, they also take unconscious action.

Chemical catalysts called enzymes digest food, and, in an automatic and unconscious way, we perspire when hot. Defined in this way, action does not always involve obvious motion. Machine motion actuators include motors that spin, solenoids that push/pull, bellows that expand/contract, and hydraulic or pneumatic cylinders that extend/retract. Converters are like actuators as they provide a useful effect from power. In furnishing heating and cooling HVAC systems make extensive use of converters. These energy converters are essentially passive actuators.

The thinking part of humans is primarily located in the brain. Its biological importance to the life process is so great that nature has fully encased it in a protective skull. Later in this chapter ideas related to protecting the microprocessor, the machine's equivalent to the brain, will be discussed. Ordinarily, in the brain and the microprocessor "something is going on." People are very complex and unique. We all seem to have minds of our own. For humans we usually think of the brain as hardware-like and the mind, that which is going on, as software-like. The machine's mind is the software program run by the MP.

Open Loop

The authors reside in a state that strictly enforces speeding laws, and automobile insurance costs are based on traffic violations. As a result, a $75 speeding ticket can mean an insurance premium increase of $300 a year. So people drive carefully. While driving on open roads, it is important to get into the habit of checking the speedometer. Quite unintentionally, the speed can gradually creep up, and there goes the money for the new stereo system. This type of speed control is called open loop. You select a safe and ticket-free speed and maintain it by checking the speedometer, and then adjusting foot pressure on the accelerator. The control resides in the human operator. It is manual, not automatic.

Many wood-burning stoves require human intervention to provide reasonable temperature control. Through the manual adjustment of the draft and fuel, the stove produces enough Btu's to keep the room comfortable. Open loop here means that control is afforded by direct human action. Most of our everyday life involves making judgments about the world around us, and exerting some control to make it as we think it should be. We live most of our lives in an open-loop or manual way.

Closed Loop (Self-Correcting)

As life became more complicated and technical, clever individuals designed systems that did not have to rely on human control. One of the first self-correcting systems was the toilet tank. Water level in the tank is controlled automatically. The valve filling the tank gets information related to the process (tank filling) from a float. The float, through a direct mechanical link, informs the valve of the tank's water level. It seems as if plumbers beat the electronics people to the punch in copying Mother Nature's closed-loop control.

This same sort of self-correcting level control system is used in humidifiers, commercial air conditioners, and boilers, and throughout virtu-

ally all of the technologies that involve the manipulation of liquids. Gasoline engine carburetors use this same float system. Figure 10-2 is a simplified diagram of the toilet water tank, which functions in the same way as do many tank level controls. The flapper valve that serves to drain the tank has been omitted for clarity.

Actuator: In this case, a valve is used to control the flow of water under pressure from the main supply. When the valve is lifted, water enters the tank and begins the filling process. When seated, the valve stops water flow. The action (filling) is achieved by the actuator valve.

Sensor: The phenomenon that is controlled in this system is water level. As a result, a float serves as the sensor. Like most sensors, it produces information, in this case, mechanical, by following the tank water level (phenomenon). It communicates water level information mechanically through a lever rod.

Feedback: An important and unique part of the closed loop is feedback. It provides specific information about the ongoing process. It's the "how am I doing?" part of the process. In this same way, it is a good idea to get feedback from customers related to jobs in progress. You have to ask for feedback to avoid costly and frustrating corrections later on. The lever rod is asking the float (sensor), "How's the tank filling proceeding?" Feedback closes the loop in this self-correcting system.

Desire: Many conflicts can be avoided by being very clear about what you want. In the HVAC field, contracts are necessary to spell out in exact detail what you will do and what the customer will pay. In this same way, the desired water level is set by adjusting the up/down position of the slotted fulcrum (pivot) of the lever rod. The location of the fulcrum expresses the desired level. Setting the temperature of a thermostat or the relative humidity of a humidistat is a technically clear way to express the climate conditions you desire. Many closed-loop control systems have the desire built into the system, while others, like the day/night thermo-

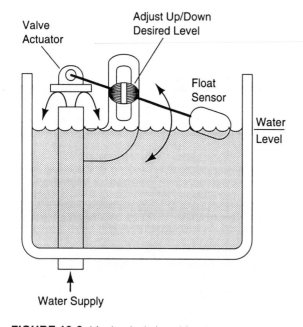

FIGURE 10-2 Mechanical closed loop.

Valve Actuator

Adjust Up/Down Desired Level

Float Sensor

Water Level

Water Supply

stat, change the desire from time to time. Process controllers recalculate desire often. For example, microprocessors used to control robots constantly produce new desired positions for the manipulator arm. In simple language, desire is "what you want."

Compare: An important idea related to closed-loop control is the comparison of desire and feedback. Desire (water level you want) and feedback (how the tank is filling) are compared. In the case of the toilet tank, the comparison is entirely mechanical. The desired level is set on fulcrum (up/down adjustment), feedback is communicated from the float sensor through the lever rod (up/down water level), and the comparison of these two results in valve actuation. Remember the op amp comparator from Chapter 9? It compared two voltages. In the same way this mechanical system compares

two levels. If the comparison of the two levels results in a difference action is taken. When the comparison results in no difference, action stops.

Toilet tank review: Moving from reality (figure 10-2, the toilet tank) to symbolic representations makes the closed-loop idea more general and useful. Figure 10-3 is a symbolic representation of the previous discussion. It shows that the desired level has been communicated to the comparator by the up/down adjustment (A). As water fills the tank through the actuator valve, the sensor float rises, sending feedback (B) to the comparator. When the comparator finds no difference between desire and feedback, the filling process is terminated by valve closure. The process will begin again if the tank is drained or the desired level is changed.

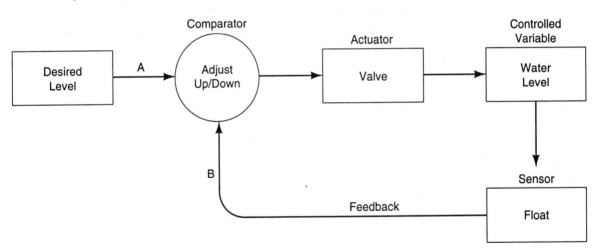

FIGURE 10-3 Closed-loop idea.

CONTROL IDEAS

The primary difference between prehistoric society and modern-day society is civilization. Civilization resulted from the desire of people to work together in an equitable way. It demands self-control and provides, in return, a better, happier, more productive, and richer life. The collapse of civilization can be observed in various countries throughout the world. Entire societies tumble into destruction and death. On

a personal basis, the authors have seen HVAC contractors lose families and businesses through drug and alcohol abuse. Although it can be difficult to maintain civil control, and self-control, their loss would make a reasonably well-ordered life impossible. All of these social ideas have parallels in machine control.

Null Control

Generally comparator circuits are designed to compare desire and feedback in an analog or digital way. If the comparison results in a zero, the actuators in the system are turned off. For example, an analog comparator might receive a +3-VDC signal from a computer (desire) and respond by actuating a heater. In return, the heat would be sensed and provide negative feedback voltages. When the feedback voltage is -3 VDC (enough heat) the comparator's inputs (desire and feedback) exactly cancel. The result of this comparison is zero or nothing. Responding to this null condition, the comparator turns off the heater. Most comparators are designed to energize actuators whenever their null or rest condition is disturbed. With a heating system, this disturbance may simply be that it gets colder (feedback) or that the thermostat is set higher (desire). The actuators controlled by the comparator are intended to bring the system back to a null condition. Comparators exert control in an attempt to keep the system at a null state. Automobile speed controllers attempt, by varying throttle position, to maintain no (null) difference between the selected speed (desire) and actual speed (feedback). This idea is true of humans as well. Standing up may be the present null setting of your balance system. If you are shifted off of that null by someone bumping into you, quite automatically your muscles adjust their tension to regain balance (null).

Response Time

How quickly a system reacts to a change in desire or a sensed phenomenon (feedback) is described as its response time. Due to their large mass, ocean freighters respond to engine and rudder changes slowly. A ship can travel many miles before engine reversal actually stops it. A great deal of anticipation is necessary to guide large ships on the open seas; in harbors, however, anticipation alone is not enough. Tugboats afford a shorter response time in the same way that heat anticipator coils in thermostats shorten the response time of the heating system. Outdoor heat anticipators can provide early warning of sharp temperature changes and shorten HVAC system response

time. Microelectronic controller circuits can permit response times that are as short as a few billionths of a second. While this is generally desirable, systems that respond too quickly can consume more energy and result in increased appliance wear. For example, constantly changing speed on the highway produces large penalties in fuel costs and mechanical wear. A heating or cooling system that responds too quickly to small temperature changes consumes additional fuel.

Error

Remember, the old saying, "learn from your mistakes?" That saying is true. If people were perfect, they would have nothing to correct, and as a result would learn very little. Closed-loop control systems attempt to maintain a variable temperature, for example, at a selected level. If the temperature were controlled perfectly, no error would be permitted. However, without error (difference between desire and feedback), the comparator could not provide control for the actuators. Is seems that both humans and closed-loop controllers require error to function.

Errors made by control systems can be separated into two major subcategories. The steady-state error illustrated in figure 10-4 arises from a difference between thermostat setting and actual room temperature. This type of error is called steady state since it is consistent. We often become adjusted to steady-state error and compensate for it. For example a "slow oven" takes longer to cook a roast. To compensate, the chef increases cooking time. Scientifically, there is no such thing as a slow oven; rather, the thermostat requires calibration. When the thermostat reads 350 degrees the actual oven temperature is 320 degrees. This slow oven has a steady-state error of −30 degrees. In the same way, speedometers, bathroom scales, tire pressure gauges, and electrical meters can be subject to steady-state error. Correction is accomplished by comparing the suspected device with a known standard. For example, an accurate thermometer permits recalibration of room thermostats.

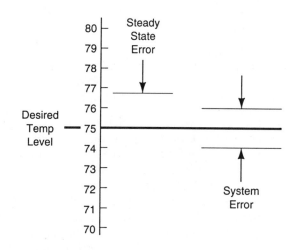

FIGURE 10-4 Control error.

Dynamic or changing, errors (non-steady state) may be collectively referred to as system error. Since these errors change in both degree and time figure 10-4 displays an error range of 2 degrees. In the field of control technology, a large number of errors fall into this subcategory. Some dynamic errors are unintended. Loose steering, for example, permits a car's direction to be influenced by wind, road conditions, and tire wear. In this same way, loose temperature sensors may move about and introduce dynamic errors in appliance operation. Water and resulting rust in air-conditioner sensors can make their functioning unpredictable. Rather than measuring refrigerant pressure and responding smoothly, they develop a "sticky" erratic action. These unintended dynamic system errors require careful diagnosis and often entail component replacement.

Intended dynamic system errors are residual to system functioning. In simple words, that means that a closed-loop system that regulates temperature must permit temperature to vary. If temperature did not vary, the closed-loop control would not measure error, and, as a result, would not function. Intentional errors designed into the system relate to the range, or degree, over which the phenomenon is allowed to vary before control action is initiated. For example, automobile speed controls ordinarily permit speed variations of 2–4

MPH. Home heating systems generally allow temperature to vary 3 to 6 degrees. Upper and lower limit settings permit adjustment of dynamic system error that is a result of range. Remember, some error is necessary for the system to function.

How fast the system responds to change is related to the speed with which the system corrects error. For example, a 10,000-Btu air conditioner in a small room removes excess heat fast—in some cases, so fast that the room is uncomfortable. Oversized HVAC systems effect a change in temperature too fast. Both range and speed error are by-products of system design. A proper match of these two dynamic system errors results in a control idea called damping.

Imagine an oversized heating system (fast error correction) adjusted to permit very little temperature change (narrow range). This system is capable of responding quickly (fast error correction) to a thermostat that, due to narrow temperature range settings, requests heat too often. Figure 10-5 shows in graphic form the rapid on/off cycle of the heating system. Far too much change is permitted in this system, which results in high fuel consumption. An underdamped system permits excessive change. Examples are large differences in day/night temperature setting, constantly adjusting the thermostat, curtains blocking window air conditioners, and keeping time with the music on your car radio with your foot on the accelerator pedal.

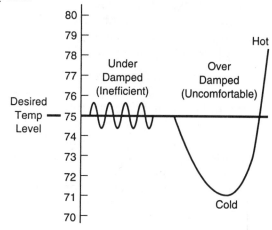

FIGURE 10-5 Control system damping.

HVAC systems with wide range settings and undersized appliances result in a system that is overdamped. The design, in this case, permits temperatures to fall below the comfort level. The heating system, once energized, stays on too long, and often results in too high a temperature. Proper settings of a correctly sized HVAC system provide reasonably comfortable conditions with efficient fuel consumption.

INTERFACING

Microprocessor chips run programs that are stored in memory. The results of program calculations are also stored in memory. This is a bit like a person who has great daydreams, but lives in isolation. Interfacing is the part of the microprocessor controller that enables information input and output, thereby connecting it to the real world. Without interfacing, microprocessors are isolated.

Interfacing is a complicated technical topic that requires extensive study. However, some of the general considerations of this topic may be easily grasped. As an example, let us see how a sensor might be interfaced with a microprocessor so that its data could be incorporated into the microprocessor's calculations. The process would proceed as follows:

1. *Signal level adjustment.* Assuming the sensor is a thermocouple its output is a weak analog electrical signal, which results from temperature. Often an op amp (discussed in Chapter 9) would be used to amplify the feeble signal.

2. *A/D converter.* The increased analog signal output of the op amp would serve as the input to an A/D converter. This conversion is necessary since microprocessors work with numbers (not analog signals). As the number is represented by 0/5 volt (binary 0/1) wires may be used to carry the number some distance. A small drop in voltage due to wire length would cause error in an analog voltage, but does not affect the

definite nature of the 0/5 volts representing the number.

3. *Latching.* Digital output of the A/D converter may be stored in a memory-like chip that temporarily holds the number that represents temperature.

4. *Data strobe.* When the microprocessor reaches the part of its program where it requires the number representing temperature, it processes a program instruction that electronically connects the latched data (temperature number) to memory. For a very brief instant, a signal called a strobe (pulse) disables sensor input to the latch and reads the latched data into memory. This prevents data from changing while the microprocessor is reading the data into memory.

In this way, the program causes the microprocessor to check the value of all system sensors. As the program proceeds, it arrives at numerical results that serve as output to various actuators. The output process is like the input process in reverse:

1. *Data strobe.* As a result of the program a number representing heat requirements resides in memory. It is "strobed" into an output latch.

2. *D/A converter.* The number representing heat requirements is converted to an analog voltage.

3. *Amplification.* The feeble voltage from the D/A converter may be increased by an op amp.

4. *Power Control.* The op amp's output could be used to provide a gate signal for a triac (discussed in Chapter 9). The triac provides power to the heater.

The previous discussion was intended to present just a quick glimpse of interfacing so don't be disturbed if you missed some of the idea. See if you can draw a simple block diagram to help "think it through." Imagine what would happen to this microprocessor controller if the heater (240 VAC) should short-circuit back through the output circuits to the microprocessor. Poof . . . brain dead!

Opto-Isolators (Brain Protection)

Input and output interfacing circuitry failures are often prevented from destroying the microprocessor by opto-isolators. These tiny devices encapsulate an LED (input) with a light-sensitive photo-triac or phototransistor (output). When an electrical signal energizes the LED, its light illuminates and energizes the photo devices. Figure 10-6 shows the internal arrangement of the opto-isolator. Since light carries the signal no direct electrical connection is made between input and output. Isolation between input and output is often as high as 10 kV. Control systems with many sensors and actuators would require a great deal of interfacing cir-

cuitry and many opto-isolators. However, there is a common alternative.

Multiplexing and Demultiplexing

Multiplexing is a fancy name for sharing. It means that many sensors, or, for that matter, actuators, can share input or output connections to a microprocessor. This is achieved by switching from one device to another in a regular and synchronized fashion. The microprocessor's input time is divided via *time division multiplexing*. For example numbers representing outdoor temperature, indoor temperature, relative humidity, and plenum temperature could be latched into the microprocessor, one after another. In a synchronized and coordinated way, the microprocessor controller could copy each number into appropriate memory locations for later program processing. Figure 10-7 shows simplified multiplex and demultiplex chips.

Computers do the same thing when they send out data to printers. Some printers connect to a parallel port that transfers eight binary places at once. This parallel port has a wire for each number place (eight plus ground and strobe). Each of the eight wires carries a binary 0 (0 volts) or a binary 1 (5 volts). Just like the interface discussed previously, the data from the computer are strobed into the printer latch. Some printers connect to a serial port, which, like time division multiplexing, sends one number place after another, in a series. In the printer, the serial data are demultiplexed to numbers. Again, refer to figure 10-7.

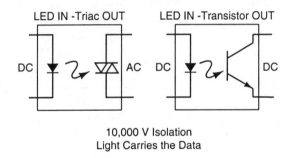

10,000 V Isolation
Light Carries the Data

FIGURE 10-6 Opto-isolators.

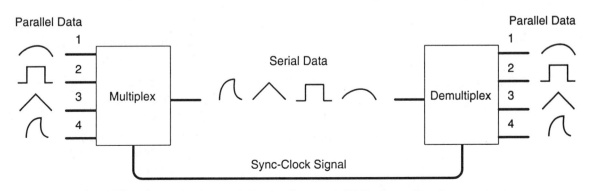

FIGURE 10-7 Multiplexing and demultiplexing.

SENSORS

Humans, as well as machines, derive an awareness of their environment from information provided by sensors. Sensors keep the closed-loop control system in touch with the outcome of its actuators, as well as the surrounding environment. Sensor technology is a high-growth, specialized area. In the past few years tremendous improvements in accuracy and reliability have been achieved, along with price reductions. Of special note is the increased use of microelectronic circuit fabrication techniques called micromachining. Through application of this photographic etching process, both sensors and actuators the size of a human hair are fabricated. During your working life, expect to find a large increase in the use of sensors. HVAC microprocessor controllers will increasingly rely on these sensors to provide ideal climate control with reduced fuel consumption and fewer environmental consequences. The brief survey of sensors that follows is intended to provide the specific ideas for common HVAC sensors and a general idea of the scientific principles involved in sensor technology. Many types of sensors not ordinarily used in HVAC systems—for example, tactile, chemical, and audio—have been omitted.

Temperature

With the exception of the capillary tube, most of the temperature sensors presented in figure 10-8 have been described in previous chapters. The most common example of a capillary tube is the thermometer. Thermal expansion of a liquid contained in a bulb results in the liquid's being forced up the thermometer's bore or capillary tube. The same principle is often used in remote temperature sensing. A liquid contained in a bulb, usually elongated, is connected to a thin capillary tube. The tube carries the thermally pressurized liquid to an actuating device that can expand and contract, for example, a bellows, a diaphragm, or a Bourdon tube. The expanding device can operate a microswitch that provids thermostatic action for refrigerators, air conditioners, hot plates, and many other devices. The expansion device, in this case, a Bourdon tube, may be connected to a needle to indicate temperature.

Name or Action	Sensor Output	Brief Description
Bi-Metal	Bending Motion	Unequal expansion of two layered metal strips.
Capillary Tube	Liquid Expansion	Expanding liquid provides force to bellows or bourdon tube.
Thermocouple	Electricity (current)	Dissimilar metals provide a small current when heated.
RTD (Temperature Dependent Resistor)	Resistance Change	With a temperature increase most materials increase resistance.
Thermistor	Resistance Change	With a temperature increase the thermistor decreases resistance.

FIGURE 10-8 Temperature sensors.

Solid-state pressure sensors may be used to provide electronic signals that are temperature dependent. Care needs to be exercised while handling the capillary tube to avoid kinking or rupturing it.

Humidity

Humidity sensors depend on the water vapor dissolved in the air to function. Unfortunately, both the air and the water vapor carry impurities, that results in relativly short sensor life and large inaccuracies with aging, especially in dirty environments. Humidity sensors should be kept away from kitchen grease and smoking, and from areas such as bathrooms that are subject to sharp humidity changes.

The psychrometer shown in figure 10-9 is an automated version of the wet- and dry-bulb temperature, relative humidity measuring method. Since the maintenance of this equipment is high, it is seldom used in HVAC appliances. In the future, micromachining may provide more practical psychrometers.

Both the organic and resistance hygrometers depend on the absorption of water vapor to sense humidity. This process is slow and response times are long. Furthermore, impurities are also absorbed, which leads to inaccurate readings. The organic version of this sensor depends on the effect of moisture on hair, while the resistance version is an example of the tendency of many substances to attract water. Before refrigeration, salt was commonly used to preserve meat. The high salt content absorbed water, preventing microorganisms from entering the meat and, in this way, prevented spoilage. Capacitive hygrometers, when calibrated, can provide very accurate measures of humidity; however, they require additional electronic support circuits, and are also subject to corruption by impurities.

Name or Action	Sensor Output	Brief Description
Psychrometer	Electronic	Through electronic sensors wet and dry bulb relative humidity may be preformed automatically.
Organic Hygrometer	Small Expansion	The length of a stretched hair changes with relative humidity variations. The addition of a micro-switch provides humidistat action. In some instances treated polymer materials are also used.
Resistance Hygrometer	Resistance Change	Lithium chloride deposited on a substrate changes resistance with humidity variations. Additional circuits provide RH measure and/or switching action.
Capacitive Hygrometer	Capacitance Change	The insulating nature of air changes with humidity. As a result an air insulated capacitor varies its capacitance with relative humidity. Additional circuitry can provide very accurate measures of RH.

FIGURE 10-9 Humidity sensors.

Position

Simple position sensors are used to indicate limits of motion. For example, some hydronic heating zone valves use limit switches to indicate valve open/close. Figure 10-10 presents an overview of position sensors. Notice that most of the sensors are limit switches that employ mechanical, photo, and/or magnetic devices to provide the switching. Switches that depend on actual contact closure are subject to dirt accumulation and wear.

As a result there is a general trend toward "contactless" electronic, magnetic, and photo switches. Some position sensors can give very accurate indication of specific position location.

Linear variable differential transformers (LVDTs), optical encoders, and magnetic resolvers are sensors designed to provide electronic indications of specific position. Encoders and resolvers will be discussed later in connection with servo motors. The LVDT is a widely used and highly accurate position sensor. Although

Name or Action	Sensor Output	Brief Description
Limit Switch	Switch Closure	Any mechanical action can, by the installation of stops, operate a micro-switch.
Photo Switch	Electronic or Relay Circuit Interruption	A light beam may be interrupted by a guillotine-like action. Photo-sensitive semi-conductors can control circuits directly or through relays.
Magnetic Reed Switch	Switch Closure	When a magnet is moved close, thin metal reeds sealed within a glass tube are caused to contact and complete a circuit.
Magnetic Hall Effect	Semiconductor Resistance Change	When a magnet is moved close to special semiconductor devices a change in resistance occurs. Additional circuitry can be used to provide a switching action.
LVDT (Linear Variable Differential Transformer)	Electronic imbalance of two equal but opposite AC waves	Basic LVDTs are transformers with two secondaries and a moveable core. By shifting the core the proportion of electricity of electricity induced into the secondaries is changed. This proportion may be interpreted with additional circuits to provide very accurate position information.

FIGURE 10-10 Position sensors.

its name makes this device sound complicated, its function is really simple.

As figure 10-11 shows, the LVDT is a special transformer with a single primary coil. When this coil is energized with AC, an expanding and contracting magnetic field is produced. This field is directed equally toward two secondary coils by the highly permeable magnetic core. The secondaries are wound in opposite directions, which results in AC outputs that are of

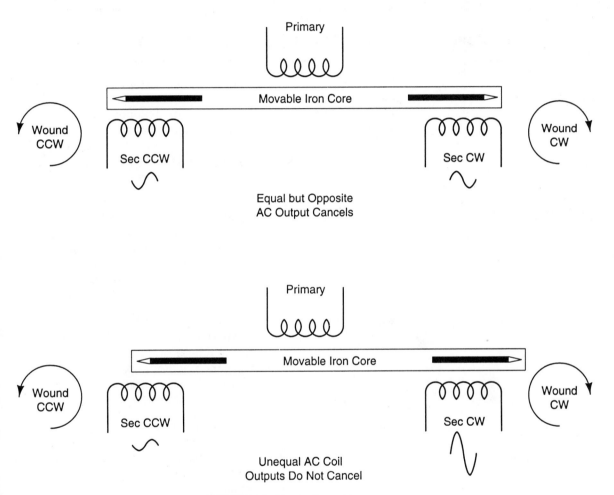

FIGURE 10-11 LVDT position sensor.

opposite polarity. While the core remains centered, as shown at the top of figure 10-11, the outputs of the two coils cancel exactly cancel each other. If the core is shifted left or right, as shown at the bottom of figure 10-11, the intensity of magnetism producing current in the secondaries is disturbed. The left coil produces less AC than the right coil. This imbalance can be electronically measured and provide accurate core position information.

Liquid Level

Answers to the questions "Is the oil tank full?," "Do we have enough gas to make it home?," and "Is the boiler full?" are provided by liquid level sensors surveyed in figure 10-12. The float level sensor is by far the most commonly used sensor. The buoyancy of a float provides the leverlike action previously discussed in relation to the toilet tank. All of the common mechanical, electronic, magnetic, and photo switching devices may be actuated. The float arm may also be used to control a variable resistor. In that case, the resistance can be interpreted as tank level.

In applications where liquid levels frequently cycle from empty to full, a pressure tube level sensor is often used. As the fluid level rises, pressure builds in a vertically mounted tube. A diaphragm microswitch pressure sensor is

Name or Action	Sensor Output	Brief Description
Float	Float Level (switching and resistance change)	Floats may provide switching action by mechanically operate micro-switches, magnetic reed switches, hall devices and photo switches. Float operating levers may be connected to variable resistors and provide level measurement.
Pressure Tube	Switch Closure	A vertical tube with a diaphragm micro-switch at the top is inserted in the liquid. As the liquid rises pressure is produced within the tube and against the diaphragm micro-switch.
Direct Probe	Switch Closure	Conductive metal probes are inserted in the liquid. Depending on the nature of the liquid resistance and/or capacitance may be sensed. Additional circuits are necessary.
Optical	Switch Closure	A light beam may be used in a transmissive or reflective way to sense liquid level. Photo semiconductors provide signal for additional circuits.
Thermistor	Resistance Change	Self-heating thermistors can use the heat sinking or cooling nature of a liquid. Additional circuits are required.

FIGURE 10-12 Liquid level sensors.

mounted at the top of the tube. When the level provides enough pressure, the switch opens and terminates the filling cycle. When the tank drains, any air that has been dissolved out of the tube is naturally replaced. Washing machines frequently use this type of sensor.

The electrical nature of two probes may sense tank level. Some liquids are conductive enough to use resistance sensing, while non-conductive liquids depend on capacitive change of the probes. In both instances, probe oxidation and coating can be a problem. The transmissive, refractive, and reflective nature of some liquids enables the use of optical level sensors. Again, problems may be encountered with sensor coating.

A self-heated thermistor can provide liquid level information. With this type of sensor, current flow heats the thermistor. If the liquid level is below the thermistor, its resistance remains low and current remains high. When liquid rises, contacts, and cools the thermistor, its resistance increases and current flow decreases. The change in current flow is an indicator of tank level.

Pressure and Vacuum

Pumps, blowers, fans, boilers, and compressors used in HVAC appliances produce pressures that, for proper system functioning, as well as safety, must be monitored. In instances where pressure falls below atmospheric pressure (15 pounds per square inch or, PSI), this condition is referred to as vacuum. Most pressure sensors are relative in that they compare one pressure with another. Often, the other pressure is atmospheric. For example, 30-PSI tire pressure is 30 PSI above 15 PSI atmospheric pressure. Imagine being inside the tire and measuring outside air pressure. In that instance, relative to the tire dweller's world (30 PSI), the atmosphere would be a vacuum (−15 PSI). One way or another, these devices are actually pressure differential sensors.

Pressure sensors employ some flexible or movable mechanism to actuate mechanical, electronic, or photo switches. These same types of mechanisms can be used to vary resistance, capacitance, and optical effects, or in conjunction with position sensors. These devices provide an analog signal that infers actual pressures. Therefore, these sensors may operate like switches or provide actual pressure infor-

mation. In this same way, automobile oil pressure may be indicated by a light (low/O.K.) or a gauge (actual pressure).

The bellows pressure sensor shown in figure 10-13 is usually used with low pressure. It provides a relatively long movement stroke, which can actuate sensing devices that require large motion. In many instances, this same mechanism is used as an actuator. At somewhat higher pressures, the diaphragm mechanism is useful. It provides less movement; however, it lasts long and is inexpensive. The diaphragm, like the bellows, is also used as an actuator. All of these devices may develop leaks, which can be detected with a simple suction test. For reference a summary description of these sensors is presented in figure 10-14.

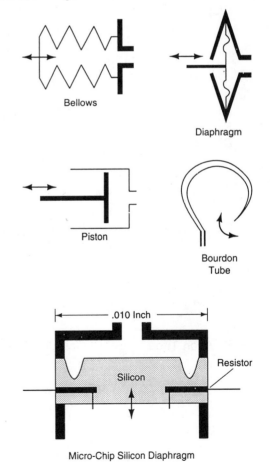

FIGURE 10-13 Pressure sensor mechanisms.

The microchip silicon diaphragm sensor shown in figure 10-13 is only 10 thousandths of an inch wide. Pressure is applied to the diaphragm through the opening at the top. It does produce a slight diaphragm deflection that results in stress. Embedded in or deposited on the silicon diaphragm are silicon resistor strain gauges that change resistance when distorted. In this way, high pressures result in resistance changes. Many silicon pressure sensors have op amp

Name or Action	Sensor Output	Brief Description
Bellows	Expand-Contract Motion	Bellows can be open on one side to pressure or connected to pressure source by tubing. The expansion resulting from pressure can operate a micro-switch, vary a resistor, change capacitance and/or operate mechanical devices.
Diaphragm	Buckling-Bulging Motion	Pressure differentials on opposite sides of a diaphragm can operate a micro-switch, change capacitance, vary a resistor, and bend a silicon resistor. A silicon resistor varies its resistance when bent or twisted.
Micro-Chip Silicon Diaphragm (IC)	Electronic Signal	Integrated circuit pressure sensors employ a photographically etched microscopic diaphragm that has pressure differential expressed on opposite sides. Often the necessary circuits are integrated into this small and reliable IC sensor.
Piston	Extend-Retract Motion	A piston-cylinder sensor has pressure applied to the piston which results in linear motion. This motion can operate micro-switches or operate other valves directly. Position sensors may also be used to infer pressure.
Bourdon Tube	Coil-Uncoil Motion	The bourdon tube is a metallic coiled tube that resembles a New Years party favor. Pressure tends to uncoil the tube, which can operate mercury tilt switches, micro-switches, gauge needles, and variable resistors.

FIGURE 10-14 Pressure and vacuum sensors.

and A/D converters combined on the same chip to provide signal conditioning.

The piston sensor mechanism at the top center of figure 10-13 is used when high pressure are encountered. Often high pressures in air conditioners place opposing pressures on opposite sides of the piston, or opposing pressure may be provided by a spring. Pistons may then directly actuate flow valves or provide sensor movement. Frequently "O" rings are used to provide piston sealing. Again, the chart in figure 10-14 describes this device.

The French hydraulic engineer Eugène Bourdon (1808-84) devised a coiled tube sealed at one end. When pressure is applied to the open end it tends to straighten. The most common use of the Bourdon tube illustrated in figure 10-13 is the pressure gauge. In that application, a Bourdon tube provides enough motion to tilt

mercury switches, open/close contact points, and actuate electronic sensors. The most common application is the New Year's Eve party favor. Review the pressure sensors shown in figure 10-14.

Flow Sensors

One theory proposes that trees on the perimeter of a forest grow thicker, as they are exposed to more wind. They biologically sense to-and-fro movement and "actuate" thicker trunk growth. The vane flow sensor illustrated at the top left of figure 10-15 is literally blowing in the wind. It responds to flow phenomena like a sail. Through lever action, it can provide movement that is sensed in a variety of ways described in figure 10-16. These vane sensors are often used to monitor air flow, and may be referred

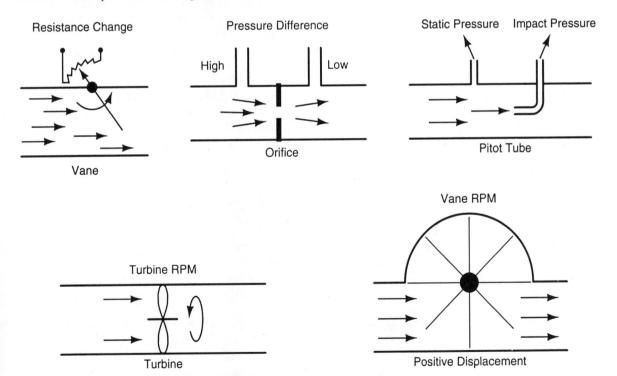

FIGURE 10-15 Flow sensor mechanisms.

to as a sail switch. It is common for them to become "stuck" due to dust and resinous air impurities.

One of the common flow sensors depends, for its functioning, on the principle that flow through a restriction provides a pressure drop

Name or Action	Sensor Output	Brief Description
Vane	Pendulum Motion	A free hanging or spring dampened sail-like vane is rotated less than 90 degrees. Its rotation can operate a micro-switch, photo switch, and/or mercury tilt switch. Flow measurement is achieved with variable resistors or position sensors.
Orifice Pressure Differential	Pressure Sensor Type Dependent	Flow provides pressure differences on opposite side of a restriction (orifice). Pressure sensors placed before and after the orifice can infer flow in the same way that resistance drops voltage.
Venturi Pressure Differential	Pressure Sensor Type Dependent	A narrowed restriction called a venturi causes the fluid flowing through it to speed-up, which reduces its pressure. Pressure sensors in the throat of the venturi infer flow. Carburetors employ this principal to vaporize fuel.
Pitot Tube	Impact Pressure	An L shaped tube is inserted in the fluid stream. The fluid impacts the open end provideing pressure for a sensor. Impact pressure is compared to static fluid pressure to imply flow.
Turbine	Spin Rate	A pinwheel-like turbine is placed in the stream of flow and its RPM provides counting information that is electronically or optically sensed.
Positive Displacement	Rotary Sensor Dependent	Vane, rotating piston and any water wheel-like device may be used to count units of flow. These devices are the fluid equivalent of ammeters.
Thermistor	Resistance Change	A self heated thermistor is placed in a gas stream. The cooling causes a change in resistance, which infers flow. Since gasses vary in cooling effects, with additional circuits gas content may be sensed. CO_2 is common.

FIGURE 10-16 Flow sensors.

for its functioning. Recall that current flow through a resistor produces a voltage (pressure) drop. In this same way, the orifice pressure differential sensor (see figure 10-15) restricts flow and results in a pressure difference. Two pressure sensors, one before the restriction and one after, provide a measure of the difference, which is proportional to flow. Figure 10-16 gives a brief description of this action.

If the restriction in the orifice flow sensor is smoothed so that the pipe diameter is "necked down" another scientific principle is responsible for pressure differential. As fluid flows through a smoothed restriction called a venturi, it speeds up. This increase in speed yields a pressure drop. A drilled port in the venturi area can be used with a pressure sensor to indicate this pressure drop. In the automobile carburetor, this action draws gasoline into the airflow stream, resulting in a combustible mixture. A venturi flow sensor compares pressure drop in the venturi with static (nonventuri) pressure, and, in this way, provides an indication of flow. Figure 10-16 shows this action.

In addition to HVAC applications, a flow-sensing device called the pitot tube is used to indicate the speed (flow) of both aircraft and ships. You may have seen this L-shaped tube near the front of a plane. As figure 10-15 shows, it depends on fluid impact pressure as a mechanism to sense flow. To provide an accurate measure of flow, impact pressure is compared with static (nonimpact) pressure. Again, refer to figure 10-16 for a description.

The pinwheel toy screws its way through air that flows through it, almost as if the air were an invisible threaded nut. For this reason, the term screw propeller is commonly used. The turbine flow sensor in figure 10-15 depends on the screw propeller idea to measure flow. Mechanical, magnetic, electronic, or optical sensors may be used to provide RPM indications, which also imply flow. This effect may be observed when a turned-off window fan spins as a result of wind passing through the fan blades.

Water and gas meters are required to provide accurate counting information related to units (gallon, liters, etc.) that have passed through the meter. They are part of a general group of flowmeters referred to as positive displacement meters. Though these flow sensors are less related to HVAC, a simplified illustration of their function is presented in figure 10-15. Suspected inaccuracies should be reported to the utility company, which ordinarily installs the meter. It is interesting to note that many of the utilities furnish these meters with remote telephone data links that eliminate the necessity for on-site reading of the meter.

Thermistor flow sensors are, like the thermistor liquid level sensors, self-heating. When used to sense flow, the flow stream cools the thermistor. This results in lower temperature and higher resistance (negative temperature coefficient). Flow may then be indicated, with additional circuitry, by resistance. Since gases differ in their ability to absorb heat, pairs of thermistors may be used to indicate gas content. In that case, one thermistor is placed in air and the other in the gas.

ACTUATORS

As we grow, we learn to focus our energy and force. Action is taken on things that provide us with the greatest benefits. In a general way, we become smarter and learn how to fight life's important battles, while ignoring the fruitless ones. In a parallel way, the history of technology is marked by inventions that provide better and more economical ways to convert electric and fluid (hydraulic and pneumatic) power into action. Actuators are devices that provide action, usually supervised by a control system. The control system, like the mind, decides what is to be done and the actuators, like muscles, take action.

Electrical Actuators

The first actuator in figure 10-17 is the solenoid. In its simplest form, it is basically a coil of wire, ordinarily wound on a bobbin, that surrounds a magnetic plunger. The plunger is usually centered within the coil, and is free to move in and out. Often a spring is employed to hold the plunger partially out of the coil. When the coil is energized its resulting magnetic field pulls the plunger into the coil. In this way, electric energy is converted into linear motion. The amount of current a solenoid requires to "pull in" is usually more than it requires to "hold in." Furthermore, the solenoid's mechanical force is directly related to energizing current.

The solenoid's short linear motion, rings doorbells, engages automobile starters, locks doors, punches holes, clamps brake bands, and actuates all kinds of mechanical devices. Perhaps the most common use of the solenoid in the HVAC field is for valve actuation. Valves that employ a flat or tapered seat covered by a washer, ball, or tapered plunger are commonly solenoid actuated. Common solenoid valves fail due to coil opens or short circuits (ohmmeter and current draw test), sticking plungers (inspect and shake for plunger freedom), and seat-washer leakage (flow test). Smaller valves are replaced as a unit, while larger ones may be repaired.

Name	Input Power	Resulting Action	Common Applications
Solenoid	DC or AC	In/Out, Back/Forth Linear Motion	Open/close valves, engage/ disengage motors and brakes
Brush Motor	Primarily DC	Speed and Torque Controllable Rotating Power	High power small motor in applications that require speed control
Induction Motor	Primarily AC	Very Reliable Rotating Power	Work horse of hvac for pumps, compressors, fans, blowers, etc.
Stepper Motor	Pulsed DC	Angular Rotational "Step" Per Pulse	Position control when feedback not required
Servo Motor	Primarily DC	Highly Controllable Rotary Position and Torque	Automated equipment in which human-like feed back is required, precise regulation of rotary motion for valves, robot arms, etc.

FIGURE 10-17 Electrical actuators.

By far the largest amount of electricity is converted to motion by motors. Chapter 7 presented the most common DC and AC motors. During that discussion motors were separated into two categories, brush motors and induction motors. Both of these motor types are commonly used as actuators. As figure 10-17 shows the brush motor is smaller and easier to control, while induction motors offer higher reliability, but greater weight and size. Two new motors are presented in figure 10-17 that require explanation.

Stepping Motors

The name stepper motor implies this motor's action. In simple terms the stepper motor rotates a fixed amount (angle) for each electric-power pulse. Institutional clocks, which commonly use a central time controller, are mechanical stepper motors. They usually step 1 minute (1/60 of circle or 6 degrees) for each electrical pulse. Figure 10-18 provides the idea behind a simple stepper motor.

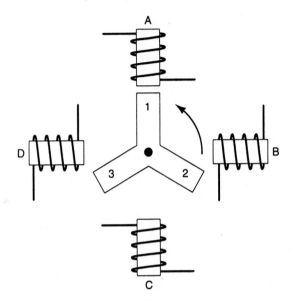

CCW Rotation 30 Degrees per Step
Stator Coil Energized A B C D A B C D A B C D
Rotor Pole Attracted 1 2 3 1 2 3 1 2 3 1 2 3

FIGURE 10-18 Basic stepper motor.

Notice that the motor of figure 10-18 contains a four-coil stator and a three-pole rotor. This unequal stator–rotor design is common to stepper motors. The stator coils are energized one at a time with this type of stepper. Whenever a stator coil is energized, it attracts the nearest rotor pole. In the position shown, the stepper in figure 10-18 has power applied to coil A, which results in a holding force exerted on rotor pole 1. If power is removed from coil A and applied to coil B the closest rotor pole (2) is attracted. In this way, the rotor turns 30 degrees counterclockwise. Follow the coil sequence at the bottom of the illustration to understand how this stepper can rotate through 360 degrees. How might the rotor be caused to turn clockwise?

Steppers are generally microprocessor or computer controlled. Although they provide accurate rotational motion, they have one

major drawback. If the stepper's position is mechanically changed, for example, the rotor is forced two poles off position, the controller has no way of knowing this. As a result, the motor will be controlled, but all motor rotations will be in reference to the incorrect starting position. You probably have seen a stepper finding its correct starting position. Computer printers often use steppers to move the print head left and right. When the printer is first turned on, the printer's microprocessor instructs the print head to move full left. The full left position is sensed by an electrical or optical switch that signals the printer's circuits that the print head is in the correct starting position. The stepper is useful for relative movement, but it requires additional sensors to indicate its "home" position. Though the stepper is not yet commonly used in HVAC equipment, its application in robotics and computer-controlled machines is extensive. As HVAC equipment becomes increasingly based on microprocessor controllers, expect to see many stepper-motor–related applications.

Servomotors

The last actuator shown in figure 10-17 is the servomotor. The servomotor is a common brush type, often with self-contained reduction

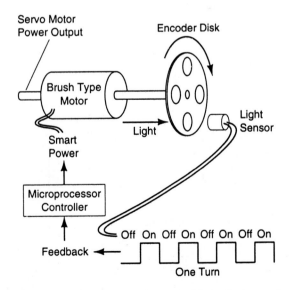

FIGURE 10-19 The servomotor and its optical encoder.

gears (gear head motor). Its speed and torque are dependent on power, while its direction is DC polarity dependent. Microprocessor controllers can, with solid-state transistors and triacs, easily control these power variables—all that is required is feedback. The controller needs to know what the motor is doing in response to power. Optical encoders provide the solution.

Optical encoders "window" light. Fastened to the rear of the servomotor is an opaque disk with holes or slots. Figure 10-19 illustrates how a light source, often an LED, is chopped by the rotating encoder disk. The flashing light is

converted to an electrical signal by a light-sensitive device, a phototransistor, for example. The resulting on/off pulses of electricity, shown at the lower right of figure 10-19, provide feedback to the microprocessor controller. With this information, the controller can count motor rotations.

Optical encoders are available with many light sources and light sensors. They are arranged from the center of the encoder disk outward. More complex encoder disks with many windowed tracks, like record bands, are used. Devices with eight tracks are common,

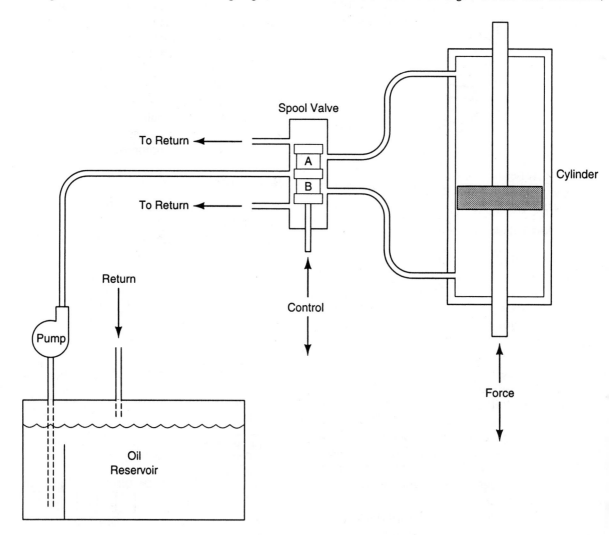

FIGURE 10-20 Hydraulic system.

and they provide information, that indicates the direction of the motor, as well as its rotation, to an accuracy of 1/256 of a circle (1.4 degrees). In this way, accurate feedback provides positive motor control.

Servomotors are, at present, the most humanlike of actuators. They are widely used in robots and robotlike manufacturing systems. A similar motor, which uses a magnetic encoder called a resolver, is used in VCRs and CD players to "track" recorded signals. Guidance and navigational equipment also make extensive use of servomotors. Again, like the stepper, the servomotor is a newcomer to the HVAC field. With increased emphasis on accurate, environmentally clean systems, expect to see an increased in the use of these motors.

Hydraulic Systems

Fluid force is used to great advantage with hydraulic systems. Hydraulic fluid (oil with special additives) is pumped from a reservoir to a high pressure, and then it passes through spool control valves to actuators. Within a hydraulic cylinder, fluid pressure is exerted on a movable piston. Depending on fluid flow direction, the piston moves in or out of the cylinder. Fluid on the low-pressure side of the cylinder returns to the reservoir. In the reservoir, particulate matter settles out of the fluid, while clean fluid overflows a baffle to the pickup tube. Figure 10-20 shows the basic components of a simple hydraulic system. Filters, pressure regulators, and flow rate valves have been omitted for clarity.

The spool valve is composed of a bored hole into which a spool-like cylinder is closely fitted. The spool is machined with various recesses to permit fluid flow. Fluid is directed by passages drilled through opposite sides of the bored hole. The position of the spool valve in figure 10-20 does not permit fluid to flow from one drilled passage to another. A solenoid or lever handle may be used to move the spool upwards. In that case, fluid can flow from the pump pressure feed line through recess (B) to the front of the hydraulic cylinder. This pressure moves the piston

upwards. As the piston moves up, it pushes fluid out of its top port through recess (A) of the spool valve and back to the reservoir. If the spool valve is moved below its center position, the process reverses and the piston moves down.

Hydraulic systems can produce great force. If the pump supplies fluid at 400 PSI and the piston area is 10 square inches, the resulting force developed by the piston is 4000 pounds (400 times 10). The large forces provided by hydraulic systems make them idea for earth-moving equipment, lifting devices, presses, and heavy-duty robots. They are also employed to great advantage in steering and braking systems. Though HVAC systems make limited use of this type of hydraulic system, the principles upon which they function are widely used in fluid systems involving pumps and valves in both burners and air conditioners.

Pneumatic Systems

Pneumatic systems use air in place of hydraulic oil. Air is compressible and springlike, while oil, a liquid, is not. As a result, pneumatic systems and the action they produce have "give." For example, a feeble obstruction of the hydraulic cylinder's motion will be crushed. Pneumatic cylinders generally work with less pressure and provide compliant (elastic) forces. As a result, pneumatic cylinders, bellows, and diaphragms may be used to operate duct doors. When the door reaches the end of its travel, the pneumatic actuator exerts its force with no further effect. End-of-travel sensors and adjustments are not often required. An additional benefit of air-operated systems is the elimination of fluid return lines, as the exhaust air is simply vented to the atmosphere. Where vacuum sources are available, diaphragms and bellows are commonly used for actuation. In hydraulic and pneumatic systems, similar components serve similar functions.

Hydraulic and Pneumatic Actuators

Hydraulic systems are like electrical circuits. The pump-reservoir is the source of pressure (PSI), while the battery is the source of electrical pressure (volts). The flow of amps or gallons per hour is controlled by switches or valves. And in a general way, hydraulic and pneumatic actuators are like electrical loads. Each converts its power source to useful work. Both systems are circuit-like.

Figure 10-21 presents a general summary of these actuators. Hydraulic and pneumatic motors are actually pumps operated in reverse.

Rather than producing pressure, they spin as a result of applied fluid pressure. Just as DC generators may be made to operate like brush motors, fluid pumps may operate like fluid motors. Air motors are commonly used in impact wrenches and air sanders. Dental tools are often operated by air or water pressure. Review figure 10-21 to be certain you have a good understanding of these actuators.

Name	Input Power	Resulting Action	Common Applications
Diaphragms	Fluid Pressure	Linear Buckling-Bulging Motion	Duct doors, clutches, brakes
Bellows	Fluid Flow and Pressure	Linear Extend-Contract Motion	As above with more motion available
Cylinders	Fluid Flow and Pressure	Long Distance Linear Extend Contract Motion	Construction, manufacturing equipment, large HVAC duct doors, air conditioner valves
Spool Valves	Mechanical, Electrical and Fluid Actuators	Linear Motion Valve Widely Used in Hydraulics and Pneumatics	Fluid flow to cylinders, automatic transmissions, complex fluid circuit flow control
Motors	Fluid flow and pressure	Full or Partial Rotation	Air and oil powered motors in manufacturing and construction, large HVAC system rotary power

FIGURE 10-21 Hydraulic and pneumatic actuators.

PROGRAMS

Most people ignore instructions from time to time. When an item doesn't work, we go to the troubleshooting section of the manual, which often, says "When all else fails, read the instructions." Microprocessor controllers always read the instructions. In this case, the instructions, step 1, step 2, step 3, etc., are called the pro-gram. The program says to do this, then do that, move this number; if one number (outside temperature) is larger than another number (inside temperature), then skip ahead in the program to air-conditioning instructions, and so on. Programs can perform mathematical (+, −, ×, etc.) and logical (AND, OR, NOT, etc.) opera-

tions, as well as branch operations (redirect program flow). All of these program instructions are written in numbers that represent specific instructions called operation codes (op codes). Op codes are instructions that are understood by microprocessors. Each microprocessor has its own language of op codes called the instruction set. For example, a 486 microprocessor has an instruction set that is entirely different from that of a 6800. As a result, writing programs in machine language (the instruction set of a particular microprocessor) is difficult, abstract, and highly specialized. You could do it, but you would have to change your occupation.

Many microprocessor controllers have program terminals that reduce the difficulties in writing programs. With these computer terminals, a far more understandable (English-like) language is used. Once the program is completed, a special computer program called a compiler is run to convert the simpler language to the op codes of machine language. If you ever work with large commercial HVAC systems, you may attend an industry workshop to learn a compiler language.

The important idea is that the program contains the instructions that the controller follows. Ordinarily the program is like our personalities—it stays basically the same throughout its life. Just as we remember our personalities and act like ourselves, the microprocessor remembers its program. Other things, like telephone numbers and temperature levels, change frequently. Humans and microprocessor controllers have to remember things that change frequently.

ROM and RAM

ROM chips contain instructions that are permanent, even when power is turned off. Since these chips contain permanent instructions, they are referred to as read-only memory. Even though programmers and engineers put a great deal of effort in programs stored in ROM chips, errors are made which may require new program chips, referred to as ROM upgrades. They often contain corrected and improved programs.

RAM chips may be thought of as the controller's working memory. The information these chips hold frequently changes while the MP controller is running. Random-access memory provides information storage that can be both changed and accessed in any order. This read-and-write memory comprises the megs of memory to which computer technologists refer. Read-only memory, which often contains the program, is permanent, whereas random-access memory can be changed.

Smart HVAC Systems

In the future (and to a limited degree, now), HVAC systems will depend more and more on microprocessor controllers. Microchips become more powerful and less expensive every year. With artificial intelligence programming techniques, control that is more humanlike is becoming a reality. When you are well into your profession, perhaps just a few years from now, it is hard to imagine what might be available, but let's try!

With a controller as simple as the one in figure 10-22, a well- and cleverly written program could provide amazing climate control. Suppose that the outside temperature were to drop sharply. Initially, the controller might energize the furnace at a moderate level, which would provide some heat, but, unfortunately, not enough. The controller can be made to remember this mistake and correct it the next time. HVAC control becomes better and better as the controller learns what is required in this building, given particular climate conditions. Further, the controller may be programmed to learn the preferred environment

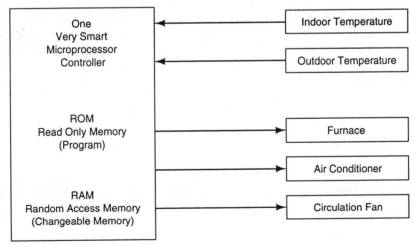

FIGURE 10-22 Very smart microprocessor controller.

of the building inhabitants. This would be a very smart controller.

With experience and an understanding of the fundamental principles that underlie HVAC systems, you will gradually become an HVAC expert.

SUMMARY

This chapter began with the idea that machines can be made to copy Mother Nature's example and become more humanlike. Humans and self-correcting machines depend on feedback. Feedback provides information about the present status of a process. Closed-loop control systems make use of feedback to bring a process closer to what is desired.

Most closed-loop control systems compare desire with the feedback provided by sensors. If the two are not in agreement, actuators are signaled to "take action." When desire and feedback agree, the comparator finds no (null) difference and takes no action. With these systems, some error is necessary. Too much error—for example, high/low temperature—results in a control system that is overdamped and responds to error slowly. Underdamped systems respond too rapidly and consume excessive fuel.

Input devices (sensors) and output devices (actuators) are connected to the brain (microprocessor) through interfacing. In the event of a sensor or actuator failure the microprocessor is often protected by opto-isolators. These micro-chips transfer signals by light, thereby eliminating electrical brain damage. In some instances, many input or output devices may share a microprocessor port by multiplexing.

Sensors provide feedback to the controller. Just about any phenomenon may be sensed. Figures 10-8, 10-9, 10-10, 10-12, 10-14, and 10-16 provide a convenient summary of common HVAC sensors. Look back and study them.

The "get stuff done" part of the control loop depends on actuators. Electrical, hydraulic, and pneumatic actuators are signaled by the microprocessor to take action. Figures 10-17 and 10-21 summarize the common HVAC actuators. Of special interest are the stepper motor, which steps a fixed amount for each electrical pulse, and the servomotor. Optical encoders provide position and speed information so that a brush-type motor may be microprocessor controlled.

Hydraulic and pneumatic systems depend on fluid circuits, which direct pressurized fluids and/or vacuum to cylinders and motors. The resulting force provided with hydraulic systems is high and definite. Since air is compressible, pneumatic systems usually afford less force and more compliant or elastic motion.

With programs based on artificial intelligence based programs microprocessors can learn in the same way that you do. If both you and the microprocessor make use of and learn from experience, one smart microprocessor controller will be maintained by one smart HVAC technician.

PROBLEM-SOLVING ACTIVITIES

Review Questions

1. Draw a simple but complete diagram representing closed-loop control. Label the parts.

2. What does the term *null* mean in reference to controllers?

3. Identify a closed-loop system that is not related to HVAC and describe its operation using correct terminology.

4. Describe the connection between HVAC system sizing and the idea of dampening.

5. Identify 10 sensors around the home or automobile that are not HVAC related. (Hint: Don't forget electronic equipment.)

6. Draw a solenoid valve. Use any valve mechanism you desire and label the parts. Describe "how it works."

7. How does the differeing nature of a liquid and gas affect the functioning of fluid power systems?

8. Draw and describe a spool valve.

9. What is the roll of the program in the microprocessor controller?

10. What function does memory serve in making humans and microprocessor controllers smarter?

SECTION TWO

Heating and Cooling Control
and
Application Technology

CHAPTER **11**

Oil-Fired Combustion Technology

The use of oil lies at the center of the United States economy. Since the first commercial oil well was drilled in Pennsylvania over 130 years ago, oil usage has grown to the point that it now influences virtually every aspect of modern technology. Formed by the decomposition of plant and animal life sealed in sedimentary layers of the earth millions of years ago, crude oils are available in grades numbered 1 through 6. Number 1 and 2 fuel oils are light oils used in residential and light commercial applications. Number 1 oil is commonly known as kerosene; no. 2 is the standard fuel oil grade and is also used as diesel fuel. The heavier grades, 4–6, are very viscous (thick), and must be heated before they can be either pumped or burned. The heating content of residential and light commercial no. 2 fuel oil is between 138,000 and 144,000 Btu/gallon. As are other common fuels—for example, gasoline and fuel gases such as propane and butane— crude oil products are also used in a wide variety of manufacturing processes.

COMBUSTION CHARACTERISTICS

All fuel oil grades contain approximately the same percentages of carbon and hydrogen: 86% carbon and 14% hydrogen. During the combustion process, assuming complete mixing with air, about 140,000 Btu of heat are released. Based on the composition of no. 2 residential fuel oil, 75,00 Btu come from the combustion of

carbon and 65,000 Btu result from the combustion of hydrogen.

To burn 1 gallon of fuel oil, 100 pounds of air must be supplied to the oil burner. Given the fact that the composition of air is 21% oxygen and 79% nitrogen, all 21 pounds of the oxygen are consumed during the combustion process. The nitrogen, which is inert due to its electron bonding, is merely heated from room temperature to several hundred degrees during the combustion process and exits the chimney chemically unchanged.

During combustion, the hydrogen in the oil unites almost totally with oxygen and exits the chimney as water vapor. If the combustion is incomplete, hydrogen forms toxic and corrosive compounds as by-products, which can cause a variety of respiratory problems if inhaled directly. Any excess oxygen from the combustion process unites with carbon atoms in the fuel oil. Complete combustion of the carbon results in carbon dioxide (CO_2). Incomplete combustion can yield carbon monoxide gas (CO), smoke, and particulate carbon (soot).

In order to maximize the efficiency of the oil combustion process taking into account minor variations in chimney draft and oil viscosity, standard industry practice is to deliver excess air to the oil burner combustion head. Too little excess air results in excessive smoke and inefficient combustion. Conversely, too much excess air lowers the temperaure of the combustion process, decreasing efficiency in the process. Under ideal conditions, about 100% excess air (twice the volume of air needed) should be made available for the combustion of the fuel oil. Figures 11-1 and 11-2 illustrate the combustion of 1 gallon of fuel oil with no excess air and 100% excess air, respectively.

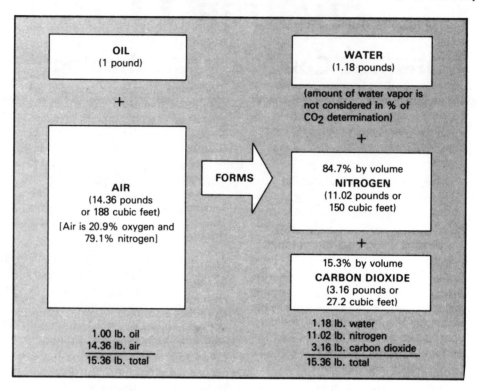

FIGURE 11-1 Products of combustion of 1 pound of fuel oil burned with no excess air. (Courtesy of R. W. Beckett, Inc.)

OIL BURNER OPERATION

All of the theoretical considerations of combustion, gas analysis, and efficiency testing come together in the design and construction of the modern oil burner. These oil burners operate far more efficiently and reliably than did their earlier counterparts, although component design and configuration remains essentially the same. A typical oil burner is illustrated in figure 11-3A. Figure 11-3B shows a cutaway offering a detailed view of the component arrangement. These units can be used to fire either hot air furnaces, water heaters, or hot water or steam boilers.

Note from figure 11-3 that the basic component arrangement consists of: the drive motor, blower and fuel pump coupled assembly, primary safety control circuitry, ignition transformer, nozzle and electrode assembly, and flame retention head. These components are encased within the burner housing. Before undertaking an analysis of separate components, we will examine how the oil burner ignition process functions. These steps, which occur in a very fast sequential order, are as follows:

1. The room thermostat calls for heat, which activates the primary relay coil, closing the electrical contacts.

2. Line voltage flows to both the burner motor and the primary coils of the ignition transformer. The burner motor

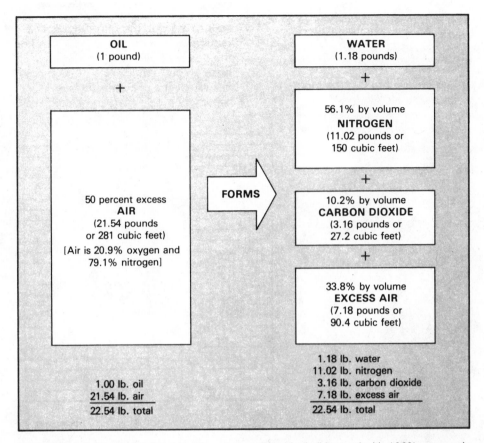

FIGURE 11-2 Products of combustion of 1 pound of fuel oil burned with 100% excess air. (Courtesy of R. W. Beckett, Inc.)

delivers air to the combustion head; fuel oil is delivered to the nozzle assembly from the fuel pump, which is mechanically coupled to the drive motor.

3. Line voltage is increased from 120 volts to 10,000 volts in the secondary coils of the transformer.

4. Current flows through the transformer contacts to the electrode assembly. The position and gap of the electrodes cause a spark to jump the electrode gap.

5. The air velocity from the blower causes the spark to move forward of the electrodes above the nozzle assembly. In this position, the spark is ready to ignite the minute oil droplets atomized by the nozzle assembly.

FIGURE 11-3A High-speed oil burner. (Courtesy of R. W. Beckett, Inc.)

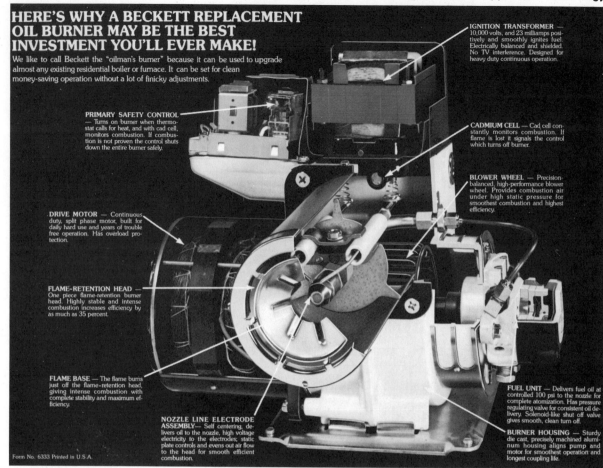

HERE'S WHY A BECKETT REPLACEMENT OIL BURNER MAY BE THE BEST INVESTMENT YOU'LL EVER MAKE!

We like to call Beckett the "oilman's burner" because it can be used to upgrade almost any existing residential boiler or furnace. It can be set for clean money-saving operation without a lot of finicky adjustments.

IGNITION TRANSFORMER — 10,000 volts, and 23 milliamps positively and smoothly ignites fuel. Electrically balanced and shielded. No TV interference. Designed for heavy duty continuous operation.

PRIMARY SAFETY CONTROL — Turns on burner when thermostat calls for heat, and with cad cell, monitors combustion. If combustion is not proven the control shuts down the entire burner safely.

CADMIUM CELL — Cad cell constantly monitors combustion. If flame is lost it signals the control which turns off burner.

BLOWER WHEEL — Precision-balanced, high-performance blower wheel. Provides combustion air under high static pressure for smoothest combustion and highest efficiency.

DRIVE MOTOR — Continuous duty, split phase motor, built for daily hard use and with years of trouble free operation. Has overload protection.

FLAME-RETENTION HEAD — One piece flame-retention burner head. Highly stable and intense combustion increases efficiency by as much as 35 percent.

FLAME BASE — The flame burns just off the flame-retention head, giving intense combustion with complete stability and maximum efficiency.

NOZZLE LINE ELECTRODE ASSEMBLY — Self centering, delivers oil to the nozzle, high voltage electricity to the electrodes, static plate controls and evens out air flow to the head for smooth efficient combustion.

FUEL UNIT — Delivers fuel oil at controlled 100 psi to the nozzle for complete atomization. Has pressure regulating valve for consistent oil delivery. Solenoid-like shut off valve gives smooth, clean turn off.

BURNER HOUSING — Sturdy die cast, precisely machined aluminum housing aligns pump and motor for smoothest operation and longest coupling life.

Form No. 6333 Printed in U.S.A.

FIGURE 11-3B Component arrangement of modern high-speed oil burner. (Courtesy of R.W. Beckett, Inc.)

6. Flame propagation takes place, spreading the flame to all of the oil droplets within the combustion chamber.

Oil Burner Components

With an understanding of this process, we proceed with an examination of the individual components of the burner mechanism.

Drive motor: Oil burner motors are rated for continuous duty to provide trouble-free service for many years. Modern oil burners use motors that spin at 3450 RPM, giving burners the designation "high-speed flame retention." This designation differentiates these units from their earlier counterparts, which used motors operating at only 1750 RPM. These high-speed motors also feature internal thermal overload protection and extended shafts to allow coupling of the motor to both the blower wheel and fuel pump. Thermal protection consists of a thermally operated switch that is wired in series with the motor windings. In the event of a motor overheat condition, the switch will open, cutting power to the motor windings. Most switches are rated to open at either 284°F or 329°F, depending on the motor design. There are two types of switches in use in oil burner motors.

1. A manual reset thermally protected switch that involves pushing a button to reset the switch.

2. An automatic reset thermal switch that resets after the motor has cooled down.

Problems commonly associated with motor failure are usually the result of overheating, which is caused either by some type of mechanical or electrical failure within the motor, or by an external condition that places undue stress on the motor. The most common of these conditions are as follows:

1. A fuel pump failure that causes the pump to seize, which overloads the motor.

2. An obstructed blower wheel causing the motor to partially or totally bind.

3. A defective internal starting switch that fails to disconnect the starting windings of the motor from the power after the motor has reached full speed, causing an overheat condition.

4. Improper pump alignment, usually caused by loose mounting bolts. This places undue stress on the blower to pump to coupling assembly, binding the motor.

5. An improperly sized motor. In most instances the load imposed exceeds the nameplate rating of the motor. Using an ammeter, make sure that the input current does not exceed the nameplate current rating by more than 10%.

The blower motor shaft is connected to both the blower wheel and fuel pump by use of a coupling assembly, as shown in figure 11-4.

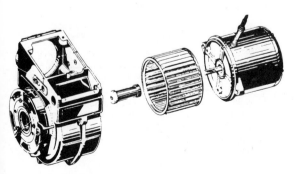

FIGURE 11-4 Burner housing, blower wheel, motor, and coupling arrangement of typical oil burner.

Off-cycle air losses can occur when the oil burner is not operating. Room air enters the oil burner air intake slots and travels throughout

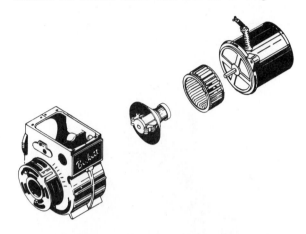

FIGURE 11-5A

FIGURE 11-5B

the appliance and out of the chimney. In this process, residual heat from the boiler or furnace, as well as heated building air, is removed. To prevent these heat losses, the standard drive coupling between the motor and fuel pump can be replaced with an automatic air inlet shut-off damper (figure 11-5). This type of shut-off is centrifugally operated. When the motor is operating, the shut-off valve opens, admitting air to the blower wheel. The valve automatically closes the air inlet passage when the motor stops spinning, reducing air travel through the burner and heat exchange passages of the appliance to a minimum.

FIGURE 11-5C Air inlet shutoff mechanism. This device replaces the standard coupling between the burner motor and fuel pump, and eliminates air flow through the burner motor during the off cycle. (Courtesy of R.W. Beckett, Inc.)

Ignition transformer: Ignition transformers are rated to produce 10,000 VAC at low current to provide the high-intensity spark for the combustion proces. These units are specially shielded to eliminate radio and television interference. Note from figure 11-6 that the transformer terminals are wound spring steel. and rest on the elelctrode extensions. This configuration both provides an effective means for transmitting the high voltage to the electrodes, and enables different transformers to be used in a variety of burner applications, since the distance between the transformer and electrodes will vary from one manufacturer to another.

Air tube assembly: The air tube assembly houses the nozzle, electrode assembly, and combustion head. During burner operation, a

FIGURE 11-6 Oil burner transformer. Most of these transformers feature universal mounting plates to fit a variety of oil burners. (Courtesy of France, a Scott Fetzer company)

stream of high-velocity air is delivered through the burner tube from the blower wheel. Air tubes are supplied in a variety of lengths, depending on the specific appliance in which the burner will be installed. Note from figure 11-

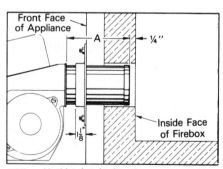

"A" = Usable air tube length.
The burner head should be ¼" back from the inside wall of the combustion chamber. Under no circumstances should the burner head extend into the combustion chamber. If chamber opening is in excess of 4 3/8", additional set back may be required.

FIGURE 11-7 Oil burner air tubes are selected based on required lengths for proper installation of the oil burner in the appliance.

7 that the length of the air tube is a function of boiler design and combustion chamber configuration, and is designed to bring the combustion head to within 1/4 inch of the inside wall of the combustion chamber. In no instance should the combustion head ever extend into the combustion chamber. In applications in which a new oil burner is to be installed in an existing furnace or boiler, the proper air tube must be selected based on the burner and/or appliance manufacturer's recommendation.

Oil burner nozzles: Figure 11-8 illustrates a typical oil burner nozzle.

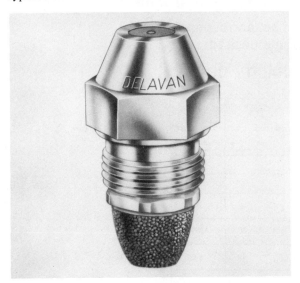

FIGURE 11-8 Oil burner nozzle. The nozzle incorporates a filtration screen in addition to the oil-line filter to eliminate nozzle clogging. (Courtesy of Delavan, Inc.)

For the nozzle to work properly, energy is required to atomize the fuel oil into small droplets. Atomization increases the surface area of the fuel oil. This enables air to combine with virtually all of the fuel oil, which results in high operating and combustion efficiencies. Fuel oil is delivered to the nozzle from the fuel pump at pressures between 100 and 150 PSI in residential applications, and at approximately 300 PSI in commercial installations. As the oil passes through the swirl chamber and swirl

slots, a cone-shaped film of oil is formed as it exits the nozzle assembly. This film, moving at a very high velocity, separates into strands that stretch past their breaking point, forming tiny droplets. A cross section of a typical nozzle assembly within the burner head is illustrated in figure 11-9.

Nozzle sizing and spray patterns: The capacity of the nozzle, listed in gallons per hour (gal/hr) of oil delivered by the nozzle to the appliance, must be matched to the heating specifications of the specific appliance. For example, a boiler rated at a net input of 144,000 Btu/hr will specify a nozzle rating of 1.0 (1 gal/hr). Individual oil burners can usually be fired over a wide range of nozzle ratings. For most residential applications, including oil-fired water heaters, these ratings fall within the .5- to 3.0-gal/hr rating. Commercial applications require higher firing rates with correspondingly larger oil burners.

While the burner mechanism itself may be capable of handling a variety of firing rates, most appliances are restricted to relatively narrow ranges. In a situation in which a boiler or furnace may not be able to deliver sufficient heat to the building, the best solution is to upsize the entire appliance. The technician should note that it is poor practice to install a larger nozzle in the oil burner in the existing appliance. In fact, installing a larger nozzle could be extremely dangerous, and could result in a meltdown or destruction of the combustion chamber of the heating appliance. In these situations, the appliance and not the firing rate of the burner needs to be changed.

Depending upon the design of the nozzle, the cone-shaped pattern assumed by the fuel oil is classified in one of three ways: solid, hollow, or combination, as illustrated in figure 11-10.

There are many design factors that determine nozzle spray pattern specifications for each boiler or furnace. These factors include air circulation patterns within the combustion chamber, the size and shape of the combustion chamber, and the heat transfer characteristics

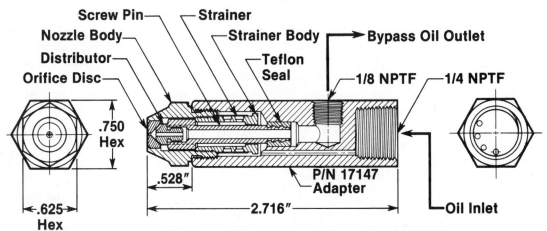

Variflo Nozzle Assembly   P/N 33769

—(see chart for dash no. corresponding
to flow & spray angle desired)

VARIFLO NOZZLE CAPACITY CHART

Dash No.	Spray Angle	Bypass Closed Calibrated Nozzle Flow–GPH	100 PSI Supply		300 PSI Supply
			Bypass Open		Bypass Closed
			Total Flow –GPH (Ref.)	Nozzle Flow –GPH (Ref.)	Nozzle Flow –GPH (Ref.)
– 1 – 2	45° 60°	.75	1.02	.20	1.27
– 3 – 4	45° 60°	1.00	1.30	.22	1.60
– 5 – 6 – 7	45° 60° 80°	1.50	1.90	.30	2.30
– 8 – 9 –10	45° 60° 80°	2.00	2.60	.38	3.30
–11 –12 –13	45° 60° 80°	2.50	3.40	.49	4.00
–14 –15 –16 –17	30° 45° 60° 80°	3.00	4.00	.57	5.00
–18 –19 –20 –21	30° 45° 60° 80°	3.50	4.60	.67	6.00
–22 –23 –24 –25	30° 45° 60° 80°	4.00	5.20	.78	6.80
–26 –27 –28 –29	30° 45° 60° 80°	4.50	6.00	.85	7.50
–30 –31 –32 –33	30° 45° 60° 80°	5.00	6.6	.97	8.60
–34 –35 –36 –37	30° 45° 60° 80°	5.50	7.40	1.02	9.10
–38 –39 –40 –41	30° 45° 60° 80°	6.00	8.00	1.17	9.90

FIGURE 11-9 Nozzle assembly in oil burner head. (Courtesy of Delavan, Inc.)

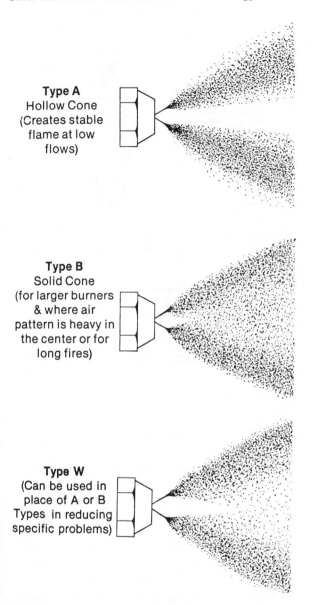

Type A
Hollow Cone
(Creates stable
flame at low
flows)

Type B
Solid Cone
(for larger burners
& where air
pattern is heavy in
the center or for
long fires)

Type W
(Can be used in
place of A or B
Types in reducing
specific problems)

FIGURE 11-10 Three common spray patterns of conventional oil burner nozzles. (Courtesy of Delavan, Inc.)

inherent in the appliance construction. Given these many factors, some generalizations can be made regarding the selection of spray patterns. Hollow spray nozzles are used most often in low firing applications ranging from .5 to 1.5 gal/hr. Hollow spray patterns tend to be very stable, a prerequisite for low firing rates. Solid cone nozzles usually work best in applica-

tions with firing rates over 1.0 gal/hr. Combination nozzles work effectively over a wide range of firing rates, yielding hollow patterns at lower firing rates and more solid spray configurations at the higher firing rates. The angle of the spray pattern, as well as the gallon-per-minute capacity of each nozzle, is stamped on the nozzle housing. Spray angles are specified by the manufacturer with an eye toward matching the shape of the flame to the configuration and air circulation patterns within the combustion chamber. In older appliances, air circulation patterns and combustion chamber designs play a crucial role in efficient oil combustion. In these applications, the only sure way to ascertain that a nozzle has been chosen properly is to conduct a full set of efficiency tests and analyze the results. In modern high-speed retention head burners, air circulation and chamber configuration play a less important role, resulting in more flexibility in the selection of burner nozzles.

Electrode assemblies: Electrode assemblies transfer high voltage from the ignition transformer and deliver a constant spark across the electrode gap. Figure 11-11A shows the arrangement of the assembly within the burner air tube. Figure 11-11B gives general gap and position measurements for a typical assembly. These measurements are general, and may vary among burner manufacturers.

Although electrode measurements may vary, the setting of both the position and gap of the electrodes is critical for proper burner firing. The position of the electrode in front of the nozzle takes into account the distance that the spark will be pulled forward of the nozzle into the airstream. This position allows for proper flame propagation and complete combustion. If the spark either leads the nozzle by too much distance or is set too far back from the nozzle tip, the burner will operate or fire poorly, or not at all. During normal service procedures, the gap and position of the electrodes, as well as the condition of the porcelain insulation, should be examined. Sooting at the tips of the electrodes is an indication of poor electrode

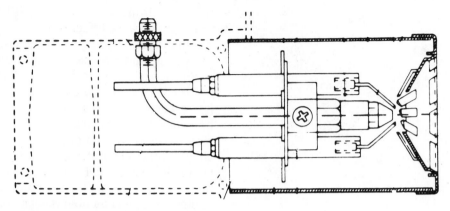

FIGURE 11-11A Arrangment of nozzle and electrode assembly in oil burner air tube.

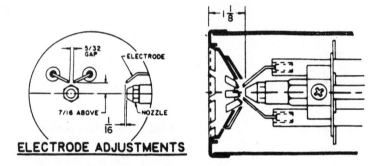

FIGURE 11-11B Position and gap of typical oil burner electrode assembly.

adjustment or nozzle fouling. There should be little wear at the tips of the electrodes if they are properly adjusted. If necessary, the tips can be filed and brought back to serviceable specifications. Electrodes normally will last for many years with few visible signs of tip degradation. Cracks in the porcelain insulation can cause grounding of the electrodes, with a resulting loss of both spark and ignition. Cracked insulation require a replacement of the entire electrode.

Combustion head: The design of the combustion head of the oil burner, along with the velocity of air delivered through the blower wheel, determines the combustion characteristics of the oil burner by creating a specific pattern of air at the end of the air tube. Combustion heads are sometimes also referred to as fire rings, retention rings, or end cones. Prior to the advent of high-speed drive motors, the flame pattern of conven-

tional, or nonretention, oil burners was of the type shown in figure 11-12.

Note both the shape of the flame and the design of the conventional combustion head in figure 11-12. The head is very simple in design, and creates the base of the flame about 2 inches off the end of the head. The flame in this design is "nonretained." Sometimes, a spinner, or turbulator, is installed inside the air tube to create a swirling pattern of air as it exits the air tube. The flame fills the entire combustion chamber and is directed against a target wall. The wall reflects heat into the flame and contributes to the air circulation pattern within the chamber. As the need for higher efficiencies arose, high-speed flame retention burners were developed. Contrast the head design and flame pattern of the nonretention burner with those of the typical flame retention burner in figure 11-13.

Conventional Combustion

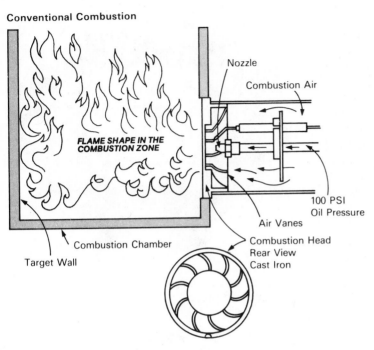

FIGURE 11-12 Flame pattern of conventional (low-speed) oil burner. (Courtesy of Delavan, Inc.)

Flame Retention Combustion

FIGURE 11-13 Flame pattern of high-speed oil burner. Note how flame is held, or retained, against combustion head of the burner. (Courtesy of Delavan, Inc.)

Flame retention heads have three basic design features: the center opening, the primary slots, and the secondary openings. The center opening directs air at the flame in a forward motion only, and tends to push the flame away from the face of the combustion head. For this reason, the center hole is usually small, forcing combustion air into the primary and secondary slots. The primary slots create a spinning motion to the air with only a slight forward motion. This results in a compact flame that is very hot and very efficient. The secondary slots both encase the flame in a surrounding envelope of air and meter the air required to achieve the proper firing rate. Air metering is accomplished either by changing the selecting a head designed for a specific firing rate, or by moving the head toward or away from a fixed ring to change the size of the slot opening.

Flame retention burners require no target wall for air circulation, and the design of the combustion chamber itself is much less important than in conventionally designed burners. In applications where a flame retention burner is to be installed in an older boiler or furnace that formerly housed a conventional oil burner, a combustion chamber liner should be installed along with the new burner as an additional safety feature to guard against the hotter flame of the flame retention burner.

Fuel pumps: Fuel pumps supply high-pressure oil to the nozzle at approximately 100 PSI. This high pressure and the resulting velocity necessary for atomization of the oil to take place as it leaves the nozzle. The fuel pump consists of a rotary gear mechanism, designed in either one or two stages, that squeezes the oil from between the gear teeth. Single-stage pumps are used in a majority of residential applications where the oil is lifted less than 8 feet from the storage tank to the burner with relatively short supply and return piping. In either residential or commercial installations where the oil must be lifted higher than 8 feet and/or the supply lines from the storage tank to the burner are long, a two-stage pump may be required. Single- and two-stage fuel pumps are almost identical in appearance and are distinguishable by their identification labels on the outside and the gearing arrangement (figures 11-14A and 11-14B).

Proper operation of fuel pumps can be checked most accurately by installing a pressure gauge in the pressure port of the pump.

FIGURE 11-14A Single-stage fuel oil pump. (Courtesy of Suntec Industries)

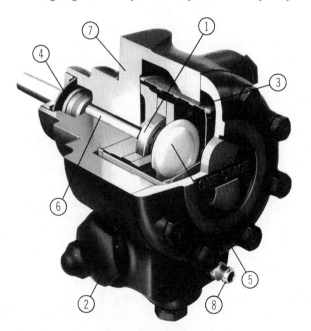

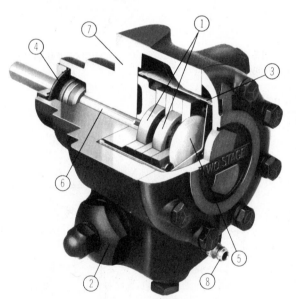

1. Gearing Assembly
2. Cut-Off Valve
3. Fuel Strainer
4. Shaft Seal
5. Anti-Hum Damper
6. Bearing
7. Pump Body
8. Bleed Valve

1. Gearing Assembly
2. Cut-Off Valve
3. Fuel Strainer
4. Shaft Seal
5. Anti-Hum Damper
6. Bearing
7. Pump Body
8. Bleed Valve

FIGURE 11-14B Two-stage fuel oil pump. (Courtesy of Delavan, Inc.)

Efficiency Testing

As we have seen in earlier chapters, during the burning of any fuel, some of the heat produced during the combustion process invariably will be lost as waste heat. In the most efficient oil-burning equipment, between 10% and 15% of the heat generated during combustion goes up the chimney. In many older oil heating systems, as well as those that were converted from coal to oil, combustion efficiencies as low as 50% are not uncommon.

With the advent of easy-to-use combustion analyzers such as the unit shown in figure 11-15, instrument testing residential of heating systems has become standard practice. One of the most important tests is the measurement of carbon dioxide as a percentage of exhaust gases.

Carbon dioxide testing: The delivery of excess air to the oil burner is a basic determinant of combustion efficiency. The carbon dioxide

The pump should normally provide oil at an exit pressure of 100 PSI. Vacuum pressure readings are also useful in determining restrictions or leaks in the oil supply lines. Vacuum readings are taken using a good-quality vacuum gauge connected to an unused intake port on the pump. Readings should be between 0 and 5 inches of vacuum. Low vacuum readings could be an indication of leaks in the suction side of the pump, while high readings usually indicate restrictions in the oil lines or clogged filters, and the like.

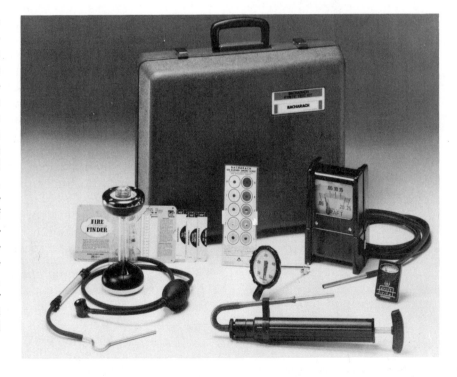

FIGURE 11-15 Combustion efficiency analyzer kit. (Courtesy of Bacharach, Inc.)

test determines in part how much excess air is available during combustion. The percentage of carbon dioxide in the exhaust gas indicates the efficiency of combustion of any specific oil-burning appliance. Properly done, this test is an invaluable tool for the service technician.

From a theoretical basis, fuel oil, consisting of pure carbon burning with the ideal amount of air, would yield 21% carbon dioxide. Assuming that 100% excess air is delivered during combustion process, carbon dioxide levels would be 10.5%. Since 21% of the air is oxygen, complete combustion would yield 21% carbon dioxide, with no excess air supplied during the combustion of the oil. With double the amount of air required for combustion, or 100% excess air during the combustion process, the percentage of carbon dioxide would drop from 21% to 10.5%. As excess air decreases, carbon dioxide levels increase. Thus the percentage of carbon dioxide present in the flue gases is an indication of the amount of excess air delivered to the oil burner during combustion. The amount

of excess air for proper oil combustion will vary from one oil-burning appliance to another. Whereas one oil burner may require 100% excess air for highest combustion efficiency, another may only require 75% excess air.

In large commercial and industrial oil-fired plants, excess air is determined by metering oxygen and monoxide levels directly while in residential installations, a carbon dioxide meter is used to measure excess air. In this procedure, a small hole is drilled into the flue pipe and a sample of the flue gases is pumped into a solution that absorbs carbon dioxide. The gas absorption is then indicated on a graduated scale reading directly in percentage of carbon dioxide (figure 11-16).

The percentage of carbon dioxide in the flue gases is an indication of the quality of the fire. With the test results, we can make some generalizations concerning the characteristics of oil fires. As we increase excess air to the fire, carbon dioxide levels will drop, and vice versa. The key element in adjusting the amount of excess air is to achieve a balance between the right amount of excess air and carbon dioxide levels. Too much excess air will actually cool the fire, reducing heat transfer to the air or water in the appliance, with resulting carbon dioxide levels of between 5% and 7%. Too little excess air, while increasing the temperature of the oil fire can also result in a "dirty" fire evidenced by large amounts of soot, with resulting carbon dioxide levels above 13%. Carbon dioxide levels of between 9% and 11% with no traces of smoke in the fire result in high efficiency operating levels. The technician should consult original manufacturer's specifications concerning the levels of carbon dioxide that apply to specific heating appliance.

FIGURE 11-16 Gas absorption carbon dioxide tester. (Courtesy of Bacharach, Inc.)

Stack temperature: The temperature of the stack gases is primarily a measurement of the efficiency of heat transfer within the lower unit of the boiler or furnace. A boiler with inadequate heat transfer surface area or one whose surfaces are dirty will transfer less heat from the passing combustion gases resulting in a higher stack temperature. During the efficiency testing procedures, the stack temperature is measured by placing a high-temperature stack thermometer in the sample hole. The temperature of the stack gases is then read directly from the thermometer (figure 11-17).

FIGURE 11-17 Stack temperature thermometers. (Courtesy of Bacharach, Inc.)

Excessively high stack temperatures are usually indicative of a problem somewhere within the appliance. As a general rule, the maximum stack temperature for conversion units is 700° F and 500° F for factory-packaged units. The most common causes for high stack temperatures can usually be traced to one of the following conditions.

1. Dirty heat exchanger surfaces.

2. Improper adjustment of draft regulator and/or excessive draft.

3. Insufficient baffling in the appliance.

4. Undersized boiler or furnace.

5. Overfired burner.

Smoke testing: Smoke testing enables service personnel to maximize the percentage of carbon dioxide just prior to increasing the smoke level of the oil fire during flame-adjustment procedures. With most oil-burning equipment, the point of highest carbon dioxide levels (the level with the correct amount of excess air for the specific appliance) is also the point at which

the fire begins to smoke (known as the smoke point). The difficulty for the technician is to adjust the air intake so that a high carbon dioxide reading can be maintained with a virtually smokeless oil fire. Smoke levels can be read from a smoke gauge chart that enables a comparison to be made between samples drawn from the flue gases and a standardized printed chart (figure 11-18).

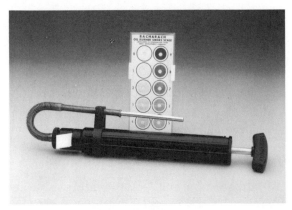

FIGURE 11-18 Smoke tester with comparison smoke chart. (Courtesy of Bacharach, Inc.)

High smoke levels indicate inefficient combustion that eventually leads to soot formation within the boiler or furnace. As soot builds up on the internal heat exchange surfaces, heating efficiency drops and stack temperatures increase. Other problems in the burner are also attributable to soot formation, including clogged fuel oil nozzles and misaligned electrode assemblies. To measure smoke levels in the combustion gases, a piece of filter paper is placed in the smoke tester. The probe of the smoke tester is then inserted into the sample hole and the sampling pump bulb's squeezed 20 times. This forces a predetermined amount of flue gas through the filter paper.

After the sample has been drawn, the paper is removed and placed on the smoke gauge chart. A visual comparison is then made between the sample paper and the samples represented on the chart in order to determine the smoke level of the fire. Adjusting an oil burner to achieve high carbon dioxide levels with a minimum of smoke level (.01 maximum) can be difficult. Some of the more common factors that can cause high smoke levels are as follows:

1. Too little excess air.
2. Too much excess air.
3. Damaged fuel oil nozzle.
4. Clogged oil nozzle.
5. Improperly sized oil nozzle.
6. Dirty or clogged air intake vents around fan.
7. Improperly sized combustion chambers.
8. Damaged burner and/or combustion head.
9. Improperly adjusted electrodes.

Draft Measurement and Adjustment

Draft, an important factor in efficient burner operation, is the evacuation pressure through the appliance and is created primarily in the chimney. Draft intensity also determines the rate at which combustion air passes through the appliance. Just as atmospheric pressure is measured by inches of mercury in a column, draft is measured in inches of water within a column. Natural draft is based on thermal currents. Since gases expand when heated, a given volume of a heated gas will weigh less than the same volume of gas at a cooler temperature. This hot gas, contained within the interior of a tall chimney, rises from the bottom to the top of the chimney. The hotter the gas, the greater is its velocity as it moves through and out of the chimney. This movement of gas as it enters and leaves the chimney is known as draft. Draft inside the chimney created entirely by the thermal currents of the exhaust gases is known as natural draft. The height, location, and materials used in the construction of the chimney all affect its draft characteristics. High chimneys generally result in a greater draft than do lower ones. Chimneys located inside a building have a greater draft than do outside chimneys, since the heat of the building keeps the chimney warm, increasing the draft. Chimney draft is measured with a draft gauge that is inserted into the flue pipe, between the draft regulator and the boiler (figure 11-19). Over-the-fire measurements can be made if the appliance is equipped with an inspection or measurement hole for this purpose.

FIGURE 11-19 Draft gauge. (Courtesy of Bacharach, Inc.)

Proper placement of the gauge is critical for accurate readings. Since the draft readings should indicate draft condition over the fire, the draft regulator must be positioned on the appliance side of the exhaust system, with the gauge downstream of the regulator. Figure 11-20 details the technique for taking draft measurements before and after a draft hood to check suction over the fire and the chimney condition, depending upon the location of the draft gauge. If used properly, this instrument will give an accurate picture of draft conditions both within the chimney and over the fire.

If natural draft is insufficient in specific applications, then induced draft fans or blowers can be installed to increase the draft. A booster fan of this type is illustrated in figure 11-21.

Too much draft can raise stack temperatures above recommended levels. This also has the effect of sucking heat out of the appliance,

FIGURE 11-20 Taking draft measurements at draft hood and checking chimney draft.

reducing heat exchange efficiency. Insufficient draft makes it almost impossible to adjust the burner for maximum efficiency, since sufficient quantities of excess air for complete combustion will be difficult to achieve. This is due to the fact that although the blower on the oil burner will supply sufficient quantities of excess air to the combustion chamber, low draft causes this air to back up and puff out of the combustion chamber. This condition is known as "back-puffing." Proper draft allows all of the air entering the combustion chamber to move

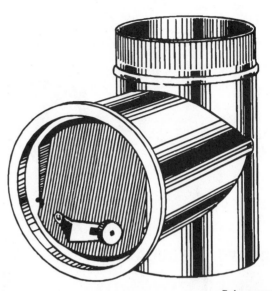

Draft controls

Deluxe control for central heating space heaters and ranges. Sizes 6" and 6"-7". Available pre-set, or adjustable. 26 guage tee and stub, 24 guage ring. Available pre-set, or adjustable to low, medium or high draft. Ring and gate revolve in collar to permit vertical or horizontal mounting. Friction-free gate mounting.

FIGURE 11-21 A booster fan for increasing chimney draft. (Courtesy of Tjernlumd Products)

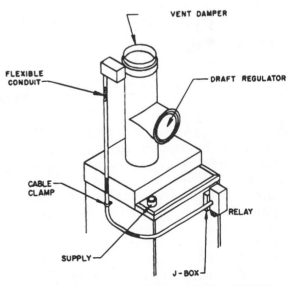

RECOMMENDED DRAFT REGULATOR LOCATION
VERTICAL PIPING

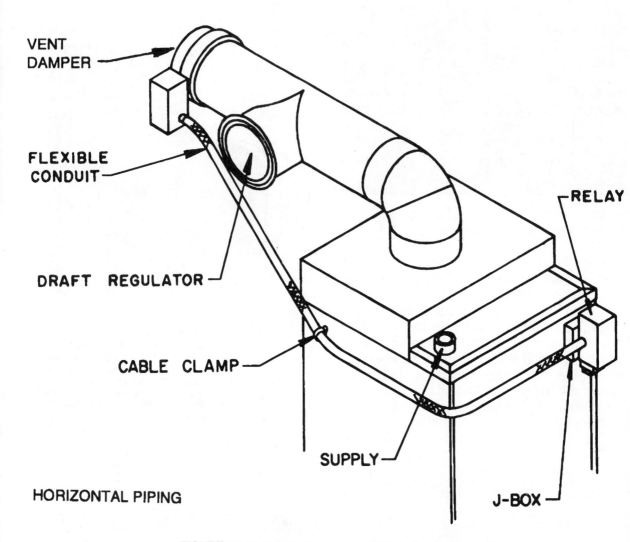

VENT DAMPER

FLEXIBLE CONDUIT

DRAFT REGULATOR

CABLE CLAMP

HORIZONTAL PIPING

SUPPLY

RELAY

J-BOX

FIGURE 11-22 Adjustable counterweight draft regulator.

through the appliance, enabling the technician to adjust the burner properly. Older oil-burning units require careful control of draft, since this influences the amount of air passing through the boiler or furnace, often occurring by air infiltration. Newer burners and appliances are tightly sealed against air infiltration and use high-speed fans to produce induced draft pressures. This makes them less dependent on the evacuation pressures produced by the chimney.

Most residential oil burners need only minimum draft for efficient operation. A draft level

of .02 inch of water column is considered ideal for most operating conditions. Check specific requirements of the appliance manufacturer for factory-specified draft settings. Chimney draft is normally regulated by the use of a draft regulator installed on the stack pipe of the appliance, depending on the piping configuration used to connect the appliance to the chimney (figure 11-22).

The draft regulator in figure 11-22 is a counterweight damper-type regulator. By adjusting a counterweight on the back of the damper door, different negative chimney pressures are required to cause the door to open. This allows boiler room air to enter the burner air intake passages, maintaining the draft over the fire at a relatively constant level. Thus the draft regulator's primary function is to maintain a constant draft level in the combustion chamber to facilitate efficient burning.

Adjustments made to the draft regulator can change the amount of excess air over the fire, as previously discussed. This can cause variations in the carbon dioxide readings. Therefore, each time a draft adjustment is made, corresponding changes in intake air regulation should be made, along with taking a carbon dioxide test, in order to maintain the proper amount of excess air for clean and efficient burning. The air intake adjustments are normally made on the fuel pump side of the oil burner, as shown in figure 11-23.

be adjusted for lowest levels at an acceptable smoke level (at or below .01). The stack temperature is then recorded. The stack temperature recorded on the efficiency calculator should be the stack temperature minus the boiler room temperature. Since stack temperature measures the heat of the exhaust gases picked up as the result of the combustion process within the boiler, it is important to remember that air entering the combustion chamber was already at room temperature (70°F – 80°F). Therefore, this temperature should be subtracted prior to making any efficiency calculations. (Some efficiency calculators may already contain this correction factor, so be sure to check prior to making final calculations). With the chimney draft adjusted to factory levels, all combustion information can then be displayed on an efficiency analyzer, as shown in figure 11-24.

FIGURE 11-23 Incoming combustion air travels through air intake slots. Adjustment is made by moving a trap over a portion of the slot to change the amount of air entering the burner unit.

Efficiency Calculations

The boiler or furnace should operate for 10 to 15 minutes to enable it to reach normal operating temperatures prior to taking final smoke, carbon dioxide, and stack temperature readings. The achievement of normal operating temperatures by the appliance is referred to as steady-state operation. Carbon dioxide should

FIGURE 11-24 Computerized combustion efficiency analyzer. (Courtesy of Bacharach, Inc.)

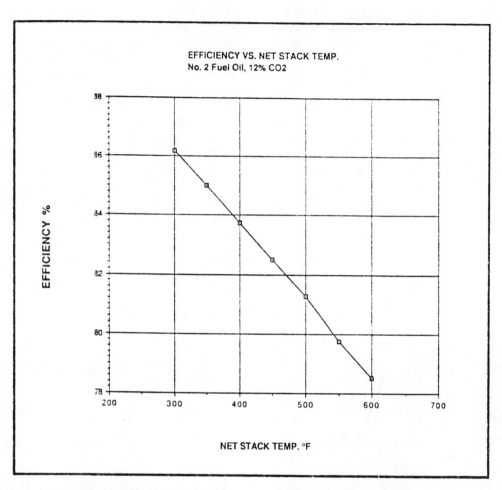

FIGURE 11-25 Oil burner efficiency as a function of stack temperature.

As a reference, consult the chart in figure 11-25, which illustrates efficiency levels as a function of net stack temperature and percentage of carbon dioxide in the flue gases. This chart is based on net stack temperatures (steady-state stack temperatures minus the room temperature), as well as on the burner's being adjusted for a maximum .01 smoke level.

With experience, the technician will develop a keen "fire sense." But while developing an intuitive sense regarding the characteristics of oil fires is important, this sense should never be substituted for carefully performed instrument testing procedures.

OIL SUPPLY SYSTEMS

Oil supply and delivery systems can vary according to the distance and height of the oil storage tank relative to the oil burner appliances. The storage tanks are constructed from either heavy gauge corrosion-resistant treated steel or fiberglass. Underground storage tanks come in standard storage sizes of 550 or 1000 gallons. Indoor oil storage tanks are usually sized at 275 gallons, and are constructed from a lighter-gauge steel than their underground counterparts. The supply system is designed in either a one- or a two-pipe configuration, depending on the requirements of the specific installation.

One-Pipe Supply System

In a one-pipe delivery configuration, a single pipe runs from the bottom of the oil storage tank to the burner appliance. A one-pipe aboveground design is illustrated in figure 11-26.

In one-pipe feed systems, any air in the oil lines must be removed, or bled, by manually opening the bleed valve on the oil pump. Since excess fuel in the one-pipe feed system remains in the oil pump, the only way to remove trapped air from the oil lines that may accumulate in the pump is to vent it from the oil pump itself. One-pipe systems are most often used in conjunction with inside oil storage tanks where

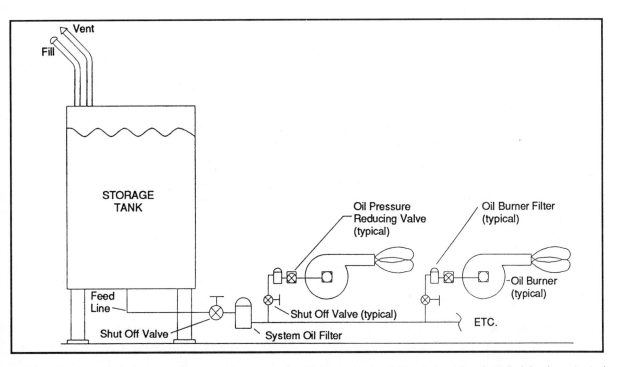

FIGURE 11-26 One-pipe aboveground oil delivery system. This system should be designed so that all piping is protected from physical damage. One-pipe systems require manual air bleeding from individual fuel oil pumps.

the pipe run from the tank to the oil burner is relatively short. If two oil burners are fed from a single-pipe system, then either check valves or electrically operated solenoid shut-off valves should be installed in each feed line, as illustrated in figure 11-27. The check valve is required to prevent oil drain-back in order to keep both pumps primed when either one or both burners are not operating. The solenoid valves are wired in parallel with their respective burner motors to open and close in response to motor operation.

Two-Pipe Supply System

A two-pipe supply system is more flexible in application than its single-pipe counterpart. In this type of installation, each appliance has both a feed and a return line that supply each oil pump. Excess oil in the pump is returned to the oil tank using a separate oil return line. A two-appliance/two-pipe feed system with an underground storage tank is illustrated in figure 11-28.

One advantage of the two-pipe system is that pump bleeding and priming are automatic—no

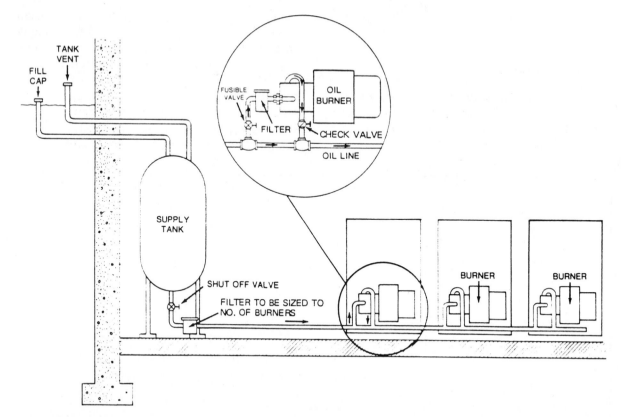

FIGURE 11-27 Check valves should be used when more than one appliance is connected to the same fuel supply system. Although this illustration features a single-pipe delivery system, check valves are recommended for two-pipe delivery systems as well.

manual bleeding is necessary, since excess oil in the pump is piped back to the storage tank via the return line. In this way, any air trapped in the oil lines is carried back to the storage tank, where it is purged and vented through the oil tank vent pipe. Also, note in figure 11-28 that when two appliances are used in this arrangement, a solenoid shut-off valve is not required on each burner since pump priming is automatic.

Tandem Tank Installation

It is sometimes advantageous to connect two oil storage tanks together in order to increase total storage capacity. This is most often done on inside storage tanks where individual tank capacities are limited, or in situations where accessibility to a particular site during the winter months is limited due to weather conditions. Local building codes may specify the maximum number of oil tanks placed inside of a residential or commercial building, and should be consulted prior to installation. A two-tank installation is illustrated in figure 11-29. Note from this installation that the two tanks are piped in series. The fill pipe runs from the first tank into the second. The system is vented from a vent line installed on the second tank. All piping in this arrangement should be 1 1/4 inches. Each tank should be fitted with its own shut-off valve and filter. The fill alarm is installed on the second tank. The two tanks are manifolded together into a common feed line to the burner appliance.

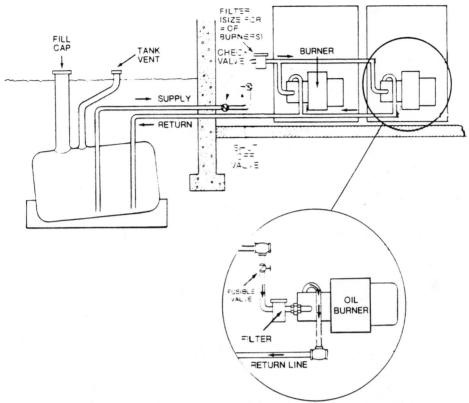

FIGURE 11-28 Two-pipe, two-appliance fuel oil delivery system. Two-pipe systems allow for automatic bleeding of fuel pumps, since excess fuel oil, along with trapped air, is returned to the fuel storage tank.

Filtration Controls

All fuel oil systems require filtration devices to eliminate impurities suspended in the fuel. Contaminants most often enter the fuel supply because of dirty storage tanks or improper handling during shipping and delivery. All residential and commercial fuel oil installations should have at least one filter assembly installed on the feed line from the storage tank to the burner. In most residential installations, the filter housing is piped into either the fuel line at the output of the storage tank or directly onto the fuel pump at the oil burner. Figure 11-30 illustrates the filter assembly connected at the feed end of the fuel oil pump. The fuel oil filter assembly in figure 11-31 uses replaceable cartridge elements, which should be changed annually under normal operating conditions. If the filter assembly is installed directly on the oil tank (for indoor tank locations), a shut-off valve should be located upstream of the filter assembly. This valve is usually installed directly onto the oil tank so that the flow of oil can be stopped when the filter element is replaced. When installing the filter assembly onto the oil pump, as in figure 11-31, the shut-off valve can be installed directly onto the filter cartridge assembly. In two-line oil feed systems, shut-off valves should be installed on both the feed and return lines.

In addition to dirt and other suspended solids, a potential problem encountered with all fuel oil tanks is that water eventually can accumulate at the bottom of the tank owing to condensation. To test for the presense of water, a water-finding paste is placed on the end of a

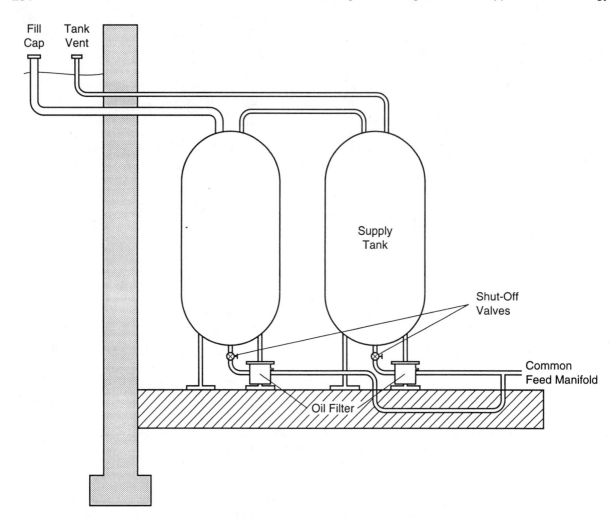

FIGURE 11-29 Two-tank tandem installation.

long stick, which is held at the bottom of the oil tank for a few seconds, and then removed. The paste, which is normally pink, turns blue in the presence of water. In indoor tanks, small amounts of water can usually be drained from the tank, since the water, which is heavier than the oil, settles to the bottom of the tank. Outdoor underground tanks generally have to be completely pumped out.

Level Indicators

Two types of gauges are used to indicate oil level within a storage tank, depending on whether the tank is aboveground or underground.

Aboveground tanks use a gauge operated by a float mechanism, as shown in figure 11-32.

This float-type gauge is incorporated into a standard housing that screws into an aboveground fuel oil tank. The level of oil within the tank causes the floating cork to move a lever, which is connected to an indicator on the top of the housing. This fitting also incorporates the tank vent fitting. The vent fitting contains an alarm that whistles when, during filling, the tank is almost full (figure 11-33). This feature helps prevent dangerous and unwanted spills from coming out the vent

FIGURE 11-30 Fuel filter and oil burner assembly.

pipe. Oil spills are not only environmentally destructive, but can result in substantial fines in many states. Oil spills from leaking indoor tanks sometimes require that the entire floor be torn up to remove any contaminated soil. The costs of a spill can be quite high both environmentally and financially. Proper operation of the float gauge and vent alarm is essential to prevent these types of spills.

FIGURE 11-31 Fuel oil filter housing with replaceable filter elements. (Courtesy of General Filters, Inc.)

Underground tanks cannot use the standard float-type level indicator. In these installations a special sending unit is piped from the tank to a remote barometric pressure-type gauge, as shown in figure 11-34.

Fire Controls

All oil systems must be designed and installed with an eye toward protection from hazardous situations, primarily the potential for fire. Design features to minimize damage in the event of fire focus on shutting off the flow of oil if the system should sense excessive heat in the vicinity of either the oil storage tank or the oil burner. A typical installation with fire protection logic is illustrated in figure 11-35. Note that shut-off valves equipped with fusible links are installed at the bottom of the oil tank, as well as in the oil line connection to the oil pump.

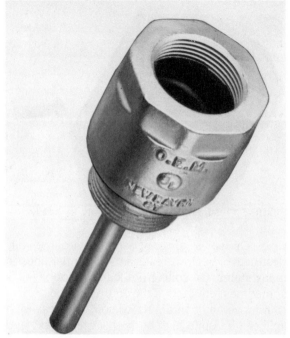

FIGURE 11-33 Combination vent/alarm fitting for fuel oil storage tank. (Courtesy of Oil Equipment Manufacturing Co., Inc.)

FIGURE 11-32 Float-type oil level gauge. The level of the fuel oil in the storage tank raises and lowers the float mechanism that indicates the level of the oil in the tank. This assembly screws directly into the top of the storage tank. These gauges are available in different float lengths, depending on the size of the storage tank. (Courtesy of Oil Equipment Manufacturing Co., Inc.)

Also, a check-valve that incorporates a fusible link should be installed between the oil filter and oil burner to prevent excessive drain-back from the oil lines to the filter housing when changing the filter element. The fusible link acts like the fuse in an electrical circuit and melts if it is exposed to high temperatures. When the link melts, the valve automatically closes, shutting off the flow of oil through the valve. The link is designed to fail at a temperature well below the ignition point of the fuel oil. Figure 11-35 also illustrates the use of an electrical thermal switch that acts as a circuit breaker,

opening the electrical circuit to the appliance in the presence of excessive heat.

ELECTROMECHANICAL AND COMPUTERIZED CONTROLS

Electromechanical Primary Controls

The oil burner primary control is the control center for all of the burner's electrical functions. On a call for heat from the thermostat, the control energizes the burner, as well as monitors flame presence to shut down the system if the flame should be extinguished for any reason. A typical primary control is illustrated in figure 11-36.

It operates in the following way. On a call for heat, the thermostat circuit closes, pulling in the starting relay. Initially, the control must have a high electrical resistance across the F-F terminals to which the cadmium sulfide (cad) cell is connected, or the relay will be unable to start the burner. After the burner is running, the cad

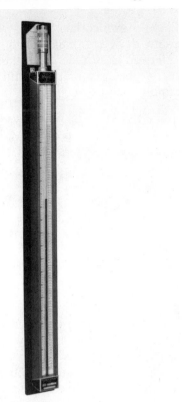

FIGURE 11-34 Barometric fuel oil level gauge for use with underground storage tanks. Connecting lines run between the tank and the gauge. (Courtesy of the Petrometer Corp.)

the oil). Intermittent ignition extends transformer life and reduces electrode wear. Both systems, properly maintained, are reliable and trouble-free.

Should the burner lose its flame for any reason, the cad cell will trigger the safety circuit to a lock-out condition cutting all power to the burner. Based on the design of the primary control, some units allow the burner to attempt to fire one additional time before going out on a safety lock-out. A lock-out condition requires the control to be reset manually by utilizing the reset button before the burner can be started again.

It is not uncommon for a service technician to respond to a no-heat call and find that the home or building owner has tried to restart the oil burner by continually pressing the manual reset button on the primary control. The technician should be aware of this potential problem. Significant amounts of liquid fuel oil that have collected in the combustion chamber can result in an out-of-control fire when the oil burner is restarted unless the chamber is checked and excess oil removed prior to performing any service procedures.

cell monitors light from the combustion flame (by changing electrical resistance in proportion to the amount of light coming from the combustion flame).

Some controls feature a constant ignition process in which the spark is continuously generated during the entire burner operating time. Other controls feature an intermittent ignition sequence in which the electrode spark is eliminated after the control goes through its safety cycle to prove flame presence. Each of these ignition patterns has advantages. Constant ignition is useful in the event of momentary disruptions in oil or air supplies or minor changes in oil viscosity (thickness of

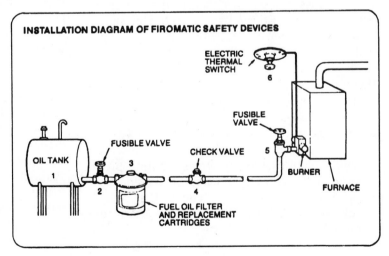

FIGURE 11-35 Recommended installation of fusible shutoff valves and heat sensors for fire safety on all oil-fired combustion systems; (Courtesy of Highfield Manufacturing Co.)

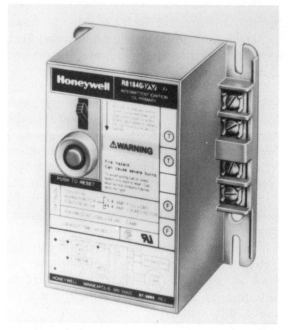

FIGURE 11-36 Oil burner primary control. (Courtesy of Honeywell, Inc.)

Computerized Primary Controls

Microprocessor-based controls are now available, in addition to the all-electromechanical controls that have dominated the field for so many years. These new controls feature a microprocessor that self-checks the control circuitry when the burner starts, and continues to monitor all burner activity during the run cycle. The status of all control functions is displayed using light-emitting diodes (LEDs). The operating characteristics of computerized controls of this type are similar to three of their electromechanical counterparts. Most of these controls incorporate intermittent ignition with a choice of 15- or 45-second lock-out times. A constant ignition spark is present during pre-lock-out. The use of diagnostic LED readouts enables a homeowner to check the relay visually and to identify the problem to a service technician over the telephone.

Flame-Sensing Controls

Light-sensitive controls (cad cells): The use of light-sensitive photoelectric cells to prove

flame presence has been a reliable oil burner safety device for many years. The photoelectric cell used in virtually all of these applications is made from cadmium sulfide and is known simply as a cad cell (figure 11-37).

FIGURE 11-37 A cadmium sulfide safety cell.

The cad cell is part of the primary control circuit and consists of the cell, the cell holder, and the wires that connect the cell to the F-F terminals of the primary control. The cell operates within the primary control by virtue of its electrical resistance in either the presence or absence of light. In darkness, the cell has a very high resistance to electricity. When exposed to light, the resistance of the cell drops.

During an initial burner start-up, the primary control must sense a high electrical resistance across the cad cell or it will not close the starting relay. If the cell either senses light or is short-circuited, the primary control will not close. After the burner starts and light from the flame is established, the resistance in the cell drops, preventing the safety circuit from locking out the control as long as the flame remains established.

Problems with a suspected cad cell can be diagnosed in the following way. If the burner will not start and a defective cad cell is suspected, one of the wires should be removed from the F-F terminals. If the burner then starts, the problem is traceable to the primary controls not sensing a high resistance (no light) setting at the beginning of the cycle. If the burner starts but continues to go off on safety lock-out, the F-F terminals should be jumped, or wired together, after the burner has started *and* before the safety lock-out time has elapsed. If the burner continues to run, the problem can be traced to the cad cell's not changing electrical resistance in the presence of light. In either case, the cell should be replaced and the burner sequenced again to verify proper performance. The low cost of cad cells warrants their replacement when any question of their reliability arises.

Pyrostat flame controls: Flame-sensing controls can also be mounted directly on the stack pipe. These devices rely on a pyrostat, which is a bimetallic coiled element that senses heat in the flue gas stream. As the element is heated by the flue gases, the rotating bimetallic coil controls an electrical safety relay. Figure 11-38 illustrates the pyrostat sensing element on a stack-mounted pyrostat control.

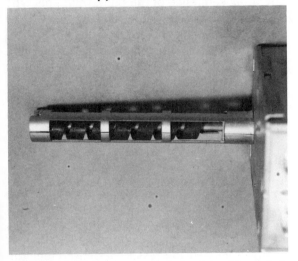

FIGURE 11-38 Stack-mounted pyrostat safety control.

When the burner is activated, heat from the combustion process causes the pyrostat to rotate, bypassing the safety lock-out. As long as heated flue gases are present in the stack pipe, the pyrostat maintains a flow of uninterrupted electric power to the burner. Should the flame be extinguished for any reason, the pyrostat rotates in the opposite direction, allowing the safety circuit to lockout the burner by opening the electrical circuit to the burner. A safety lock-out condition requires the relay to be reset manually.

VENTING REQUIREMENTS

Conventional oil-fired equipment is vented into a chimney, as illustrated in figure 11-39.

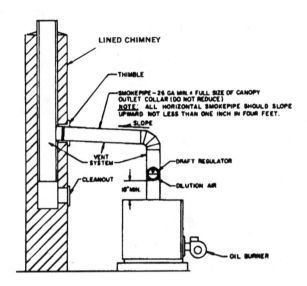

FIGURE 11-39 Venting requirements for oil-fired appliances.

For proper operation of the appliance, certain basic rules must be followed when venting any oil-fired furnace or boiler. The vent pipe should be constructed from 24-gauge black or stainless steel pipe. Many localities specify minimum pipe gauges required for venting oil-fired appliances, and should be consulted when this type of work is to be done. The connecting run to the chimney opening must be as short as possible. Note from figure 11-39 that the smoke pipe is angled upwards toward the chimney, and that there is a sufficient vertical distance between the chimney connection and the clean-out door. This distance allows for a small buildup of soot without blocking the chimney connection. The clean-out should be inspected on a regular basis during the heating season and any deposits removed.

Older chimneys, especially those located on the outside of buildings, tend to operate at lower temperatures due to a greater heat loss characteristics than is true of either insulated chimneys or of those located within a building structure. Also, outside chimneys tend to create more corrosive condensate than do either insulated or indoor chimneys. When replacing an older oil-fired appliance with a newer one, it is often found that the draft on the newer one is much less than on the previous appliance. This is sometimes caused by the older unit's higher steady-stack stack temperatures, which yield high draft levels. Newer, more efficient units operate at substantially lower stack temperatures, as well as at lower draft levels. Also, note the proper and improper locations of draft regulators in figure 11-40. Improper installation of the draft regulator will adversely affect the operation of any appliance. Consult the manufacturer's specifications for specific draft regulator applications.

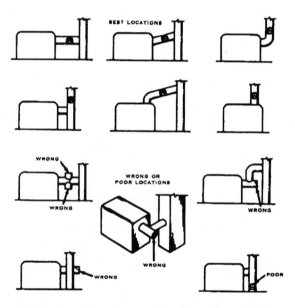

FIGURE 11-40 Proper and improper locations of draft regulators used in oil-fired appliances.

The condition of all chimneys should be determined by period inspections. Chimneys should be cleaned by professional chimney sweeps on a regular basis. Any telltale signs of chimney deterioration, such as staining of masonry or of stainless steel, should be investigated. Problems should be rectified prior to putting the chimney back into service. For a summary of common chimney problems relating to chimneys located both inside and outside of heated buildings, see Figure 11-41.

If the location of any unit will bring it into close proximity to combustible materials, the manufacturer's specifications concerning safe clearances must be consulted and followed. All minimum distances should be strictly adhered to for continued safe appliance operation.

TROUBLESHOOTING LOGIC

Given an understanding of the general operating characteristics of conventional oil burners, figure 11-42 is presented as a summary of procedures to use in determining the proper operating sequence of the oil burner. This type of troubleshooting guide is often used by service technicians in order to diagnose problems and component malfunctions. The student is advised to study this guide as a general reference tool, since it is valuable for reviewing operating sequences, as well as for pinpointing problems that will arise from time to time.

SUMMARY

This chapter has examined the theory of oil-fired combustion technology, using typical oil burner configurations as the point of reference. The oil burner brings together a number of energy inputs—high-pressure atomized fuel, turbulent air, and a high-voltage electrical spark—to accomplish efficient combustion. The efficiency of this process is determined by combustion-efficiency testing procedures combine chemical analysis with visual inspection in order to maximize delivered heat and minimize waste heat (efficiency is maximized and entropy is minimized). Oil supply and control systems offer the technician a wide variety of installation options to meet a variety of system requirements. With the fundamentals of oil combustion in mind, we move on to a discussion of gas combustion technology in Chapter 12.

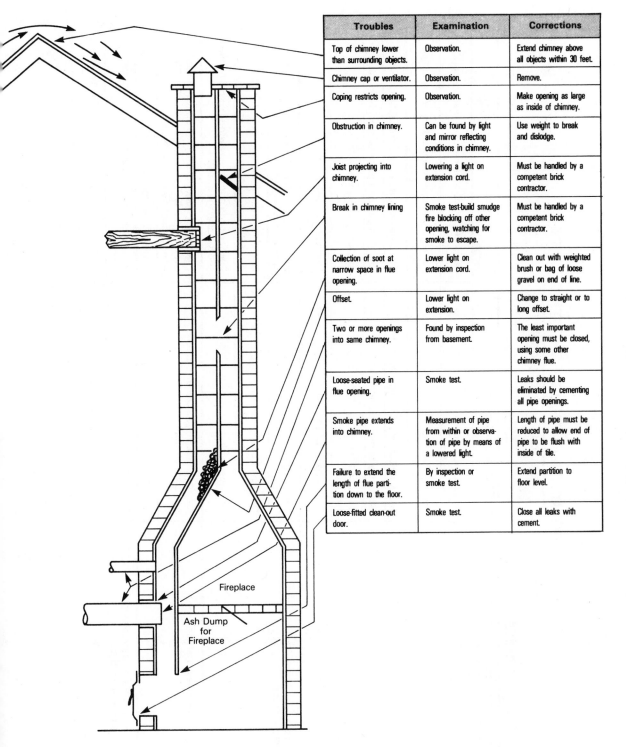

Troubles	Examination	Corrections
Top of chimney lower than surrounding objects.	Observation.	Extend chimney above all objects within 30 feet.
Chimney cap or ventilator.	Observation.	Remove.
Coping restricts opening.	Observation.	Make opening as large as inside of chimney.
Obstruction in chimney.	Can be found by light and mirror reflecting conditions in chimney.	Use weight to break and dislodge.
Joist projecting into chimney.	Lowering a light on extension cord.	Must be handled by a competent brick contractor.
Break in chimney lining	Smoke test-build smudge fire blocking off other opening, watching for smoke to escape.	Must be handled by a competent brick contractor.
Collection of soot at narrow space in flue opening.	Lower light on extension cord.	Clean out with weighted brush or bag of loose gravel on end of line.
Offset.	Lower light on extension.	Change to straight or to long offset.
Two or more openings into same chimney.	Found by inspection from basement.	The least important opening must be closed, using some other chimney flue.
Loose-seated pipe in flue opening.	Smoke test.	Leaks should be eliminated by cementing all pipe openings.
Smoke pipe extends into chimney.	Measurement of pipe from within or observation of pipe by means of a lowered light.	Length of pipe must be reduced to allow end of pipe to be flush with inside of tile.
Failure to extend the length of flue partition down to the floor.	By inspection or smoke test.	Extend partition to floor level.
Loose-fitted clean-out door.	Smoke test.	Close all leaks with cement.

Fireplace

Ash Dump for Fireplace

FIGURE 11-41 Some common chimney problems, their causes, and some corrective procedures.

VENT DAMPER TROUBLE-SHOOTING GUIDE
(Listed in order of probability)

SYMPTOM	POSSIBLE CAUSE	REMEDY
Heating required and burner will not operate. Damper open.	Thermostat is set wrong.	Reset thermostat (heat or hot water) to call for heat.
	No electrical power.	Turn on switch - replace fuse - reset circuit breaker.
	Improper wiring.	Recheck and correct any wiring errors in line and low voltage circuits.
	Stack switch or cad cell malfunction.	Check reset button: repair or replace control.
	Defective burner components.	Check, repair or replace burner components.
	Damaged or defective damper operator.	Replace damper operator.
	Damaged or defective time delay relay.	Replace time delay relay.
Burner operates normally, damper will not close.	Time delay in normal operation.	Wait at least 3 minutes for damper to close, before checking further.
	Damper is blocked open.	Check for free damper movement, and remove blockage.
	Improper wiring.	Recheck and correct any wiring errors in line and low voltage circuits.
Time delay relay chatters.	Incompatible solid state oil burner primary control.	The Robertshaw line of solid state primary controls (SJ 4000 series) are not compatible. Consult factory for modification.
	Defective time delay.	Replace time delay relay.
Burner will not operate. Damper closed and will not open.	No call for heating.	Reset thermostat (heat or hot water) to call for heating.
	Damper is blocked closed.	Check for free damper movement and remove blockage.

Symptom	Probable Cause	Corrective Action
Burner will not operate. Damper operates normally.	Improper wiring.	Recheck and correct any wiring errors in line and low voltage circuits.
	Broken return spring.	Inspect under mounting plate for broken return spring. Replace with complete drive assembly.
Burner operates normally. Damper operates normally. Bad odor is detectable.	Improper wiring.	Recheck and correct any wiring errors in line and low voltage circuits.
	Stack switch or cad cell malfunction.	Check reset button: repair or replace control.
	Defective burner components.	Check, repair or replace burner components.
	Normal time delay insufficient for system.	Open vent pipe and remove two knock-outs from damper vane. Be careful not to damage or distort vane.
Burner operates before damper is open.	Improper wiring.	Recheck and correct any wiring errors in line and low voltage circuits.
Damper vane stops in other than fully open or fully closed position.	Damper is blocked.	Check for maximum 90° damper movement. If less than 90°, remove blockage.
	Missing roll pin damper stop.	Replace stainless steel roll pin.
	Broken coupling.	Inspect and replace with complete drive assembly.
	Broken return spring.	Inspect under mounting plate for broken return spring. Replace with complete drive assembly.
	Broken spring stop.	Inspect and replace complete drive assembly.
Intermittent burner operation. Damper operates normally.	Bent or broken coupling.	Replace complete drive assembly.
	Bad ground.	Recheck and correct any wiring errors in line voltage-circuit.
	Damaged or defective switch.	Replace damper operator.

FIGURE 11-42 Troubleshooting logic diagram. This type of chart is consulted by technicians to logically identify problems and ascertain the recommended corrective procedures.

PROBLEM-SOLVING ACTIVITIES

Troubleshooting Case Studies

A. A service technician is called in to diagnose a no-heat situation in a single-family residence. In a preliminary inspection, the following conditions are observed.

1. The oil tank is full.
2. All oil valves are open.
3. There is voltage to the oil burner primary.
4. The room thermostat is calling for heat.
5. There is an accumulation of unburned oil in the combustion chamber.
6. The oil burner has a new cad cell.

What procedural steps are involved in isolating the problem?

B. After annual spring maintenance procedures have been performed on an oil burner, the burner takes a long time to ignite when it is energized, and produces a great deal of soot and smoke after ignition has taken place. Describe the procedures used to isolate the problem.

C. An oil burner ignites on a call for heat, but locks out on safety. Outline a procedure that can be used to diagnose the cad cell for proper operation.

D. An oil burner fails to fire upon a call for heat. Inspection by the service technician indicates the following conditions.

1. The thermostat circuit functions, and the burner motor turns on.
2. Visual inspection of the nozzle and electrode assembly indicates a spark across the electrodes.
3. The fan is delivering combustion air to the burner head and the coupling between the burner motor and the fuel pump is working.
4. The oil tank is full and the oil delivery valves are open.

Indicate the method you would use to diagnose the problem and a probable cause for the system malfunction.

Review Questions

1. Identify in sequence the steps that take place during the firing of a typical oil burner. For each step, identify the critical function or functions that take place.

2. Specify five factors that can cause excessively high stack temperatures. For each factor, identify a service procedure that can be used to eliminate the specific problem.

3. Differentiate between inside and outside chimneys. Which design is preferable and why?

4. Differentiate between the operation of mechanical and electronic oil burner safety controls.

5. Highlight in sequence all of the steps required to replace the oil burner nozzle and reset the electrodes. In your answer, the following conditions must be taken into account.

   a. The oil filter assembly will be replaced.
   b. A new electrode will be installed.
   c. A test of the fuel pump pressure will be performed.

CHAPTER 12

Gas-Fired Combustion Technology

The use of fuel gases in heating and cooling systems provides the installer with many options. Fuel gases either occur naturally or are manufactured. Natural gas is most often used in urban areas where utility piping distribution networks are in place, while manufactured gases are primarily rural-use fuels, delivered to the customer as bottled gas for central space- and water-heating applications. An introduction to fuel gas characteristics and burner fundamentals will be followed by a discussion of the variety of control devices associated with gas combustion systems.

CHARACTERISTICS OF FUEL GASES

The two most commonly used fuel gases for residential and commercial applications are natural gas and propane. Each has its own unique characteristics.

Natural Gas

Natural gas is composed primarily of methane, with varying smaller amounts of other hydrocarbon gases. Location and extraction techniques for natural gas are similar to those used for petroleum. In its natural state, the gas is odorless and colorless. Chemical sulfur-based odorants (such as mercaptan) are added during the processing of the gas to aid in leak detection.

Although the heating content of natural gas varies from one geographical area to another,

for purposes of heating calculations it is generally assumed that 1 cubic foot (ft^3) of gas contains approximately 1000 Btu. The specific gravity, or weight of the gas also varies slightly from region to region. The specific gravity of most natural gas is around .6 (six tenths the weight of an equal volume of air). Because natural gas is lighter than air, it disperses rapidly in open environments, such as well-ventilated basements. And because gas dispersal is an important safety factor in the event of a gas leak, it is recommended that gas-fired appliances be installed in areas that are as airy and open as possible.

Liquified Petroleum (LP) Gas

Propane is the most popular of the liquified gases; it is produced as by-products of petroleum refining. It is considered a rural fuel, since it is delivered to customers in bottles of various sizes. The familiar 25-gallon tall, thin cylinders and the 100-gallon storage bottles are the most popular residential sizes. Larger storage tanks are used in commercial and industrial applications.

Propane contains between 2200 and 2500 Btu/ft^3 (approximately twice the heating value per cubic foot of natural gas). It is also denser, with a specific gravity of about 1.5. Because it is heavier than air, it tends to accumulate on the floor near appliances and does not disperse as easily as does natural gas. As a result, explosions can result from accumulations of the gas. Chemical odorants similar to those used in natural gas are added to propane for leak detection purposes. Adequate ventilation is critical for safe use with propane, and the manufacturer's specifications regarding proper ventilating techniques should be closely followed.

COMBUSTION CHARACTERISTICS OF FUEL GASES

Yellow- and Blue-Flame Atmospheric Burners

Fuel gases can be burned, or combusted, in a number of different ways. Under atmospheric conditions, this combustion is known as either yellow flame or blue flame. The color of the flame depends upon whether the air and gas are premixed prior to combustion (blue flame) or are mixed externally at the burner ports during the combustion process (yellow flame). A yellow-flame atmospheric burner is illustrated in figure 12-1. Yellow-flame burners generally have low Btu outputs for use in such items as pilot lights and small gas lamps. Note from this

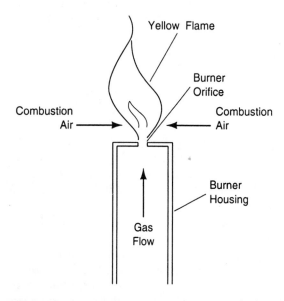

FIGURE 12-1 Yellow-flame atmospheric burner. These burners are used primarily for low-heat applications. Note that primary combustion air is drawn from around the flame.

illustration that combustion air is drawn from around the flame, where it mixes with the gas, giving the flame its characteristic yellow color.

Yellow-flame burners are low-temperature combustion devices with flame temperatures of approximately 1200°F. For this reason, the flame should not come into contact with any surface cooler than 1200 degrees or sooting will result.

In contrast to the yellow-flame burner, note the cross section of the blue-flame atmospheric burner in figure 12-2.

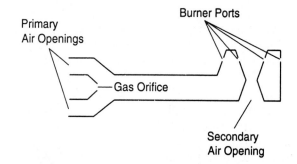

FIGURE 12-2 Blue-flame atmospheric burner. Blue-flame burners premix primary combustion air with the fuel gas prior to combustion. These burners are customarily used in domestic heating applications.

In the blue-flame burner, primary air is mixed with the fuel gas prior to leaving the burner ports. Additional secondary air for combustion is drawn as needed from around and through the burner ports. Blue-flame burners are generally used in higher Btu heating applications, such as domestic hot-water heaters, boilers, furnaces, and stoves.

Power Burners

Gas power burners function in a manner similar to that of oil burners. Fuel gas is pumped under pressure to a burner head which mixes combustion air with the gas. A spark, or hot surface ignitor, ignites the air–gas mixture. Power burners are often installed when switching a boiler or furnace from oil to gas. The operating characteristics of gas power burners are similar to those of oil burners and will be examined in more detail later in this chapter.

GAS PIPING AND PRESSURE REQUIREMENTS

System Operating Pressures

The pressures of the gas delivered to the burner ports are different for natural gas and LP gas. These pressures are a function of the different densities of the gases, as well as the differences per cubic foot in their respective heating values. Gas pressures are measured in inches of water column (w.c.) in a tube in a device called a manometer (figure 12-3).

FIGURE 12-3 A manometer used to measure gas pressures. The manometer reads out in inches of water column (w.c.) in a tube. (Courtesy of Bacharach, Inc.)

In order to take this measurement, the manometer is installed in the pressure fitting on the gas valve.

Pressure requirements at the burner ports for natural gas are between 3 and 5 inches w.c. The gas enters the building from the utility piping and distribution system gas mains at about 25 PSI. The gas meter pressure regulator reduces this pressure to approximately 7 inches

w.c. Final pressure reduction to 3 to 5 inches w.c. takes place at the individual appliance.

Propane is stored in the storage tanks at approximately 125 PSI. The pressure regulator on the storage tank reduces this pressure to about 10–12 inches w.c., which is required by individual appliances. Some appliances have their own pressure regulators so that final pressure adjustments can be made.

Propane gas requires higher operating pressures to the burner head than does natural gas, since its density is greater.

Piping Procedures

All gas piping procedures must comply with the National Fuel Gas Code, as well as local utility and state regulatory codes. A typical piping circuit to the gas valve is illustrated in figure 12-4.

Although local codes may vary regarding specific piping procedures, some general rules are offered here as guidelines.

1. An accessible manual shutoff valve must be installed upstream of the furnace or boiler outside of the appliance jacket when codes so require.

2. A moisture trap and ground joint union (see figure 12-4) must be included.

3. A pipe thread compound that is resistant to the action of LP gases should be applied to all threaded pipe joints.

4. Some utility companies may require pipe sizes larger than those shown on standard sizing charts. The installer needs to check with the local utility company regarding this, and all other, applicable piping procedures.

Pipe Sizing

Gas piping is sized to provide the required supply of gas without undue loss of pressure between the gas meter and the boiler or furnace. Figure 12-5 gives nominal pipe sizes to be used in order to deliver specific cubic-foot capacities of gas based on different pressure drops and a specific gravity of .6. Figure 12-5

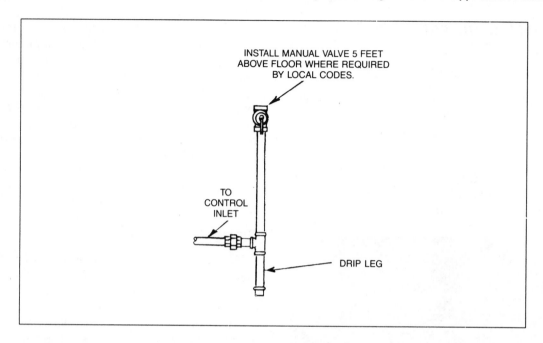

INSTALL MANUAL VALVE 5 FEET
ABOVE FLOOR WHERE REQUIRED
BY LOCAL CODES.

TO
CONTROL
INLET

DRIP LEG

1. The gas line should be of adequate size to prevent undue pressure drop and never smaller than the pipe size of the main gas control valve.

2. To check for leaks in gas piping, use a soap and water solution or other approved method. **DO NOT USE AN OPEN FLAME.**

3. Disconnect the boiler from the gas supply piping system during any pressure testing of the gas piping. After reconnecting leak test the gas connection and boiler piping before placing the boiler back into operation.)

FIGURE 12-4 Gas supply piping.

also includes multipliers that should be used if the specific gravity of the gas varies from .6. The following factors should be taken into account when consulting these tables to determine the sizing for the delivery piping.

1. Gas valve should be a minimum of 3½ inches from the appliance..

2. Length of pipe and number of fittings should be kept to a minimum.

3. Correction factors must be used for specific gravity, where required.

The gas supply line can usually be connected from either side of the appliance. The supply should be a separate line, installed in accordance with the National Fuel Gas Code, American National Standards Institute (ANSI) Z223.1, or CAN1-B149.1 or .2. After all connections have been made, or if modifications have been made to the system, a heavy soapsuds/water mixture should be used for checking all joints for gas leaks.

COMBUSTION AND CONTROL DEVICES

There are many different types of gas-burner mechanisms, each with its own set of operating procedures.

Maximum Capacity of Piping in Cubic Feet of Gas
Per Hour
(Based on a Pressure Drop of 0.3" Water and 0.6
Specific Gravity)

Length in Feet	Nominal Iron Pipe Size			(Inches)
	½	¾	1	1¼
10	132	278	520	1,050
20	92	190	350	730
30	73	152	285	590
40	63	130	245	500
50	56	115	215	440
60	50	105	195	400
70	46	96	180	370
80	43	90	170	350
90	40	84	160	320
100	38	79	150	305

FIGURE 12-5 Gas capacity of selected sizes of pipe. Properly sized piping for a particular installation ensures that the Btu capacity of the appliance can be maintained by adequate supplies of fuel gas to the burner.

Standing Pilot Ignition—Natural Vent

Almost all gas-fired systems installed up until the late 1970s were of the standing pilot ignition type. Because of this installation his-

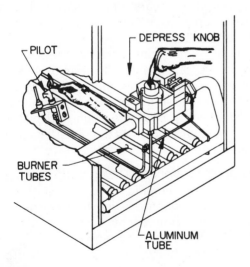

FIGURE 12-6 Standing pilot ignition. The pilot flame is supplied with gas from a separate circuit in the main gas valve. Most gas-fired ignition systems are installed with 100% shutdown capability, meaning that a loss of pilot flame turns off the gas to the pilot light circuit, as well as to the main gas burners. (Courtesy of Burnham Corp.)

tory, standing pilot units still outnumber all other types of gas-fired combustion systems. In many parts of the United States, standing pilot systems no longer can be installed under revised building codes. They are being replaced by newer units that have electronic ignition sequencing and other features to enhance burner efficiency and limit off-cycle heat losses. The configuration of a typical standing pilot system is shown in figure 12-6.

In the standing pilot arrangement, a small amount of gas is used for a pilot flame which is directed against a thermocouple. The thermocouple is a device composed of two dissimilar metals joined at either end. When heat is applied to one end of the thermocouple, a small electric current is produced—which is sufficient to operate the main gas valve as the burners cycle during normal operation. A typical thermocouple is illustrated in figure 12-7.

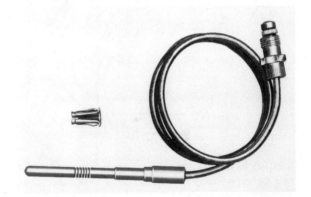

FIGURE 12-7 Thermocouple. Composed of two dissimilar metals, the thermocouple junction provides a small amount of electricity to power a relay that keeps the gas valve open during main burner operation and closes the valve upon loss of pilot flame. (Courtesy of Honeywell, Inc.)

Operating sequence: During the normal operating sequence, the flame of the pilot burner furnishes the heat that enables the thermocouple to provide electric power to the gas valve. Upon a call for heat, the gas valve energizes, allowing gas to flow to the main burners. The pilot light serves to ignite the main burners during each heating sequence. Should a loss of pilot occur for any reason, power from the thermocouple to the gas valve is interrupted, deenergizing the gas valve, which prevents the valve from supplying gas to the main burners. For the thermocouple to operate properly, the pilot flame and thermocouple control assembly should be adjusted to give a steady flame that covers from 3/8 inch to 1/2 inch of the tip of the thermocouple, as illustrated in figure 12-8.

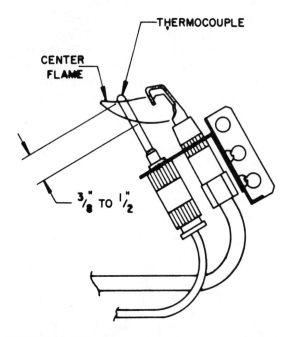

FIGURE 12-8 Pilot flame assembly. Note the position of the pilot flame against the thermocouple to ensure proper operation of the gas valve.

The components of the standing pilot ignition system are assembled into the boiler cabinet as shown in figure 12-9. Note the use of the flame rollout, or spill switch, which is located at the burner access panel. Should a blockage occur in the appliance flue passages that causes the flame to roll, or spill, out from the combustion chamber, the switch opens the circuit to the main gas valve, which shuts down the main and pilot burners.

Adjustments to the main burners are made by moving the air shutters located at the base of each burner tube. The main burner arrangement and adjustable shutters are illustrated in figure 12-10A. The main burner flame characteristics should resemble those in figure 12-10B.

A power vent damper can be added to this system to minimize off-cycle heat losses from the furnace or boiler. The vent damper is wired in series with the main gas valve. See the section on venting later in this chapter for more details concerning power vent dampers.

Power-pile generators: A modification of a conventional thermocouple system is known as a power-pile generator. Conventional standing pilot ignition systems rely on an external 24-volt power supply to energize the main gas valve. Power-pile systems generate their own electric power within the pilot ignition system. This is done through the use of a power-pile generator, which looks like an ordinary thermocouple, but is, in fact, several thermocouples connected to one another in series. This arrangement allows the electrical outputs of the individual thermocouples to be added together to generate sufficient electric energy to power the gas valve without the use of any external electric power. A power-pile appears almost identical to the thermocouple component arrangement; it is detailed in figure 12-11.

The basic operating sequence for power-pile (sometimes called millivolt) systems is essentially the same as for standing pilot systems. Power-pile generators normally use 100% shutoff valves whereby both the main and pilot gas orifices are shut down as the result of a loss of a pilot flame.

Operating sequence: The operating sequence and control schematic for a typical standing pilot ignition system are shown in figure 12-12.

This illustration contains the basic circuit diagrams for both the aquastat and main boiler control functions in a typical hydronic heating system. The operating sequence is as follows:

1. A call for heat starts the circulator through the relay in the combination

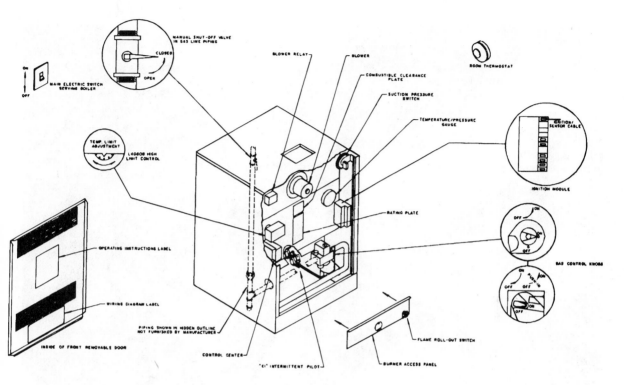

FIGURE 12-9 Assembly of burner and control elements in the boiler cabinet. (Courtesy of Burnham Corp.)

control, while opening the combination gas valve to start the main burners.

2. Assuming a pilot flame presence, both the burner and circulators operate until the thermostat is satisfied.

3. The reverse-acting switch in the combination control acts to stop the circulator if the water temperature of the system falls below that required to satisfy a domestic tankless coil.

4. If the high-limit temperature is exceeded, the high-limit control in the combination control stops main burner function.

5. The blocked vent switch stops burner operation in the event that excess blockage is developed in the venting system. The flame rollout, or spill switch, stops burner operation if the appliance flue passages become blocked.

6. The combination gas valve is equipped with a 100% shutoff provision that prevents the flow of gas to both the main burners and pilot if the pilot flame goes out for any reason.

Gas supply controls in warm-air heating systems function in essentially the same way as in hydronic applications. The major difference is that the operation of the main burners in warm-air systems is controlled by the fan limit switch, which supplies the high-limit control. Vent dampers, rollout, and blocked vent switches are identical in both hydronic and warm-air applications.

Electronic Ignition—Natural Vent

In contrast to the standing pilot ignition system, electronic ignition maintains a pilot flame only when there is a call for heat by the thermostat (or an aquastat in the case of a low-limit control for a tankless coil temperature

FIGURE 12-10A Main burner arrangement. Note the shutters used to adjust main burner flame characteristics. (Courtesy of Slant/Fin Corp.)

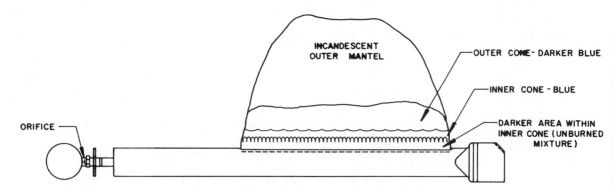

FIGURE 12-10B Main burner flame characteristics. A properly adjusted main burner flame will insure a relatively trouble-free ignition sequence throughout the heating season.

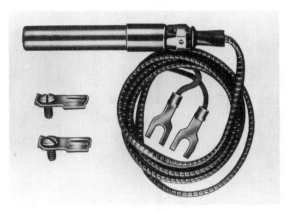

FIGURE 12-11 Power-pile generator. Power-piles are similar in appearance to the ordinary thermocouple arrangement. The ability of the power-pile to generate more power than the thermocouple allows the gas valve to operate independently of the home electricity source. (Courtesy of Burnham Corp.)

minimum). Electronic ignition systems use an ignition electrode to provide a spark in order to establish a pilot flame. Once the flame is established, the pilot serves to light the main burners. This ignition and burner assembly is illustrated in figure 12-13.

Electronic ignition uses the same safety rollout and blocked vent switches found in standing pilot ignition systems. In addition to these standard safety features, the ignition module also contains a flame sensor. Since pilot failure can occur during the start-up or operating cycle of the burners, any pilot flame failure will be sensed by the electronic control module, which is designed to close the main gas valve within 1 second of loss of flame.

When a loss of flame occurs, the ignition control module attempts to reestablish the pilot flame. Depending on the particular ignition control, one or more tries will be made to reestablish the pilot flame. If these attempts fail, the control goes into the safety lock-out mode, which must be manually reset before ignition can be tried again.

Operating sequence: Figure 12-14 shows the control schematic for an electronic-ignition, gas-fired boiler with intermittent circulator opera-

tion and a tankless heater operating on natural gas.

1. The ignition module is continuously activated at terminal 6.

2. When the thermostat calls for heat, the circulator is started through a relay in the combination control. At the same time, the ignition module is energized at terminals 2 and GR.

   OR

   When the low limit in the combination control calls for heat, the circulator is deenergized. At the same time, the ignition module is energized at terminals 2 and GR.

3. Terminal 1 and the ignition terminal are energized. Terminals 1 and GR power the pilot valve operator in the redundant combination gas valve that supplies gas to the pilot. The ignition terminal and GR supply voltage to the ignition electrode, creating an electric spark to ignite the pilot light.

4. The sensor senses the presence of a pilot flame and signals to terminal 4 that the flame is present. This electrical signal deenergizes the ignition terminals.

5. Terminal 3 is energized and supplies power to the main gas operator of the redundant combination gas valve, which allows gas to flow and ignite the main burners.

6. After the thermostat and low limit are satisfied, the circulator, ignition module, and gas valve are deenergized, which extinguishes the main burner and pilot flames.

This sequence of events is visualized by the flowchart in figure 12-15 on page 290.

The adjustment of the pilot and main burner flames is essentially the same for both standing pilot and electronic ignition appliances. Consult the manufacturer's specifications for specific pilot and burner adjustments.

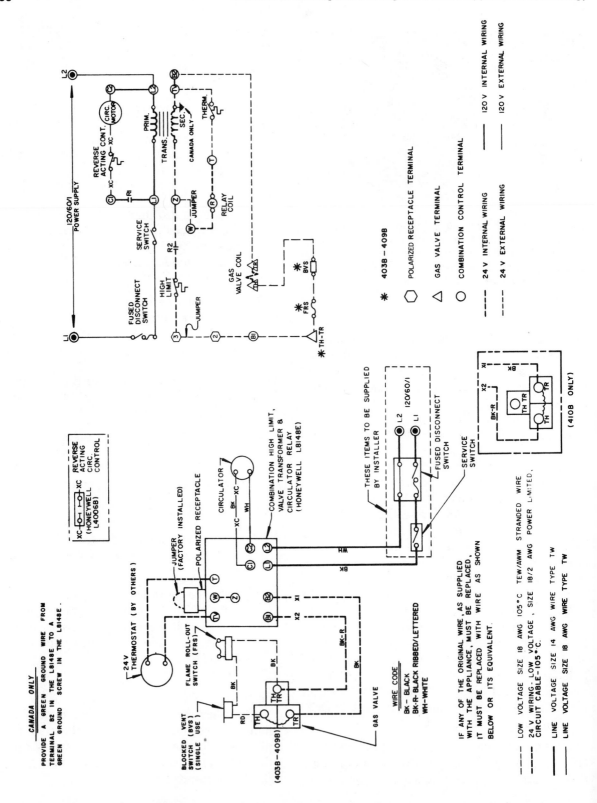

403B - 410B
WIRING DIAGRAM - CONTROL SET WB
24 VOLT INTERMITTENT CIRCULATOR OPERATION
WITHOUT TANKLESS HEATER

SEQUENCE OF OPERATION

When THERMOSTAT calls for heat, it starts the CIRCULATOR through a RELAY in the COMBINATION CONTROL and at the same time it opens the COMBINATION GAS VALVE starting BURNER operation. Where condensation of flue gas is encountered in boiler flues, a REVERSE ACTING CIRCULATOR CONTROL should be installed to stop the CIRCULATOR before the BOILER WATER TEMPERATURE drops to that at which flue gas condensation may occur. The BURNERS and CIRCULATOR will operate simultaneously until the THERMOSTAT is satisfied. In the event that excessive boiler water temperature is developed, the HIGH LIMIT SWITCH in the COMBINATION CONTROL will stop the BURNER operation only.

The BLOCKED VENT SWITCH will stop the burner operation in the event excessive blockage in the vent system is developed. The FLAME ROLL-OUT SWITCH will stop the burner operation in the event excessive blockage in the boiler section flue passageways is developed.

COMBINATION GAS VALVE is equipped with a 100% shut-off provision which prevents operation of the BURNERS and interrupts the flow of gas to the PILOT if PILOT FLAME should be extinguished.

FIGURE 12-12 Operation of a standing pilot ignition system. (Courtesy of Burnham Corp.)

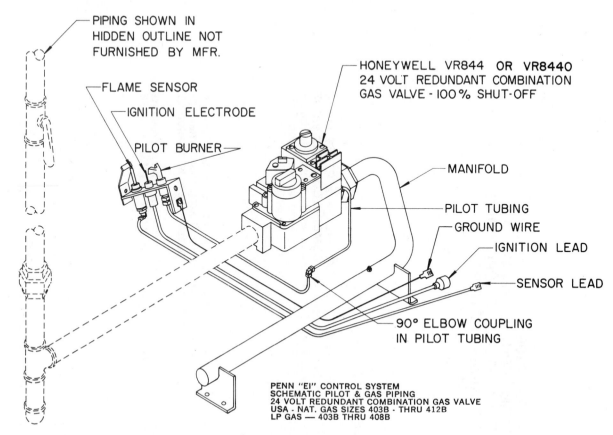

PIPING SHOWN IN HIDDEN OUTLINE NOT FURNISHED BY MFR.

FLAME SENSOR

IGNITION ELECTRODE

PILOT BURNER

HONEYWELL VR844 OR VR8440 24 VOLT REDUNDANT COMBINATION GAS VALVE - 100% SHUT-OFF

MANIFOLD

PILOT TUBING

GROUND WIRE

IGNITION LEAD

SENSOR LEAD

90° ELBOW COUPLING IN PILOT TUBING

PENN "EI" CONTROL SYSTEM
SCHEMATIC PILOT & GAS PIPING
24 VOLT REDUNDANT COMBINATION GAS VALVE
USA - NAT. GAS SIZES 403B - THRU 412B
LP GAS — 403B THRU 408B

FIGURE 12-13 Component assembly of electronic ignition system. (Courtesy of Burnham Corp.)

Induced-Draft Condensing and Noncondensing Operation

Induced-draft technology enables gas appliances to achieve high operating efficiencies through the use of an induced-draft combustion blower coupled with electronic ignition. This virtually eliminates standby heat losses during appliance off-cycles that are common with conventional atmospheric burners, since the vent passages in the appliance are sealed off against air infiltration when the unit is not operating.

Whether a boiler or furnace operates in the condensing or noncondensing mode depends upon the use of a secondary heat exchanger in condensing applications that lowers flue gas temperatures to the point where they partially condense. This condensate, which is corrosive, is removed from the appliance vent system through a condensate trap and drain system. Both a condensing and a noncondensing furnace are shown in figure 12-16.

Condensing units differ from conventional, noncondensing appliances in their use of a secondary heat exchanger, condensate trap, and connector in the condensing unit. The choice of condensing versus noncondensing operation has trade-offs. While the condensing units are somewhat more efficient given the secondary heat exchanger and high degree of heat transfer from the stack gases to the appliance, the condensate resulting from the secondary heat exchanger must be effectively removed. The condensate trap should be periodically inspected to ensure that no backup has occurred in the removal system.

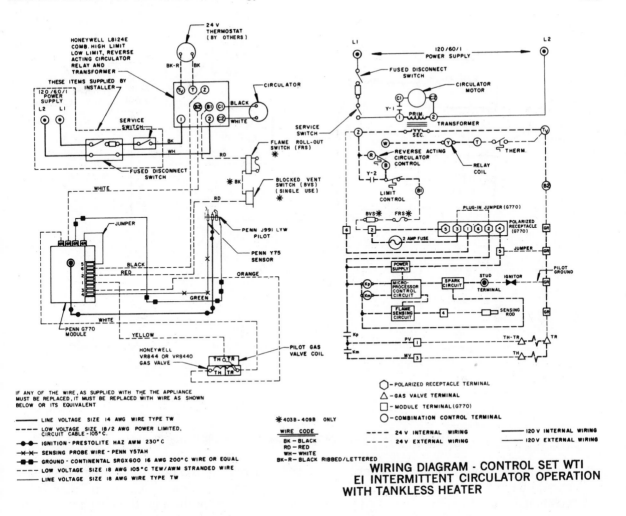

WIRING DIAGRAM - CONTROL SET WTI
EI INTERMITTENT CIRCULATOR OPERATION
WITH TANKLESS HEATER

FIGURE 12-14 Sequence of operation of electronic ignition system. (Courtesy of Burnham Corp.)

Induced-draft units require proof of operation of the blower before power is applied to the gas valve. In order to verify blower operation, one side of a differential pressure switch is connected to a pressure tap at the blower inlet using silicone rubber tubing. The other side of the switch is open to the atmosphere. When the blower rotates, inlet pressure to the blower becomes negative, closing the differential pressure switch, which establishes proof of operation. The differential suction pressure switch is located adjacent to the blower housing. Note the use of a manometer in figure 12-17 to check proper operation of the switch.

Operating sequence: On induced-draft units, the sequence of operation, using a hydronic boiler as an example, is as follows (refer to figure 12-18 to locate the specific controls listed):

1. The thermostat calls for heat. The circulator relay coil CR on the R8285D control center is energized. This circulator relay closes two sets of contacts, main pole contacts complete the 120-volt circuit to the circulator, and the auxiliary pole contacts provide 24 volts to the Y terminal and on to the safety circuits.

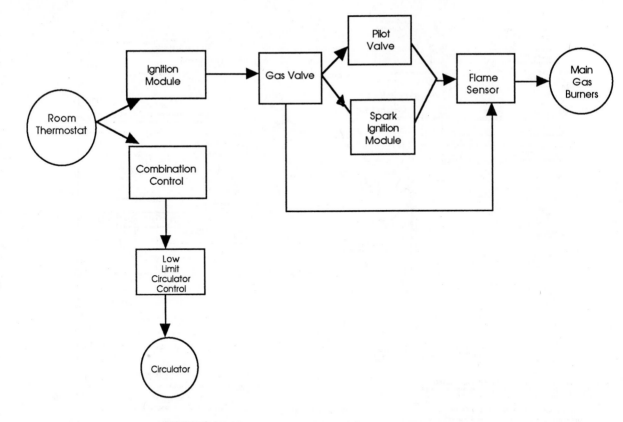

FIGURE 12-15 Logic flowchart of electronic ignition system.

2. The following safety circuit switches should be in their normally closed position:

   a. Flame rollout switch mounted on the burner access panel.

   b. High-limit switch with immersion well.

   c. Suction pressure switch connected to canopy pressure tap.

3. The blower relay coil BR is energized, which closes two sets of contacts: the main pole contacts to complete the 120-volt circuit to the blower and the auxiliary pole contacts to provide 24 volts back to the coil, which is already energized by the suction pressure switch check circuit (C to NC).

4. Sufficient blower performance activates the pressure switch contacts to switch from the NC position to the NO position. The blower relay then feeds 24 volts back through the check circuit to provide the suction pressure switch C to NO with electric power.

5. The suction pressure switch now provides 24 volts to the vent switch contacts (NV and nonmetallic PV vent systems) or through a jumper (stainless steel vent systems), which will then energize the ignition module.

6. Power to the ignition module energizes the purge timer. This allows the blower to clear all air from the boiler After 30–50 seconds, the delay timer energizes the ignition electrode and pilot valve coil in the redundant gas valve. Gas flows to the pilot burner, where it is ignited.

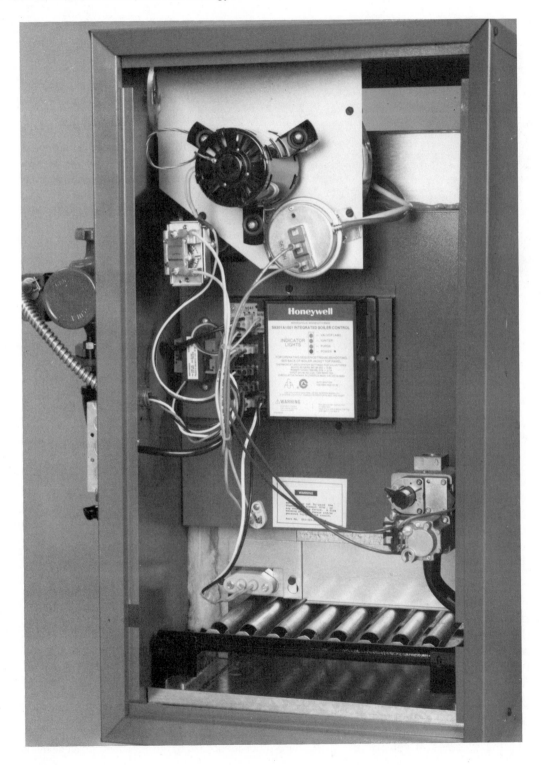

FIGURE 12-16A Induced-draft noncondensing boiler. (Courtesy of Weil-McLain, a Marley company)

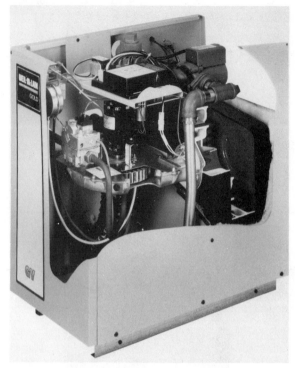

FIGURE 12-16B Induced-draft condensing boiler. (Courtesy of Weil-McLain, a Marley company)

7. When the pilot flame sensor indicates the pilot flame, the ignition module allows power to flow to the main gas valve. This valve is a step opening valve, in that only a limited amount of gas is admitted to the main burners for low fire ignition. After a short time, the valve fully opens to allow full flow for a high fire condition.

8. Should a loss of pilot flame occur for any reason, the main gas valve closes and a spark reoccurs within .8 second. If the pilot fails to ignite within 90 seconds, the ignition module goes into a 100% lockout mode. Five or six minutes later, the ignition module restarts the ignition sequence. This sequence continues until either the pilot lights or the room temperature rises above the thermostat set point.

9. The main burners, circulator, and induced-draft blower continue to operate until the thermostat is satisfied.

10. If the high-limit setting is reached, the limit switch opens, which causes the main gas valve to close and the induced-draft blower to stop. The circulator continues to operate. Upon a drop in temperature, the ignition sequence begins after a 30–50 second purge sequence.

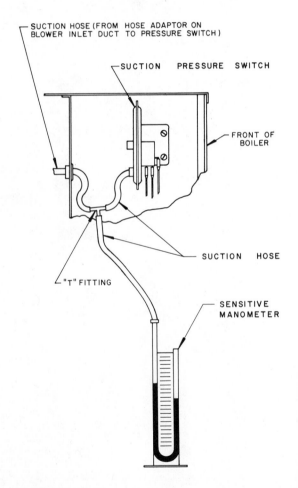

FIGURE 12-17 Use of a manometer to check operation of a differential pressure switch. The minimum suction pressure depends on the size of the boiler. This information is supplied by the individual appliance manufacturer.

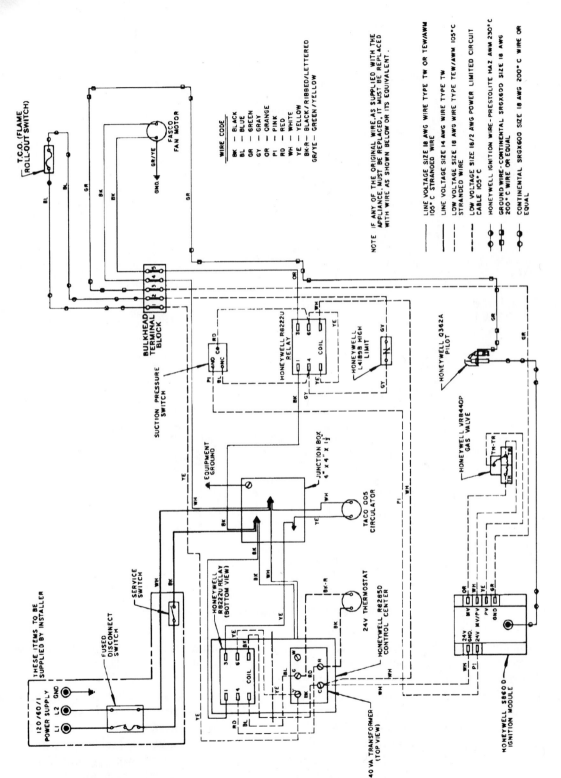

FIGURE 12-18 Sequence of operation and control circuitry of induced-draft noncondensing boiler. (Courtesy of Burnham Corp.)

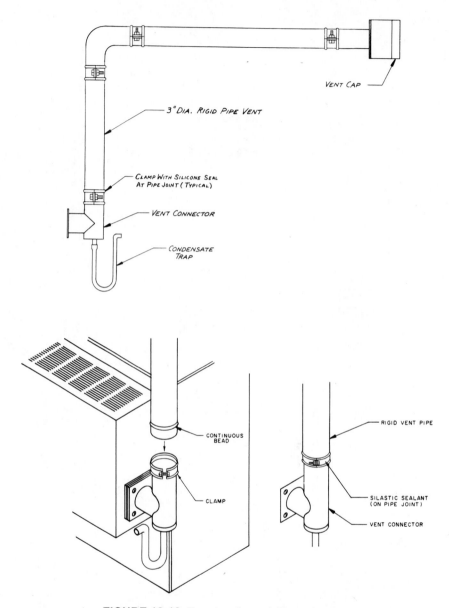

FIGURE 12-19 Trap for disposal of condensate.

11. If for any reason the flame rollout switch is subjected to temperatures above its setting, the switch will open, the gas valve will close, and the main and pilot burners will shut down. The circulator continues to operate. The flame rollout switch is a one-time use fusible link. In the event of flame rollout shutdown, the cause of the problem should be identified and rectified before installing a new rollout switch.

12. A blocked vent switch (with natural vent systems) or thermal vent switch (with power vent systems) will open in the event of either excessive pressure or blockages in the vent system. The suction pressure switch opens in the event of

induced-draft blower failure or blockage in the appliance vent passages. The gas valve cannot be powered on if any of these three safety switches are open.

Condensate formation and removal: Boilers and furnaces operating in the condensing mode use a secondary heat exchanger to remove additional heat from the flue gases prior to their venting. This reduction in flue gas temperatures is often below the dew point of the exhaust gases, causing condensate to form. In order to dispose of this condensate, a trap is incorporated into the base of the venting system, as illustrated in figure 12-19.

The trap is connected to a drain using silicone rubber tubing. Drain lines that rise above the trap at any point should be avoided. If a suitable drain is not available, the condensate should be collected in a container, which must be periodically emptied. The technician should note that this condensate is corrosive. Connections to septic systems should be made only after checking with local building code officials and appliance manufacturers to ensure compatibility of the condensate with septic system operation.

Pulsed Combustion Technology

Pulsed combustion boilers and furnaces rely on operating principles similar to those of the automobile internal combustion engine. The combustion process involves a series of small, controlled explosions of air–gas mixtures. In the process, heat from the hot gases is transferred to the surrounding air or water in the heat exchange passages of the appliance. Heat transfer in pulsed combustion systems utilizes condensing technology to capture an additional 9% to 10% of heat from the exhaust gases from the latent heat of vaporization. The pulsed combustion process is illustrated in figure 12-20A–D.

In the initial combustion cycle, a blower forces outside air into a sealed combustion chamber, where it is mixed with gas (figure 12-20A). A spark plug, similar to the type of plug used in an automobile, provides a 20,000-volt

spark to ignite this mixture of air and gas on the first cycle only (figure 12-20B).

After each successive cycle, the air and gas mixture is ignited by residual heat from the previous ignition cycle. Neither the spark plug nor the blower operates after the first ignition sequence. During a normal heating cycle, the ignition sequence continues at a rate of between 60 and 70 cycles per second. The pressure of the exploding gases in the combustion chamber forces the hot gases through the heat exchanger tubes where heat is transferred to the surrounding water or air (figure 12-20C). During this heat transfer process, the exhaust gases are cooled below their dew point and the water vapor in the flue gas condenses. The latent heat of vaporization of the gas is released to the surrounding air or water. After this initial combustion cycle, the spark plug is deenergized. Subsequent mixtures of air and gas are ignited by the residual heat within the combustion chamber (figure 12-20D).

The installation of pulsed combustion appliances in hydronic or warm-air heating systems follows normal procedures. Installation guidelines highlight features that are unique to these systems. Pulsed combustion units use 100% outside air for combustion and incorporate unique air supply and venting techniques, which are described in the "Venting Requirements" section of this chapter.

Power Burner Technology

Gas power burners combine induced-draft and oil burner technology for burning natural or propane fuel gas. The power burner is similar in appearance and operation to a conventional oil burner, and is illustrated in figure 12-21.

These units are available in several firing ranges for use with either natural or propane gas. Although most power burners are used in commercial heating applications, the lower firing range units are ideal for residential heating, especially when switching from oil to gas. The power burner can be installed in place of the conventional oil burner along with the associated gas piping.

The blower motor pressurizes the air cushion chamber, purges the combustion chamber of any unvented gases, and draws in 100% outside air for combustion. In pressurizing the air cushion chamber, it also puts a pressure on the diaphragm of the regulator.

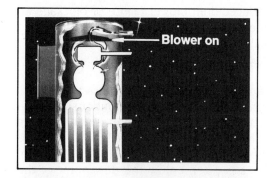

Thirty-four seconds after fan is energized both spark plug and gas valve are energized. Gas flows through the air-gas valve into the combustion chamber, where the proper mixture of air and gas is ignited by the spark plug.

When ignition occurs in the combustion chamber, the expanding gases close the air-gas valve discs. This forces the gas through the heat exchanger passages, which creates a vacuum in the combustion chamber. This vacuum draws open the air-gas valve discs to admit a fresh charge of air and gas for the next ignition cycle.

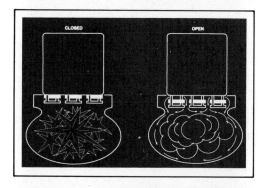

After the first ignition, the spark plug is de-energized, as the second and subsequent air-gas mixtures are ignited by the residual heat or flame left over from the previous ignition cycle. The blower motor is also de-energized, as the vacuum caused by the velocity of the exiting exhaust gases now pulls in the outside air for combustion.

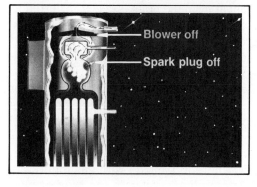

FIGURE 12-20 Operating sequence of a pulsed combustion boiler. (Courtesy of the Hydrotherm Corp., a Mestek company)

FIGURE 12-21 Gas power burner. (Courtesy of the Adams Corp.)

Several methods are used for burner ignition in these systems. In a hot surface ignition system, a specially constructed surface element provides the heat for gas ignition. Other systems employ either a conventional pilot burner to ignite the gas, or a spark ignition mechanism similar to those used on oil burners.

The reader is referred to the exploded-view drawing of a spark ignition power burner in figure 12-22 to assist in understanding the operating sequence of these units.

On a call for heat from the room thermostat, the thermostat contacts close, which energizes the coil in the motor relay, starting the motor. When the motor reaches full rotational speed, centrifugal switch contacts located on the motor close, which supplies 24 volts to the prepurge timer. After approximately a 30-second time delay, the timer delivers low voltage to the primary control relay, which closes, supplying

120 volts to the ignition transformer. An arc is established across the ignition electrodes, while simultaneously the low-voltage gas valve is opened. After the flame has been established, a low-voltage signal is sent back to the primary control, deenergizing the ignition arc. If the flame cannot be proved within 4 seconds, the primary control goes to a lock-out condition, turning off both the ignition assembly and the gas valve; however, the burner motor continues to operate if the thermostat is still calling for heat.

After a lock-out sequence, the primary control must be reset either by turning off the power to the burner or by lowering the thermostat below room temperature, and then raising it again. If during the heating cycle a loss of flame condition occurs, the ignition electrodes are reenergized and the 4-second try-for-ignition sequence is reestablished.

Power burners with pilot ignition establish main burner flames in the same manner as in conventional gas burners. Pilot-equipped power burners are supplied with either a push-button spark igniter, an automatic standing pilot, or an intermittent spark ignition pilot system. On all pilot ignition systems, the gas valve is opened only after the burner motor has come up to full operating speed during a purging sequence.

VENTING REQUIREMENTS

The requirements placed on chimneys have changed significantly over the years. Up until recently, heat loss up the chimney coming from low-efficiency boilers was of little apparent concern. With relatively low fuel prices, excessive heat losses were largely ignored. Also, chimney heat losses helped prevent the formation of damaging condensates within the chimney, and the warmth helped to provide and maintain a good draft. Newer, high-efficiency boilers and furnaces operate at much lower stack temperatures than their earlier counterparts and put little off-cycle heat into the

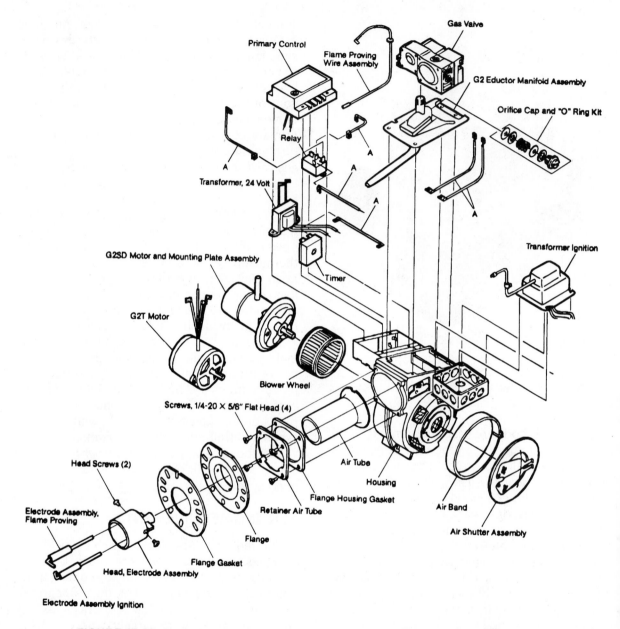

FIGURE 12-22 Component arrangement of gas power burner. (Courtesy of the Adams Corp.)

chimney. Before beginning a discussion of venting systems for the several types of gas-fired appliances examined in this chapter, some basic venting and chimney rules are offered as general guidelines.

1. No appliance should ever be vented into an unlined masonry chimney.

2. The chimney used to vent a specific appliance must be of the proper height and size for the firing range of that appliance.

3. The vent connector on the chimney must not be smaller than the vent connector on the appliance.

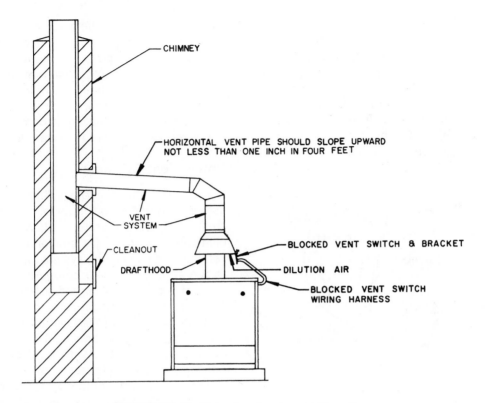

FIGURE 12-23 Natural vent system configuration.

4. All chimneys should be inspected at least once each heating season and cleaned on a periodic basis.

5. Chimney construction should follow one of the approved procedures (see figure 11-37 in Chapter 11) as recommended by the appliance manufacturer in compliance with all applicable building codes.

6. Where the appliance and a water heater vent into a common chimney, the chimney must have a cross-sectional area equal to that of the largest venting appliance plus 50% of the area of the second appliance.

7. Vent pipe should be installed into the chimney thimble or connector so that it is flush with the inside wall of the chimney. In no instance should the pipe extend beyond the inside surface of the chimney.

8. All horizontal pipe runs should be as short as possible. All pipe seams should be facing up. The pipe should be supported at least every 3 feet of run.

9. In no instance should a vent damper or any other obstruction be installed in the appliance vent pipe.

10. All installation should conform to the latest edition of the National Fuel Gas Code, ANSI Z223.1.

Natural Vent Systems

A natural vent system is shown in figure 12-23 which illustrates the use of single- or double-wall stove pipe venting into a lined masonry chimney. Note the slope of the pipe from the appliance vent outlet to the chimney thimble connection. For best results, single-wall pipe should be a minimum of 24-gauge steel.

Figure 12-23 illustrates the use of double-wall gas pipe as a total chimney installation.

The use of a draft hood (figure 12-24) is required on most natural-draft venting systems. The draft hood acts to deflect momentary downdrafts away from the flue passages, preventing these air currents from extinguishing the pilot flame. Note the installation of a blocked vent switch, or spill switch, on the outside of the draft hood.

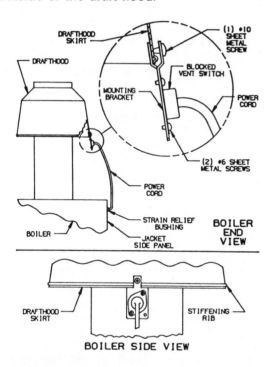

FIGURE 12-24 Draft hood and spill switch assembly. The spill switch will detect a blocked vent system. Draft gases that "spill" out of the draft hood will activate the switch to shut down the main burners. (Courtesy of Burnham Corp.)

A vent damper can be installed on these systems to minimize off-cycle heat losses. In natural vent systems, the damper is installed above the draft hood and connected with a wiring harness to the appliance ignition control module (figure 12-25). In a normal operating sequence, a call for heat from the room thermostat deenergizes the vent damper through the main boiler control. This causes the damper

blade to open. When the blade reaches the fully open position, an end switch in the damper is closed which energizes the main gas valve to begin the burner ignition sequence. After the heating cycle has been completed, the gas valves are deenergized, which extinguishes the main burner flame. The vent damper motor is energized, which closes the damper blade.

Power Vent Systems

Induced-draft technology incorporates the use of venting systems made of either stainless steel or high-temperature plastic pipe. This eliminates the need for conventional chimney venting. The power vent system piping connects to the appliance fan discharge box (figure 12-26), which can also incorporate a liquid trap for use with condensing appliances.

All power venting systems must be installed according to the National Fuel Gas Code, ANSI Z223.1, in compliance with local code regulations. Some generalizations concerning the installation of power venting systems are offered as guidelines to assure a quality installation.

1. Vent pipe should not pass through interior walls or floors. If this must be done, special fittings and adaptations to the venting system are required. Consult local applicable codes.

2. When passing vent pipe through either combustible or noncombustible exterior walls, consult the guidelines offered in figure 12-27. The distance of the vent cap to the ground must be such as to maintain adequate clearance above average snowfalls for the specific geographical area.

3. There are minimum and maximum vent lengths, depending on the specific appliance. Generally, the minimum vent length is around 2 feet, with a maximum length of approximately 30 feet. Consult the manufacturer's recommendations concerning individual appliances.

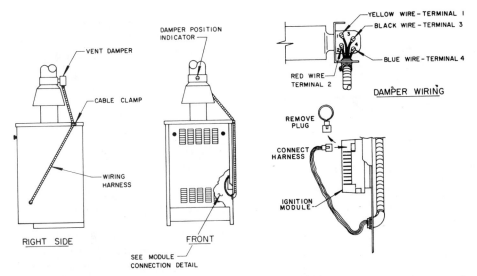

FIGURE 12-25 Installation of motorized vent damper in vent system. (Courtesy of Burnham Corp.)

4. The use of power saws to cut vent pipe should be avoided since personal injury, as well as damage to the pipe, may result.

5. Minimum clearances must be maintained to a variety of surfaces and obstacles. Consult the manufacturer's specifications for distances to doors, windows, public walkways, and so on.

Since moisture and ice may form on the surfaces surrounding the power vent terminal, these surfaces should be inspected frequently and kept in a good state of repair.

Pulsed Combustion Venting

The venting requirements of pulsed combustion appliances differ from those of conventional natural- and induced-draft combustion systems. Both combustion and exhaust air are piped to and from the appliance. Most pulsed combustion boilers and furnaces can be vented using an existing dormant chimney. In this instance, the chimney acts as a vent pipe chase. It is most important that the chimney used to house the air supply and exhaust pipes not be connected to any operating heating appliances. This type of installation is shown in figure 12-28.

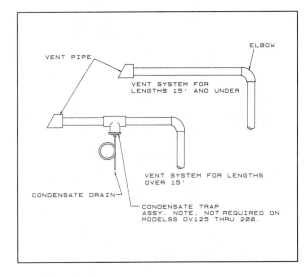

FIGURE 12-26 Typical power vent arrangement with and without condensate trap. (Courtesy of Utica Boilers, Inc.)

Note from figure 12-28 that the exhaust pipe must be a minimum of 12 inches taller than the fresh-air intake pipe to eliminate the possibility that the boiler or furnace will inhale the exhaust gases.

If a through-the-roof/chimney venting installation is not possible, then the appliance must have fresh-air and exhaust pipes installed as shown in figure 12-29.

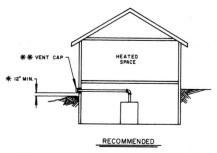

RECOMMENDED

* INCREASE MINIMUM HEIGHT ABOVE GRADE TO MAINTAIN ADEQUATE CLEARANCE ABOVE AVERAGE SNOW FALL FOR GEOGRAPHIC AREA WHERE BOILER IS INSTALLED.

** RECOMMENDED CLEARANCE BETWEEN WALL SURFACE AND BACK OF VENT CAP IS 6". MINIMUM ALLOWABLE CLEARANCE IS 3".

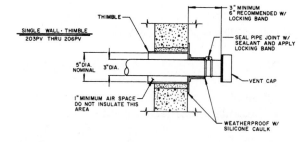

FIGURE 12-27 Vent pipe requirements through combustible and noncombustible exterior walls. (Courtesy of Burnham Corp.)

Following are some general rules to heed when either installing or inspecting pulsed combustion vent and supply piping.

1. Pipe specifications supplied by the manufacturer must be closely followed. These specifications include the maximum length of pipe runs based on the diameter of the pipe being used in the installation.

2. There must be no low spots in either the supply or exhaust pipe.

3. Since a certain amount of vibration is produced as a by-product of the pulsed combustion process, vibration-eliminating hangers are required to support the piping.

4. In geographical areas where outdoor temperatures can fall below 15°F, special venting procedures may be required to eliminate frost. Consult the manufacturer's recommendations for special fittings and installation procedures that must be followed in these instances.

5. Where possible, always vent pulsed combustion units through either the rear of the building or the roof. If exhaust noise is objectionable, supplemental mufflers should be installed.

Venting systems on all types of gas-fired appliances should be periodically inspected. Any indications of deteriorating piping or connections should be remedied. The proper switching action of blocked vent and spill switches should be verified. If there is any question or doubt as to the reliability of these safety devices, they should be replaced. Remember, the venting system removes potentially lethal substances from inside the home or building. If the venting system fails for any reason, the consequences can be deadly. So, when in doubt—inspect, and fix or replace.

DORMANT CHIMNEY INSTALLATION

1. When running piping up through an existing chimney, be sure chimney and flue are dormant; that is, not connected to a fireplace, water heater or any other heating appliance. In Canada, the exhaust vent shall not be run through interior part of an open chimney unless the ex-haust vent is insulated.

2. Air intake pipe termination must be as shown in Figure 4.2, with a 180° ell installed on air intake pipe.

3. Exhaust pipe termination must be 12" longer than air intake pipe to prevent exhaust recirculation.

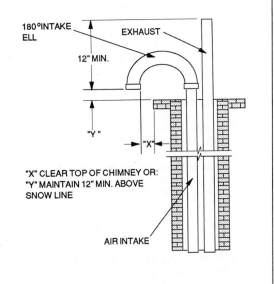

180°INTAKE ELL   EXHAUST

12" MIN.

"Y"   →"X"←

"X" CLEAR TOP OF CHIMNEY OR:
"Y" MAINTAIN 12" MIN. ABOVE
SNOW LINE

AIR INTAKE

FIGURE 12-28 Venting a pulsed combustion appliance into an unused chimney. (Courtesy of Hydrotherm, a Mestek company)

CHECKING GAS INPUT AND PRESSURE RATES

For the proper and safe operation of all gas appliances, it is important that the gas input rate and maximum inlet pressure shown on the rating plate of the appliance not be exceeded. When making these measurement adjustments, be sure that all other appliances that operate from the same gas meter are turned off. All adjustments are made while the appliance is firing, except where noted below.

Pressure Regulation

To adjust the gas input pressure, a water manometer or water column gauge must be connected to a shutoff cock, which is installed upstream from the manometer in the 1/8-inch pipe tapping in the boiler or furnace manifold (figure 12-30).

The manometer must be installed while the appliance is off. Installation of the gas cock upstream of the manometer or water column gauge allows the gas pressure to be increased gradually. This helps to avoid blowing any water out of the manometer due to a sudden rush of gas into the instrument. Before making any adjustments, the manufacturer's specifications should be consulted.

For LP gas, the regulator on the gas valve is adjusted to 10 inches w.c. For most gas valves, turning the adjustment screw clockwise increases the gas pressure, and vice versa.

The procedure for natural gas pressure adjustment is similar to that for propane, except that the manifold pressure is usually adjusted to 3 1/2 inches w.c. Minor pressure

THROUGH-THE-WALL INSTALLATIONS

CONCENTRIC VENT TERMINAL

NOTICE: Concentric vent terminal may be used for through-the-wall venting only. It cannot be used for dormant chimney or through-the-roof venting.

NOTICE: Concentric vent terminals are for residential installations only.

For through-the-wall venting for residential applications only, a concentric vent terminal is recommended. Follow installation instructions packaged with Concentric Vent Terminal Kit.

However, where the concentric vent terminal will not fit, use separate air intake and exhaust vent terminals, as shown below.

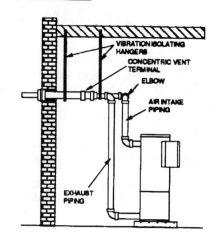

SEPARATE VENT TERMINALS

Air intake/exhaust terminations must be separated 18" minimum to 8-ft maximum ON THE SAME WALL. Exhaust outlet must be installed ABOVE or DOWNWIND from air intake to prevent exhaust recirculation. If mufflers are required, follow installation instructions packaged with the muffler sets.

FOR INSTALLATIONS USING 2" & UNDER PIPE

For exhaust, install vent pipe, plus vent terminal plate, plus 12" exhaust extension. For air intake, install vent pipe to within 36" of outside wall...add appropriately-sized reducer coupling...complete run with 3"x36" air intake pipe...and install vent terminal plate.

FOR INSTALLATIONS USING 3" PIPE

For exhaust, install vent pipe, plus vent terminal plate, plus 12" extension. For air intake, install vent pipe plus vent terminal plate.

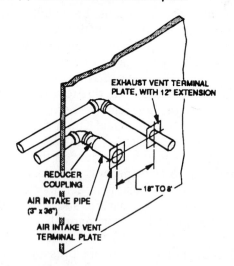

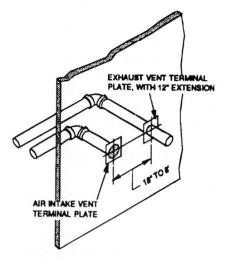

FIGURE 12-29 Venting a pulsed combustion appliance through an exterior wall. Note the use of vibration dampers to reduce noise during system operation. (Courtesy of Hydrotherm, a Mestek company)

adjustments may have to be made to ensure proper gas flow, a description of which follows.

Determining Gas Input

Since gas input to any appliance is a function of the specific gravity, heating value, and volume of gas delivered, all three factors must be known before making any calculations. Specific gravity and heating value figures may be obtained from the local utility company. After initial pressure adjustments have been made to the gas valve, the appliance is fired and the gas meter should be clocked for 3 minutes. Once the volume of gas flowing to the appliance, as determined from the gas meter, is known, the following formula is used to compare the actual gas reading with the theoretical flow of gas required by the appliance.

$$\text{Gas input ft}^3/3 \text{ min} = \frac{\text{Btu/hr.}}{\begin{array}{c}\text{Heating} \\ \text{value of} \\ \text{gas,} \\ \text{Btu/ft}^3\end{array} \times 20 \times \begin{array}{c}\text{Multiplier} \\ \text{from table} \\ \text{below}\end{array}}$$

Specific Gravity	Multiplier
.50	1.10
.55	1.04
.60	1.00
.65	0.96
.70	0.93

To illustrate the use of this formula, let us assume that a warm-air furnace carries a plated rating of 120,000 Btu/hr. Information from the gas company tells us that the utility-supplied gas has a Btu rating of 1050 Btu/ft^3 and a specific gravity of .65. To determine the proper volume of gas to be delivered to the furnace, we substitute these numbers into the equation as follows:

$$\text{Gas input} = \frac{120,000}{1050 \ \times \ 20 \ \times \ 0.96}$$

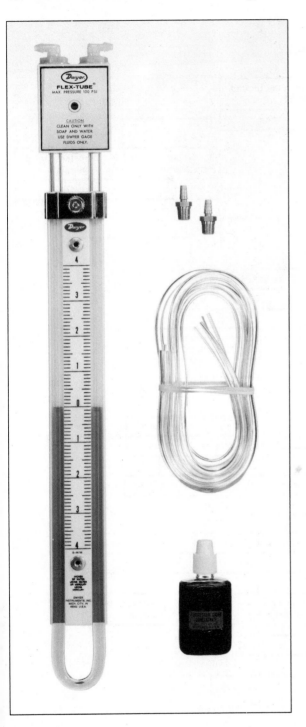

FIGURE 12-30 A manometer of the type used to measure main gas valve pressures. (Courtesy of Dwyer Instruments, Inc.)

NO SPARK

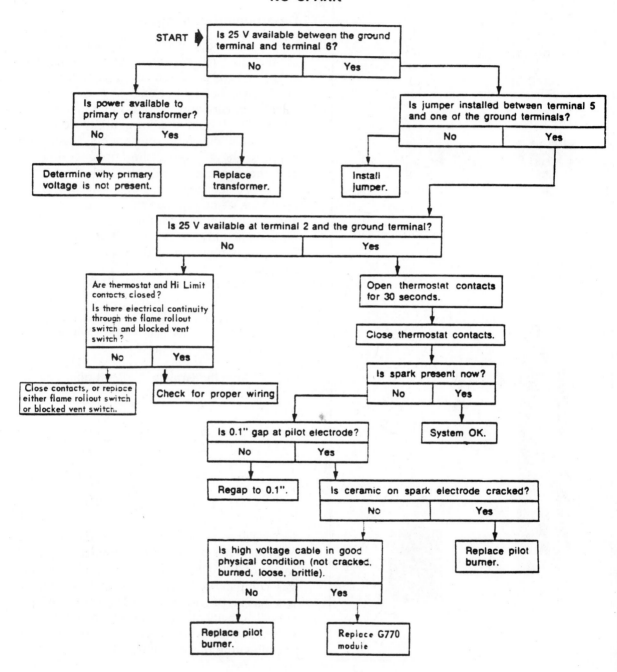

FIGURE 12-31 Troubleshooting analysis chart. (Courtesy of Burnham Corp.)

Type of Gas	Heating Values, Btu/ft³	Specific Gravity	Orifice/Drill-Size No.
Natural Gas (Manifold presssure 3.5 inches w.c.)	800	.6	40
	900	.6	41
	1000	.6	42
	1100	.6	43
Propane gas (manifold pressure 10 inches w.c.)	2500	1.53	54

$$\text{Gas input} = \frac{120,000}{21,000 \times 0.96}$$

$$\text{Gas input} = \frac{120,000}{20,160}$$

$$\text{Gas input for 3 minutes} = 5.95 \text{ ft}^3$$

Thus, for proper operation, this furnace requires 5.95 ft³ of natural gas for a 3-minute period if the gas flow to the furnace is to be exactly calibrated to the factory Btu output rating of the unit. This figure can now be checked with the actual gas reading obtained by clocking the gas meter. Minor adjustments to change gas volume can be made to the pressure regulator as long as the total pressure difference is less than .3 inch w.c. For larger changes, it will probably be necessary to change the gas orifice, by either physically replacing the orifice or drilling it out with the next largest drill size if appropriate. Consult the manufacturer's recommendations regarding specific procedures to use in this instance. The orifice/drill-size numbers appropriate for both natural and LP gas based on a specific gravity of .6 are listed at the top of this page.

TROUBLESHOOTING LOGIC

The troubleshooting logic to use in gas-fired appliance manipulation follows a sequence-of-operation approach. Sequence A takes place, which enables B and C to happen, and so on. Many different types of analysis charts are available to assist the technician in pinpointing control and operating problems. Figure 12-31 is one type of flowchart used for troubleshooting gas-fired furnaces with electronic spark ignition.

Regardless of the type of troubleshooting chart used to locate and remedy operating problems, the technician will find it a valuable diagnostic aid. Most gas-fired appliances come with some reference to troubleshooting techniques, usually packaged with the owner's operating manual. They can also be obtained from the original manufacturer of the appliance. When requesting this information, the Btu rating, model number, and serial number of the unit must be included.

SUMMARY

The material covered in this chapter dealt with the characteristics of fuel gases and the conventional methods of gas combustion. Gas is normally combusted under atmospheric conditions by using a pilot flame to ignite the main burners in a combustion chamber, or by using a power burner assembly that is similar in operating characteristics to that of fuel oil burners.

Gas-fired appliances are categorized according to their combustion characteristics: standing pilot ignition with conventional thermocouple and thermopile generators, electronic ignition in both the noncondensing and condensing modes, pulsed combustion mechanisms, and power burner devices. While some of these combustion technologies, such as standing pilot ignition, are gradually being replaced by electronic ignition units, each offers unique operating characteristics within the overall range of gas-fired devices. The venting systems, like the combustion technologies, are designed for use with specific types of firing applications. Knowledge of gas pressure and input rates is also a link in the chain of understanding the theory and practical applications of gas-fired combustion.

PROBLEM-SOLVING ACTIVITIES

Troubleshooting Case Studies

A. A homeowner complains that a furnace with a standing pilot fails to ignite the main burners. An inspection of the unit indicates that gas is flowing to the furnace and that the pilot light is burning against the thermocouple. Indicate in proper sequence the troubleshootiong steps to be used to isolate the problem.

B. A technician suspects that a defective blocked vent switch is turning off the boiler circuit. Devise a troubleshooting strategy, listing the sequence of procedures required to isolate the blocked vent switch in order to determine whether the switching is causing the operating problems.

C. A family is faced with the necessity of purchasing a new furnace and has been reading newspaper advertisements regarding new, high-efficiency condensing warm-air furnaces. They feel nervous about switching from the old, standing pilot system to this new type of furnace. Prepare a presentation that highlights the advantages of electronic ignition, as well as the efficiencies inherent in condensing appliances, as compared with the family's old, standing pilot system.

Review Questions

1. Explain the differences in characteristics between natural and propane fuel gas. Which gas is considered safer under normal operating conditions, and why?

2. What is the difference between a thermocouple and a thermopile? In what situation would a thermopile have an advantage over a thermocouple ignition system?

3. List three distinct advantages that electronic ignition systems have over standing pilot ignition systems.

4. How do induced-draft appliances differ in design and operating characteristics from natural-draft systems? Which system is inherently more efficient, and why?

5. Identify two advantages and two disadvantages of condensing appliances as compared with noncondensing units.

6. A gas-fired steam boiler has a plated rating of 140,000 Btu/hr. Information supplied by the utility company indicates that its natural gas has a specific gravity of .65 and a heating capacity of 975 Btu/hr. From this information, calculate the proper gas flow to the boiler for a 3-minute period in order to match the plated rating to the proper gas flow.

CHAPTER 13

Hydronic and Steam Heating Controls

In a hydronic or hot-water heating system, water is heated in a boiler and circulated through a piping system that incorporates a series of finned baseboard heating convectors. (See figure 1-3B for a sectional view of a finned baseboard convector.) In the baseboard units, heat from the water is transferred by conduction to the baseboard fins. Room air passing over the fins convects this heat into the room. The water continues to circulate back to the boiler, where it is reheated. This process continues as long as there is a call for heat from the room thermostat. The operation is essentially the same whether the heating system is installed in a residential or a commercial application.

BASIC SYSTEM COMPONENTS

Basic system components, which will be discussed in greater detail throughout this chapter, include the following.

Boiler: Boilers to heat water are manufactured from either cast iron or steel and are fired by conventional fuels, such as natural gas, propane, fuel oil, wood, and/or coal. Boilers are classified as either wet base or dry base. Wet-base boilers have water chambers surrounding both the top and bottom of the combustion chamber, whereas in dry-base boilers, the water chamber is usually above the combustion chamber. Pressure in residential hydronic boilers is limited to a maximum of 30 PSI.

Commercial boilers operate at higher pressures, depending on the application.

Pressure/temperature relief valve: All boilers are required by both national and local codes to have a protective device to vent excess pressure and temperature when and if they occur. These valves are installed on the top of the boiler and are piped to within 6 inches of the floor to prevent personal damage when and if the valve discharges.

High-limit controls: The high-limit control is a safety device to limit the maximum temperature of the water in the hydronic boiler (250 degrees is the absolute maximum!). These controls operate in direct response to water temperature, and are wired to turn off the fuel burner mechanism if the preset high-temperature limit is reached.

Expansion tank: In all hydronic systems, the water expands as it is heated. An expansion tank provides space for this increased volume of water by offering a cushion of compressed air that expands and contracts during the normal operating cycles of the heating system. Expansion tanks are sometimes referred to as compression tanks.

Pressure-reducing fill valve: All hydronic systems are connected to a source of cold water. Since normal building pressure is much higher than the operating pressures of most hydronic systems, a pressure-reducing fill valve is installed in the water line to reduce the incoming pressure of the feed water to that of the cold-start pressure of the heating system. Most fill valves contain either a back-flow preventer or a check valve to prevent heating system water from flowing back into the building water supply. Fill valves either can be stand-alone valves or can be incorporated into the expansion tank assembly.

Flow control/check valves: These valves prevent the flow of water in a heating circuit when the circulator is not running. They are used most often with multiple circulator systems where water would circulate through a stopped circulating pump by thermosiphon action .

Air vents: Air vents are installed at the high points in the heating system as well as in conjunction with an air scoop to remove trapped air from the system. The vent operates on a float-valve principle. As the valve fills with air, the float level drops, opening a needle valve that allows the air to be vented from the system. As the air escapes, the float valve rises, sealing the needle valve against a valve seat. The operation of this valve is fully automatic as air is trapped and vented from the system. Manual air vents are also installed on many systems to remove trapped air from the high points in the piping system.

Air scoop: An air scoop is generally used in conjunction with an air vent to removed trapped air from the heating system. The air scoop is a cast iron fitting that is installed at the top of the boiler. All water in the heating system must flow through the air scoop, which contains a series of internal baffles that catch and direct the trapped air bubbles up into a chamber in the scoop. If an air vent is installed at the top of the air scoop, the air is automatically purged from the system when enough has accumulated to operate the automatic air vent.

PIPING AND SYSTEM CONFIGURATIONS

Hydronic heating systems are arranged in a variety of different piping configurations, de-

pending on the type and size of the building construction, as well as the flexibility required of the heating system. The basic piping designs used in hydronic applications follow.

Gravity Feed and Return

Gravity systems rely on the fact that water changes density and gets lighter as it is heated. This changing density enables the circulation of the water throughout the heating system. No circulating pump is used in this arrangement, therefore, pipes used for both the feed and return lines are oversized in order to minimize internal frictional circulating losses (minimum pipe size on all lines is 1 1/2 inches to 2 inches). Note the use of an expansion tank in the typical gravity system in figure 13-1.

Forced-Circulation Supply Loop

The supply loop configuration utilizes a main supply pipe that carries heated water to and from the boiler. A circulating pump is used to move the water throughout the piping system. Terminal heating units, each with its individual shutoff and air bleed valves, tap off this main supply pipe. The supply loop system and system circulator are shown in figures 13-2A and 13-2B.

Note from figure 13-2A that the supply pipe also acts as a direct return line to the boiler. In the supply loop system, the terminal heating units at the end of the piping circuit tend to be the coolest, making it difficult to balance the system to deliver even heating. To overcome this design deficiency, a two-pipe return system is usually employed. Two-pipe systems can be of either the direct or reverse return configuration, both of which are shown in figure 13-3.

In the direct return configuration (figure 13-3A), the terminal heater that is closest to the boiler also has the shortest return line. Balancing the heating system becomes difficult in this arrangement since the heating circuits have different internal frictional losses because the length of the pipe runs are different. To overcome these shortcomings, the reverse return circuit illustrated in figure 13-3B is gener-

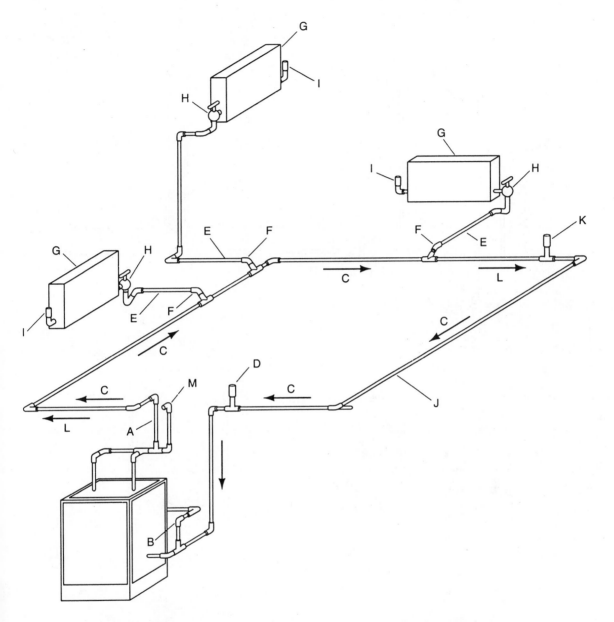

A Supply pipe
B Return pipe
C Pipe pitches downward in direction of
   arrow not less than 1/4 inch in 10 feet
D Main vent (See Figure 27 for details)
E Branches (See Figure 25 for details)
F 45 degree el (See Figure 25 for details)

G Heat distributing units
H Radiator valve
I Air vent
J Return line
K Main vent located here
L Direction of flow of water
M Make-up water line

FIGURE 13-1 Gravity feed hot-water heating system. All piping maintains a positive slope back toward the boiler. Oversized piping minimizes friction losses to allow for free flow of water. (Courtesy of the Hydronics Institute)

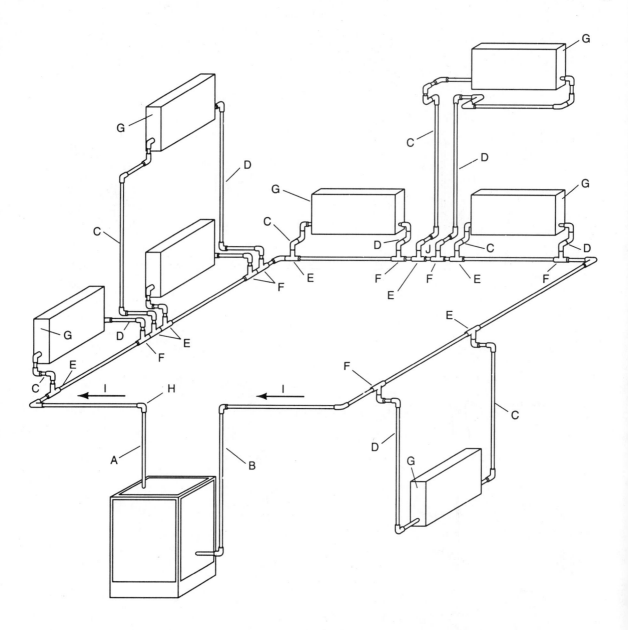

A   Supply pipe
B   Return pipe
C   Supply branches
D   Return branches
E   If one pipe fitting is designed for supply
     connection to heat distributing units,
     install here

F   If one pipe fitting is designed for return connection from
     heat distributing units, install here
G   Air vent on each unit
H   Flow control valve required if an indirect water heater
     is used and optional if an indirect water heater is not used
I   Direction of flow of water
J   Not less than 6 inches

FIGURE 13-2A One-pipe forced hot-water supply loop heating system. The supply pipe serves as both the boiler feed and return line. (Courtesy of the Hydronics Institute)

FIGURE 13-2B (Courtesy of the Hydronics Institute)

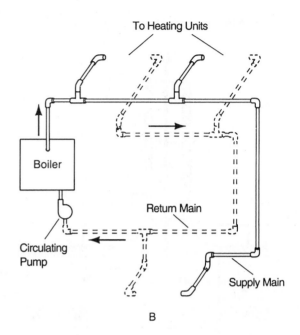

FIGURE 13-3B Two-pipe reverse return hot-water system. (Courtesy of the Hydronics Institute)

ally used. In this design, all pipe runs are approximately the same length. This enables even heating to be delivered from each terminal heater. Both direct and reverse return supply loop systems feature minimum pipe sizes of 1 inch to 1-1/4 inches on both the supply and

return trunk lines. Individual terminal heaters take off from the supply lines pipes in appropriate sizes for the heaters, usually 1/2 inch to 1 inch.

Special tee fittings are used for the feed and return taps from the supply loop. These fittings are designed to supply each terminal heater with the proper amount of water, while minimizing frictional losses by taking advantage of the forced-circulation pattern in the supply pipe. These tee fittings are only used in supply pipe systems and not in two pipe piping circuits. The construction of the fitting allows most of the water to flow normally through the straight leg of the tee, while diverting the proper amount of water up the tee and into the heating unit. A fitting of this type is illustrated in figure 13-4.

Forced-Circulation Series Loop

Series loop heating systems are the most popular type of forced-circulation installation, primarily because of their ease of installation. A typical series loop system is shown in figure 13-5.

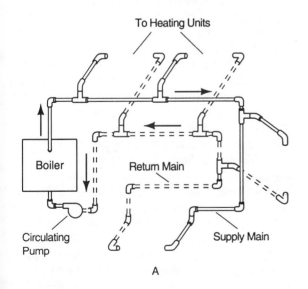

FIGURE 13-3A Two-pipe direct return hot-water system. (Courtesy of the Hydronics Institute)

Feed to Baseboard from Supply Loop   Return to Supply Loop from Baseboard

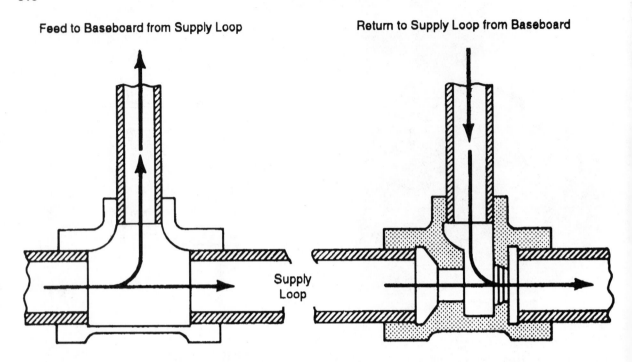

Supply
Loop

FIGURE 13-4 Flow diversion fittings used in feed and return lines of a supply loop heating system. The design of the fitting ensures an adequate supple of water to the loop along with complete retrun to the loop. (Courtesy of Taco, Inc.)

If the heating system is designed for a single zone, it may become necessary to split the piping into one or more circuits, depending on the heating capacity of the pipe sizes being used and the length of the pipe run. Should a second circuit become necessary, the design of the series loop system would resemble the piping in figure 13-6.   Note that both circuits empty into a common return line, which should be properly sized to accommodate the flow rates from both heating circuits.

Forced-Circulation Zoned Systems

As the need for additional circuits arises, it becomes advantageous to utilize a multiple-zone configuration in the heating system. Zoning adds a great deal of complexity to the system owing to the use of multiple thermostats, switching relays, circulating pumps, and/or zone valves.

Commonly, a hydronic system is zoned either by installing zone valves that open or close the piping path in each zone, or by using separate circulators that control the flow of water through each separate zone. A typical zone-valve boiler configuration is shown in figure 13-7. Figure 13-8 illustrates a zoning design that employs multiple circulators.

Both zoning methods accomplish the same design objective, but each is distinctive in terms of individual components and control applications. Multiple circulator installations are more expensive than their zone-valved counterparts because switching relays and circulators must be duplicated for each heating zone in the system. However, this duplication offers more flexibility than does a single-circulator/multiple–zone-valve installation. If one of the circulators fails, heat will still be delivered to the remaining zones through the remaining operating circulators.

Given an understanding of the basic piping configurations available in typical hydronic installations, we proceed to an examination of the control functions associated with these various configurations.

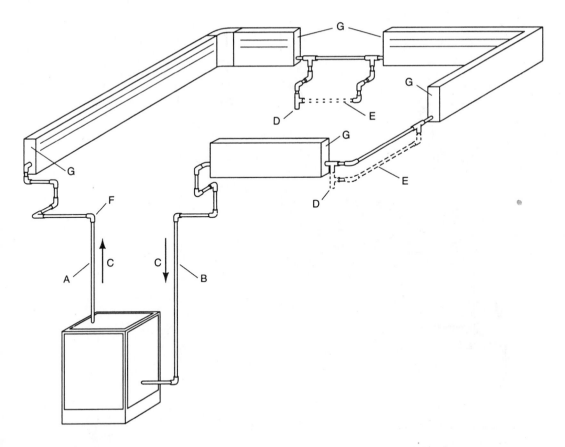

Note: Allow for expansion of baseboard units in accordance with the manufacturer's recommendations.

A Supply pipe
B Return pipe
C Direction of flow of water
D Nipple and cap installed in tee to provide for drainage

E Alternate connection between units, when required
F Flow control valve required if an indirect water heater is used and optional if an indirect water heater is not used
G Air vent on each unit, if required

FIGURE 13-5 Series loop baseboard heating system. In this single-circuit loop, water travels from the boiler through each of the terminal heating units in series and returns to the boiler for reheating of the water. (Courtesy of the Hydronics Institute)

CONTROL OF PRESSURE AND TRAPPED AIR

When a heating system is first filled with either water or antifreeze, air from within the system initially as well as air mixed in with the water itself, is trapped in the system. This air, if not removed, will accumulate, forming sizable air pockets that can prevent the system from operating normally. Also, as this charge of water will expand as it is heated, provision must be made for accommodating this volumetric expansion of water, as well as for a mechanism to relieve the pressure if it should reach an unacceptably high level. Peripheral control components are designed to handle each of these operating conditions.

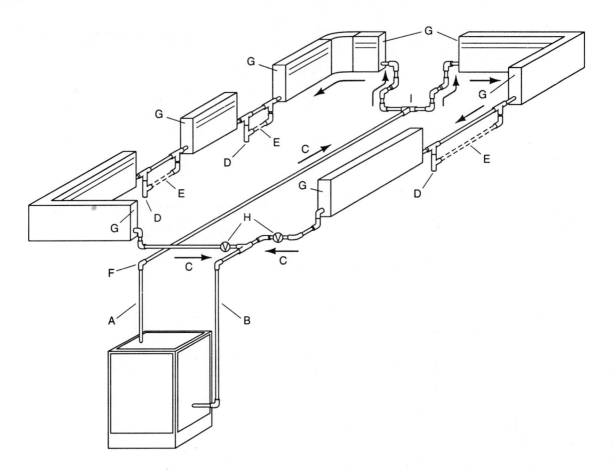

Note: Allow for expansion of baseboard units in accordance with the manufacturer's recommendations.

A Supply pipe
B Return pipe
C Direction of flow of water
D Nipple and cap installed in tee to provide
   for drainage
E Alternate connection between units, when
   required

F   Flow control valve required if an indirect water
    heater is used and optional if an indirect water
    heater is not used
G Air vent on each unit, if required
H Balancing cocks
I   Circuit splits here

FIGURE 13-6 Two-circuit series loop heating system. Additional circuits are designed into the series loop system when the heating requirements of the system exceeds the heating capacity of the piping in a particular circuit. (Courtesy of the Hydronics Institute)

Air-Purging Controls

As the heating system water is heated, trapped air is released from the water, and travels throughout the system until it either collects at a high spot in the piping circuit or is removed from the heating system. An air purger assembly (figure 13-9) is used to trap this air in one location, and then vent it through a float vent.

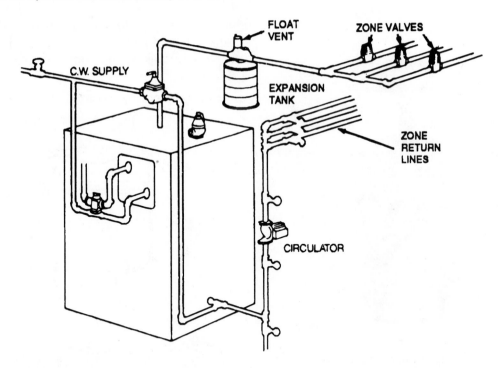

FIGURE 13-7 Zoning with electric zone valves. (Courtesy of Taco, Inc.)

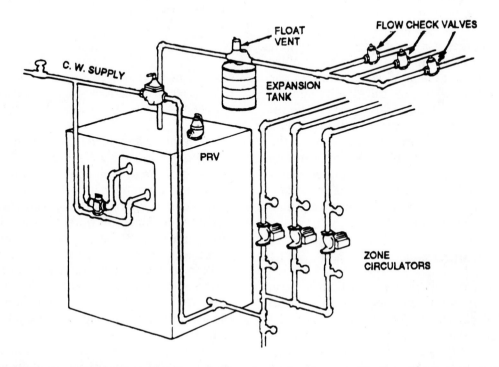

FIGURE 13-8 Zoning with circulators. Note the use of flow check valves to prevent the flow of heated water in those zones not calling for heat. (Courtesy of Taco, Inc.)

FIGURE 13-9 Air purger and float valve assembly. The expansion tank connects into the bottom of the air purger. The assembly directly between the air purger and expansion tank is an automatic fill valve for the cold-water supply. (Courtesy of Amtrol, Inc.)

The air purger is a one-piece cast assembly with an internal baffle that diverts air bubbles up to the float assembly, while allowing the major flow of water to travel continuously throughout the piping with a minimum of resistance. When enough air collects in the float vent, the float drops, opening the valve on the top of the vent assembly. System pressure forces this trapped air out of the vent. As water is forced into the vent assembly, the float rises, sealing the escape valve. This action is continuous during system operation.

Pressure Controls

Note that the air purger and float vent in figure 13-9 are shown connected to a pressurized expansion tank. This tank contains an expandable flexible diaphragm that separates water in the heating system from an air charge in the pressure tank. Figure 13-9 also shows an automatic pressure-reducing and fill valve in-

stalled between the air purger and expansion tank. Figure 13-10 details the operation of the expansion tank during system operation.

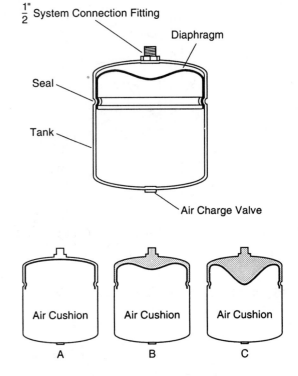

FIGURE 13-10 Operation of a diaphragm-type expansion tank during normal heating system operation. (Courtesy of Amtrol, Inc.)

When the system is first filled with water (A), no water enters the pressure tank, since the pressure of the system equals the air pressure in the cushion in the expansion tank. When the system comes up to operating temperature, the expanded volume of water pushes into the expansion tank (B). If the water temperature increases, the diaphragm flexes against the air cushion under the expanded volume of water. As the system water cools and the volume of water is reduced, the air cushion in the expansion tank pushes its water back into the system.

Figure 13-11 shows a cross section of the air purger, float vent, and expansion tank assembly during normal continuous system operation.

In conjunction with the expansion tank, a pressure-relief valve, set to discharge at 30 PSI, is installed in an appropriate tapping on the boiler as an added safety feature. This valve must be piped to within 6 inches of the floor to prevent personal injury if the valve should discharge during heating system operation.

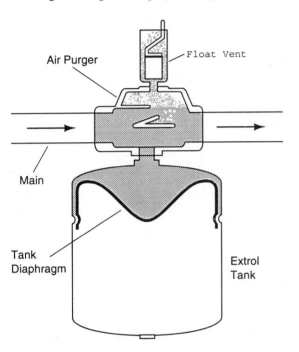

FIGURE 13-11 Continuous operation of float vent, air purger and expansion tank assembly. (Courtesy of Amtrol, Inc.)

FIGURE 13-12 Pressure-relief valve used in hydronic boilers. The lever on top of the valve is used for manual release of the valve. The outlet must be piped to within 6 inches of the floor to prevent personal injury should the valve discharge. (Courtesy of Watts Regulator Co.)

BOILER OPERATING SEQUENCES

Controls for hydronic heating systems that include zone valves, circulating pumps, and multiple circulators can be looked at separately from those controls that relate to specific burner functions. Burner controls are discussed in Chapter 11 (oil-fired technology) and Chapter 12 (gas-fired technology). References to specific fuels are made where appropriate.

Operating Logic

The block diagram of figure 13-13 illustrates the control logic of a single-zone hydronic heating system (oil is used for illustration purposes only).

This logic forms the basis of all conventional hydronic systems. Additional zones and peripheral components are modifications of this basic logic. In all instances, the room thermostat initiates the heating process. Assuming proper water temperature in the boiler, the burner will be energized and heat will be delivered to the building as long as it is called for by the thermostat.

All control functions operate from the main boiler aquastat. This switching device integrates the control of boiler water temperature, the

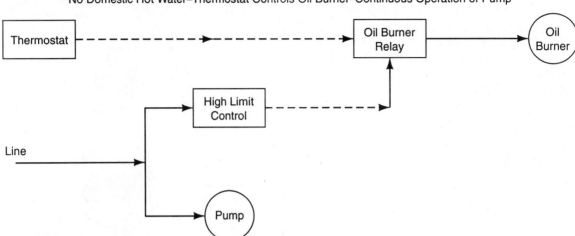

No Domestic Hot Water–Thermostat Controls Oil Burner–Continuous Operation of Pump

1. Thermostat controls only the oil burner.

2. If boiler water temperature reaches High Limit Control setting, the oil burner is shut off. A drop of approximately 20 degrees in boiler water temperature will start the oil burner again. Local ordinances may require a second High Limit Control (set 10° F higher).

3. Pump operates continuously.

FIGURE 13-13 Control logic of single-zone hydronic heating system. (Courtesy of the Hydronics Institute)

thermostat circuit, and electric power flow. Boiler aquastats are generally of the immersion type that utilizes an oil- or gas-filled capillary tube for sensing water temperature. The liquid or gas in the tube expands or contracts in response to changing water temperatures in the boiler. The expansion and contraction in the capillary tube actuate the switching mechanisms, which is often a bellows and snap switch unit in the aquastat. The capillary tube is installed in an immersion well, which is threaded into a tapping in the boiler. The well is filled with a heat-conducting grease that maximizes heat transfer from the boiler water to the capillary tube. An aquastat and immersion well are illustrated in figure 13-14.

Oil and Gas Control Aquastats

Although oil- and gas-fired boilers sometimes have different aquastats to control burner function, depending on the voltage require-

ments of either the oil burner or gas valve, the aquastats function in essentially the same way with either fuel. Figure 13-15 shows a typical hydronic aquastat both with and without the exterior case. Oil burner aquastats usually have built-in F-F (cad cell) terminals. Gas output terminals are low voltage (24 volts). Oil-fired output terminals are 110 volts for typical oil burner motors.

The remaining primary components of the heating system are connected to the aquastat as shown in figure 13-16. This illustration depicts a multiple-circuit zone-valve configuration that uses only one circulating pump. Based on the heating requirements, a single thermostat may replace the multiple-zone thermostats as illustrated. Terminals F–F are for connections to the cad cell in the oil burner. In order to understand the special features of this type of control, consider the operating sequence of the aquastat:

FIGURE 13-14 Aquastat relay with immersion well. (Courtesy of Honeywell, Inc.)

1. The room thermostat calls for heat, closing the relay across the T–T terminals.

2. Assuming that the proper resistance is present across the F–F terminals from the cad cell and the temperature of the boiler water is below the fixed differential (usually between 10 and 15 degrees below the high limit), the relay initiates burner operation.

3. If the water temperature is above the low–limit circulator setting, the relay contacts that supply power to the

circulating pump will close, initiating circulator operation.

4. The burner will continue to run until the water reaches the high-limit setting. At this point, the burner will turn off.

5. The circulator continues to operate as long as the thermostat continues to call for heat and the water temperature remains above the low-limit setting.

6. The burner will turn on again after the water temperature has dropped below the high-limit differential. For example, if

FIGURE 13-15 Aquastat of the type used for oil- and gas-fired appliances. (Courtesy of Honeywell, Inc.)

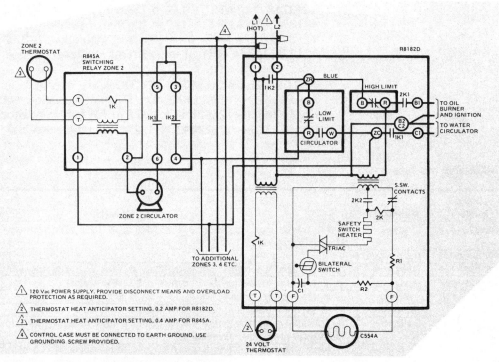

FIGURE 13-16 Multiple-circuit zone-valve configuration showing connections to aquastat controller. (Courtesy of Honeywell, Inc.)

the high limit is set at 200°F and the aquastat has a high-limit fixed differential of 15°F, then the burner turns off at 200°F, and, assuming a continued call for heat, turns on again when the water temperature drops to 185°F.

Domestic Hot-Water (Tankless Coil) Heating Controls

The use of a tankless coil for heating domestic hot water in a boiler offers both advantages and disadvantages for the building owner throughout the year. The tankless coil is made up of a long length of tubing (usually between 1/2 and 3/4 inch in diameter in residential boilers, with larger coils in commercial units depending on the amount of water to be delivered) that is fabricated into a housing and inserted into an opening in the boiler. A tankless coil installed in a residential boiler is illustrated in figure 13-17A. If the boiler lacks an internal tankless coil, an external coil can be

plumbed into the system as an additional heating zone. This type of installation is shown in figure 13-17B.

During the heating season when the boiler is maintaining a heated reservoir of water most of the time, the use of a tankless coil for indirect water heating offers efficiencies of approximately 50–60%. This efficiency figure is comparable to that for a conventional gas-fired water heater, and higher than that for electric water heaters (oil-fired water heaters usually average efficiencies in the neighborhood of 75% to 85%). This operating efficiency, along with the advantage of not having to deal with a separate water heating appliance, has resulted in the widespread use of tankless coils.

The disadvantage of a tankless coil configuration arises from the fact that the boiler must continue to operate during the nonheating season solely for the purpose of providing domestic hot water. The off-season efficiency of this process is only in the neighborhood of 15–20%. The use of the coil is, therefore, an

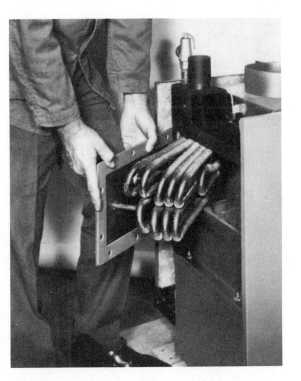

FIGURE 13-17A Tankless coil installed in a residential hydronic boiler. (Courtesy of Slant/Fin Corp.)

operational trade-off in most instances, with the savings realized in the heating season offset by additional costs during the months when the boiler would otherwise be idle.

The aquastat controller in figure 13-16 has a built-in switching capability, through the low-limit controller, to maintain a fixed low-limit setting throughout the year. This would be needed for indirect domestic hot-water heating via a tankless coil. In this way, even when the boiler is not used for space heating, the water temperature remains sufficiently high to provide hot tap water. The low-limit dial in figure 13-16 provides for an adjustable differential that is measured from 10 degrees below the low-limit setting. This is a fixed differential cut-in point built into the aquastat, and is active even when there is no call for space heating from the room thermostat. For example, assume that we adjust the control to a low-limit setting of 140°F with the adjustable differential dial set at 25 degrees. This setting provides two operating

platforms or levels, one for the circulator and one for the fixed low-limit differential. In this example, the circulating pump will not be energized unless the boiler water temperature is above 140°F, the temperature setting used during the normal heating season. Also, if the boiler water temperature falls to 130°F (the low-limit set point minus the 10-degree fixed cut-in point), the burner will be energized. In this instance, the burner will continue to operate until the water in the boiler has been heated to 155°F, which incorporates the +25-degree differential measured from the cut-in point of 130°F.

It should be noted that tankless coils are not normally installed in conventional gas-fired boilers, since the hot-water recovery rate of gas-fired boilers is considered by many heating contractors to be too low to assume the additional heating load imposed by the installation of the coil.

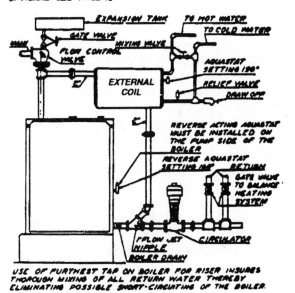

FIGURE 13-17B External shell and tube heat exchanger used to provide domestic hot water when a boiler lacks an internal tankless coil. (Courtesy of Everhot All Copper, Inc.)

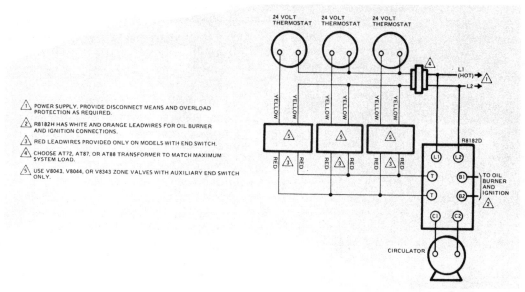

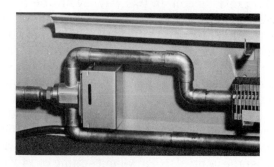

1. POWER SUPPLY. PROVIDE DISCONNECT MEANS AND OVERLOAD PROTECTION AS REQUIRED.

2. R8182H HAS WHITE AND ORANGE LEADWIRES FOR OIL BURNER AND IGNITION CONNECTIONS.

3. RED LEADWIRES PROVIDED ONLY ON MODELS WITH END SWITCH.

4. CHOOSE AT72, AT87, OR AT88 TRANSFORMER TO MATCH MAXIMUM SYSTEM LOAD.

5. USE V8043, V8044, OR V8343 ZONE VALVES WITH AUXILIARY END SWITCH ONLY.

FIGURE 13-18 Multiple-circuit control schematic using zone valves for each heating circuit. (Courtesy of Honeywell, Inc.)

Zone Valve Controls

Figure 13-18 illustrates multiple-circuit control using zone valves to control the flow of water in each heating circuit. Note from this schematic that a separate low-voltage (24-volt) transformer is used to power the multiple transformer/zone-valve circuit. Low-voltage controls are normally outside the range of code inspections, thus enabling factory-wired boilers to be installed without the need for additional electrical inspections. Low-voltage controls also offer a degree of safety to the building owner or operating engineer that high-voltage controls would lack. All connections to normal house current, including supply lines to the appliance, fall under existing electrical wiring codes.

A zone valve of the type used in most residential and commercial applications is shown in figure 13-19. These zone valves are used to control water flow through conventional hydronic systems. Most zone valves feature replaceable power heads, which allows the removal and replacement of defective motors while leaving the main body of the valve and the piping intact. The operating sequence for a typical zone-valve installation is as follows:

FIGURE 13-19 Zone valve used for water flow control. (Courtesy of Flair International, Inc.)

1. The room thermostat calls for heat, closing the specific electrical circuit in a heating zone and energizing the zone valve motor, which opens the zone valve.

2. When the zone valve motor opens all the way, a set of electrical contacts in the valve close, which completes the thermostat circuit and energizes the main aquastat relay.

3. Assuming that the proper resistance is present across the F-F terminals from the cad cell and the temperature of the boiler water is below the aquastat's high-limit fixed cut-in differential, the relay initiates burner operation.

4. The circulator relay is energized if the boiler water temperature is above the low-limit setting. The circulator operates as long as there is a call for heat from the thermostat, circulating water from the boiler through the open zone valve and throughout the heating circuit.

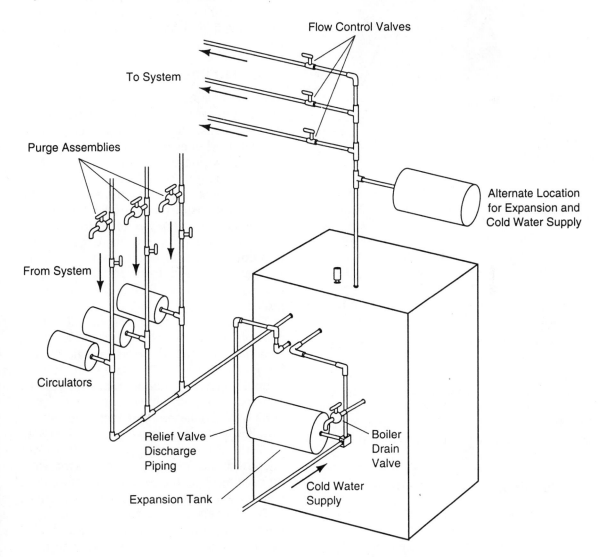

FIGURE 13-20A Multiple-circulator zoning configuration.

FIGURE 13-20B Circulator switching relay. (Courtesy of Honeywell, Inc.)

5. Water circulates through circuits in which the zone valves are opened. Zones in which there is no call for heat have no water circulation, since these zone valves remain closed. When the room thermostat is satisfied, the thermostat circuit opens and deenergizes the zone valve, which automatically closes under spring tension.

In multiple–zone-valve systems, multiple zone valves can be open in any combination. As long as one zone is calling for heat, the boiler will cycle through its normal heating sequence. When all of the room thermostats in the system have been satisfied, the boiler automatically reverts to a standby mode, cycling off the low-limit set point on the aquastat.

Multiple Circulator Zone Controls

A multiple circulator zoned system schematic is shown in figure 13-20A.

These systems use separate circulators, each controlled by a separate switching relay, as

illustrated in figure 13-20B. Note that the switching relay draws 120-volt power from the main boiler aquastat. In this illustration, the circulator switching relay draws power through the ZR and ZC terminals. This wiring arrangement makes use of the low-limit circulator control in the main boiler aquastat. This type of wiring enables the switching relay to energize the burner when the thermostat circuit is closed, but prevents the circulator motor from running unless the boiler water temperature is above the low-limit set point. Each switching relay contains a 24-volt transformer that powers the individual zone thermostat and relay circuitry.

Cold Start Design

In a cold-start configuration, which is sometimes factory configured on boilers without tankless coils, the boiler relay energizes the circulator whenever the thermostat calls for heat. The wiring diagram for the controller is shown in figure 13-21. Note from this illustration that some cold-start controls require that a separate aquastat or controller be installed if low-limit circulator protection is to be provided. The low-limit control prevents circulator operation until the boiler water reaches a minimum set temperature. However, some cold-start aquastats do contain built-in low-limit circulator protection.

The high-limit controller in the aquastat controls the burner circuit only. The circulator runs as long as there is a call from the thermostat for heat. The cold start design has certain limitations if additional zones are to be added to the heating system. For example, if additional circulators and switching relays are installed with no adaptations to the cold-start aquastat, then any time an additional zone is energized, the original zone circulator will also be running. This happens because some cold-start aquastats normally do not have thermostat (T-T) terminals, and so are energized from the thermostat terminals on the oil burner primary control. All zone circulators have their thermostat circuits wired in parallel with the primary

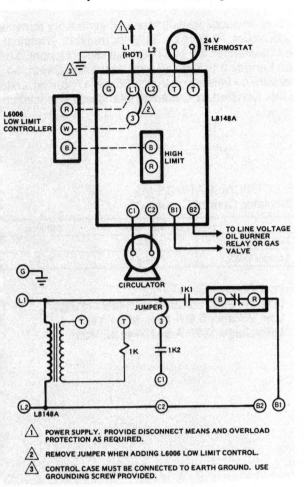

1. POWER SUPPLY. PROVIDE DISCONNECT MEANS AND OVERLOAD PROTECTION AS REQUIRED.

2. REMOVE JUMPER WHEN ADDING L6006 LOW LIMIT CONTROL.

3. CONTROL CASE MUST BE CONNECTED TO EARTH GROUND. USE GROUNDING SCREW PROVIDED.

FIGURE 13-21 Cold start wiring circuit. (Courtesy of Honeywell, Inc.)

control as well. In order to isolate the original zone circulator from the boiler aquastat, the circulator must be disconnected from the cold-start aquastat. It is then connected to a separate switching relay that is energized by the original zone thermostat. This configuration is shown in the schematic in figure 13-22.

COMMERCIAL BOILER SEQUENCING AND CONTROLS

Newer commercial installations utilize modular boiler installations for heating large buildings. In older systems, one large boiler was generally used at full output for space heating all year long, even though the full firing capability of the unit was needed less than 20% of the year. To overcome these inefficiencies, modular heating systems consisting of two or more smaller compact boilers are installed in series with one another. Using computerized controls, the modules are step-fired, which allows the system to use boiler capacity in proportion to outside temperatures and building heat loss. Figure 13-23 shows the heating load curve for a typical commercial building.

Note from this illustration how a modularized heating system can offer significant savings over a single large boiler. Also, the failure of one boiler does not mean loss of all building heat. The remainder of the modules are capable of meeting partial or almost complete heating requirements, depending on the outside temperature.

Figure 13-24A is a pictorial of a multiple commercial boiler design. These boilers feature heavier castings and gasketing material than used in residential units in order to stand up to the demands of the higher operating temperatures and pressures in commercial heating systems. A commercial modular boiler of this type is shown in figure 13-24B. Figure 13-24C illustrates the end section of commercial finned radiation. This type of convector features heavier fins and enclosures and a larger water pipe than do residential baseboard units.

Computerized controls are required for the efficient sequence firing of modular boilers. A control of this type, as shown in figure 13-25, continuously monitors outdoor temperatures, as well as heating system water temperatures, to determine how many boilers must fire in order to maintain building comfort. These controls are also capable of rotating the firing order of the boilers in the module (the first boiler on is the first one off). This feature equalizes the operating time of all of the boilers, and permits the planning of routine cleaning and maintenance procedures. Such controls can usually be added to so that if more boilers or features are required in the future, the control technology can accommodate these additions.

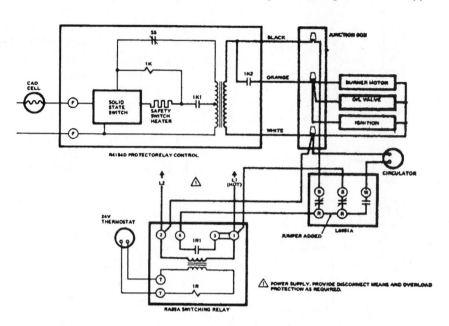

FIGURE 13-22 Multiple-circulator zoned wiring, including switching relays and burner primary. (Courtesy of Honeywell, Inc.)

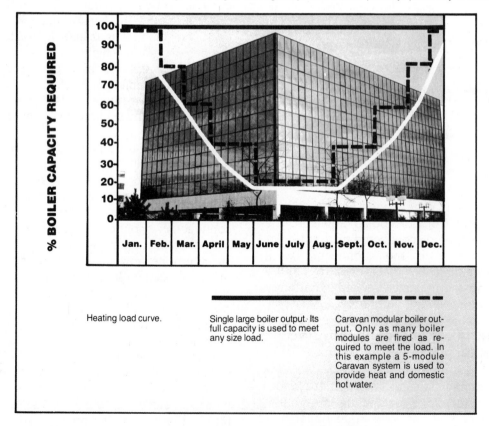

Heating load curve.

Single large boiler output. Its full capacity is used to meet any size load.

Caravan modular boiler output. Only as many boiler modules are fired as required to meet the load. In this example a 5-module Caravan system is used to provide heat and domestic hot water.

FIGURE 13-23 Heating load of a typical commercial building. (Courtesy of Slant/Fin Corp.)

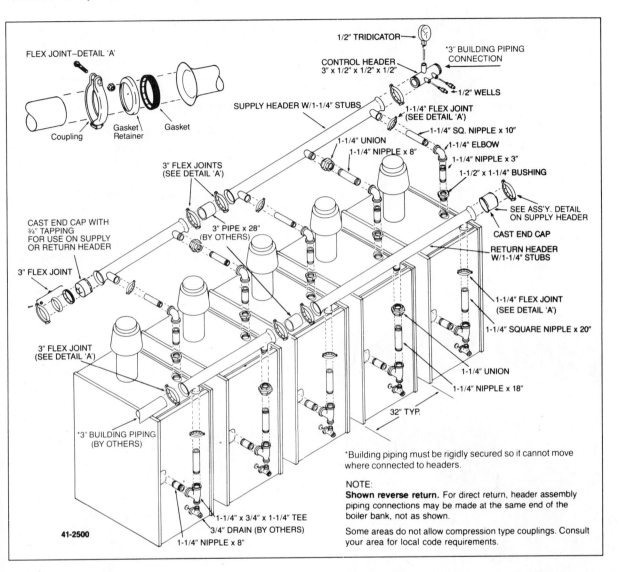

FLEX JOINT–DETAIL 'A'

Coupling   Gasket Retainer   Gasket

1/2" TRIDICATOR
*3" BUILDING PIPING CONNECTION
CONTROL HEADER
3" x 1/2" x 1/2" x 1/2"
1/2" WELLS
SUPPLY HEADER W/1-1/4" STUBS
1-1/4" FLEX JOINT (SEE DETAIL 'A')
1-1/4" SQ. NIPPLE x 10"
1-1/4" UNION
1-1/4" ELBOW
1-1/4" NIPPLE x 8"
1-1/4" NIPPLE x 3"
1-1/2" x 1-1/4" BUSHING
3" FLEX JOINTS (SEE DETAIL 'A')
SEE ASS'Y. DETAIL ON SUPPLY HEADER
CAST END CAP WITH 3/4" TAPPING FOR USE ON SUPPLY OR RETURN HEADER
3" PIPE x 28" (BY OTHERS)
CAST END CAP
RETURN HEADER W/1-1/4" STUBS
3" FLEX JOINT
1-1/4" FLEX JOINT (SEE DETAIL 'A')
1-1/4" SQUARE NIPPLE x 20"
3" FLEX JOINT (SEE DETAIL 'A')
1-1/4" UNION
1-1/4" NIPPLE x 18"
32" TYP.
*3" BUILDING PIPING (BY OTHERS)

*Building piping must be rigidly secured so it cannot move where connected to headers.

NOTE:
Shown reverse return. For direct return, header assembly piping connections may be made at the same end of the boiler bank, not as shown.

Some areas do not allow compression type couplings. Consult your area for local code requirements.

41-2500
1-1/4" x 3/4" x 1-1/4" TEE
3/4" DRAIN (BY OTHERS)
1-1/4" NIPPLE x 8"

FIGURE 13-24A Modular boiler installation typical of commercial hydronic installations. (Courtesy of Slant/Fin Corp.)

VENT DAMPER CONTROLS

To reduce burner off-cycle heat losses, electric vent dampers can be installed on most atmospheric-type gas-fired furnaces and boilers equipped with draft hoods (figure 13-26). Many factory-equipped boilers and furnaces feature electrical control centers that accept plug-in wiring harnesses for these dampers. The dampers incorporate a safety interlock that allows the burner to fire only when the vent damper blade is completely open.

THERMOSTATS

Thermostats, whether made up of simple bimetallic metal elements or of solid-state circuit components, all perform the same basic function: sensing changing temperatures, and opening or closing an electrical circuit in

FIGURE 13-24B Commercial modular boilers. These boilers are installed in tandem for sequential firing. In this way, the boilers offer the flexibility to meet varying heating loads of the building without the need for all boilers to fire at the same time. (Courtesy of Slant Fin/Corp.)

FIGURE 13-24C Commercial finned radiation. Commercial radiation features heavy fins, large-diameter water tubing, and heavy-gauge steel enclosures. Note the use of hangers in the cabinet that support the fin tube. Where additional heating capacity is required, two or more rows of tubing are hung beneath the enclosure. (Courtesy of Slant/Fin Corp.)

response to these temperature changes. The principle of bimetallic thermostats was detailed in Chapter 10. Note the thermostat shown in figure 13-27 for detailing normal operation.

Principle of Operation

As the room temperature changes, the bimetallic element coils or uncoils, forcing the electrical contacts to open or close in response to the temperature changes. This is due to the uneven expansion and contraction rates of the different metals used in the thermostat's electrical contacts.

Some thermostats utilize a small ball of liquid mercury inside a glass tube to make electrical contact when the bimetallic element rotates and tilts the glass. A heat anticipator circuit is usually incorporated into the thermostat design. This is an adjustable wire coil that produces a small amount of heat in response to the electric current that turns the boiler or furnace on. The heat generated in this coil will cause the thermostat to close before the room reaches the set temperature of the thermostat. Thus, this circuit anticipates the remaining heat that will be produced by the heating system prior to its final heat cycle. The heat anticipator, if properly adjusted, gives precise control over the total amount of heat delivered by the heating system and enables close maintenance of temperature levels in the home or building. This adjustment is made by consulting the inside cover of the main boiler or furnace control. Heat anticipator settings are usually given in amps of electric current that correspond to calibrations on the heat anticipator coil in the thermostat housing. For thermostats to work properly, they must be perfectly level. This can be checked and adjusted by removing the outer cover of the thermostat, installing a small level, loosening the set screws, and rotating the base or subbase of the thermostat housing in either direction.

Multistage and Set-Back Thermostats

In many instances, more than one thermostat set point is required if different heating and/or cooling control functions must be maintained

FIGURE 13-25 Computerized firing control for modular boiler installations. (Courtesy of Slant/Fin Corp.)

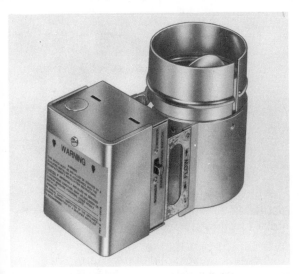

FIGURE 13-26 Motorized vent damper of the type used to control off-cycle heating losses through the boiler. (Courtesy of Honeywell, Inc.)

throughout the day or week. Clock-based thermostats, sometimes called set-back thermostats, are inexpensive and can yield significant savings in reduced fuel consumption. Some thermostats, such as those used to control both winter space heating and summer cooling, have contacts that close on both temperature increase and decrease. An indicator lever can be set to switch control to the heating or cooling function. Multistage thermostats are also used in dual-fuel heating installations or in applications where one set point controls hot water circulation and the other set point controls burner operation. A typical two-stage thermostat is illustrated in figure 13-28.

Solid-State and Programmable Thermostats

Solid-state electronic control technology is used in the production of programmable ther-

FIGURE 13-27 Bimetallic thermostat. (Courtesy of Honeywell, Inc.)

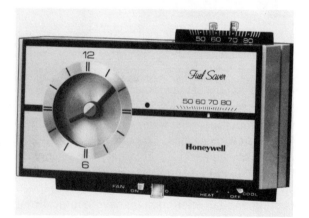

FIGURE 13-28 Multistage thermostat. (Courtesy of Honeywell, Inc.)

FIGURE 13-29 Set-back thermostat. Pins inserted into the clockwheel determine specific set-back times, and are adjustable. Several set-backs can be set on this thermostat during a 24-hour period. (Courtesy of Honeywell, Inc.)

The thermostat in figure 13-30 enables the home or building owner to program a variety of temperature settings throughout the 24 hours of the day, as well as the seven days of the week.

FIGURE 13-30 Multistage solid-state programmable thermostat. (Courtesy of Honeywell, Inc.)

mostats for temperature sensing and circuit control. Figure 13-29 illustrates a simple set-back thermostat. These thermostats allow a lower temperature level to be set during times when a home is unoccupied or a building is not being used. The use of the set-back thermostat instead of the familiar mechanical thermostat without a set-back feature can save from 10% to 15% of heating costs during a typical heating season.

Many of these solid-state thermostats can control both heating and cooling systems in the building from one single thermostat control. The microprocessor that these thermostats rely

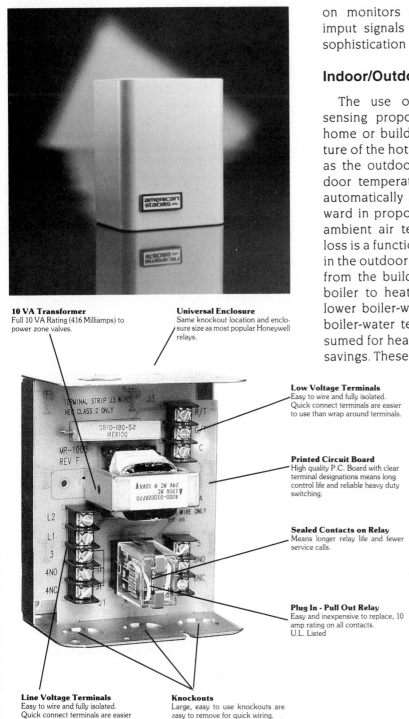

10 VA Transformer
Full 10 VA Rating (416 Milliamps) to power zone valves.

Universal Enclosure
Same knockout location and enclosure size as most popular Honeywell relays.

Low Voltage Terminals
Easy to wire and fully isolated. Quick connect terminals are easier to use than wrap around terminals.

Printed Circuit Board
High quality P.C. Board with clear terminal designations means long control life and reliable heavy duty switching.

Sealed Contacts on Relay
Means longer relay life and fewer service calls.

Plug In - Pull Out Relay
Easy and inexpensive to replace, 10 amp rating on all contacts. U.L. Listed

Line Voltage Terminals
Easy to wire and fully isolated. Quick connect terminals are easier to use than wrap around terminals

Knockouts
Large, easy to use knockouts are easy to remove for quick wiring.

FIGURE 13-31 Indoor/outdoor proportional controller. (Courtesy of American Stabilis, Inc.)

on monitors general time and temperature imput signals simultaneously, to achieve this sophistication in one single thermostat control.

Indoor/Outdoor Proportional Controllers

The use of indoor/outdoor temperature-sensing proportional controllers enables the home or building owner to vary the temperature of the hot water in the boiler automatically as the outdoor temperature changes. As outdoor temperatures rise, these controllers will automatically reset the boiler high limit downward in proportion to the rise in the outdoor ambient air temperature. Since building heat loss is a function of outdoor temperature, a rise in the outdoor temperature lowers the heat loss from the building. This condition enables the boiler to heat the building effectively with a lower boiler-water temperature. The lower the boiler-water temperature, the less fuel is consumed for heating purposes, with resulting cost savings. These controllers incorporate two sensing bulbs, one for measuring indoor temperature and the other for outdoor temperature sensing. A controller of this type is illustrated in figure 13-31.

STEAM HEATING SYSTEMS

Although their popularity in new construction has diminished, steam heating systems have proved to be a durable heating option over the years, and many are still in use. Steam heating was the heating system of choice during the late 19th and early to middle 20th centuries. The effectiveness of steam heating systems comes from the nature of steam as a heating medium. As water in a boiler is heated to steam, the steam contains the latent heat of vaporization. This heat is given off during the

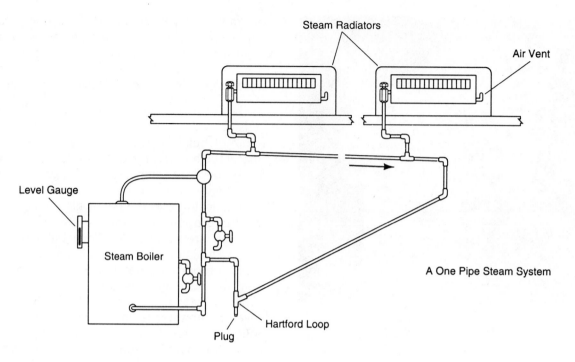

FIGURE 13-32A One-pipe steam heating system.

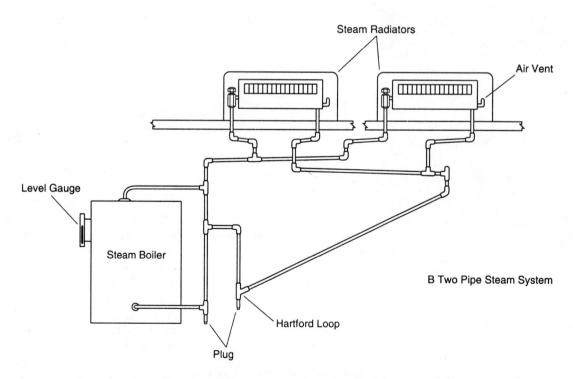

FIGURE 13-32B Two-pipe steam heating system.

heating process when the steam vapor comes into contact with the relatively cool surfaces of the steam radiators. The steam condenses, releasing the latent heat of condensation in the process, which is transferred to the building through the radiator surfaces. The condensate then travels back to the boiler, where the heating process is repeated. The latent heat that is trapped in the steam is extremely dangerous to anyone coming into accidental contact with the vapor. Burns result not only from the sensible heat of the steam (approximately 220°F), but also from latent heat released as the steam condenses on the relatively cool surfaces of the skin. The homeowner, building owner, and technician should be extremely careful when dealing with any type of steam-based appliance.

The rising popularity of hydronic, electric, and heat pump systems has relegated steam use primarily to larger commercial installations where either the steam is available as a utility, or the complexity of steam generation is worth the economies that it provides. An explanation of the piping configurations is required before examining the controls associated with steam heating systems.

Steam Piping Configurations

Steam systems are designed in either one- or two-pipe configurations, as shown in figures 13-32A and 13-32B. In one-pipe arrangements, a single pipe serves a dual function as both the steam delivery pipe and the water condensate return line to the boiler. In a two-pipe system, separate pipes are used for the steam feed and water return lines.

Steam Radiators

Steam radiators differ from hydronic baseboard convectors in both design and function. Note the design of a typical steam radiator in figure 13-33.

Steam enters the radiator through the supply pipe, driving most of the air from within the radiator out through the radiator vent. The air that remains trapped within the flutes of the

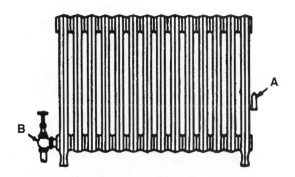

A. Air vent. Use lower opening in radiator
B. Radiator valve

FIGURE 13-33 Steam radiator (typical).

radiator acts as an insulator to prevent the radiator from overheating. Note that the radiators are made up lengthwise of separate tubular columns that are bolted together, resulting in a unit with a large surface from which to radiate the heat. The length of the radiator depends on the amount of heat that is required in a particular room. Although the radiator in figure 13-33 is a one-pipe unit, most radiators have threaded ports in both ends to enable their use in either one- or two-pipe steam systems.

Most residential steam systems are equipped with manually operated radiator valves to open or close individual radiators during the heating season. Note that these valves are not intended to serve as regulators by being turned to the half-open position. When used in this way, the valve seats quickly erode and ruin the valve. Commercial steam systems utilize thermostatically operated high-capacity valves for this purpose. These valves are available either as self-contained units or with remote sensors, depending on the installation requirements. Figure 13-34 shows a variety of thermostatically operated valves that are suitable for either steam or hot-water commercial heating applications.

Steam Heating Controls

Steam controls for both residential and commercial boilers are designed to maintain safe pressure limits and water levels in the boiler. Maximum pressures for residential steam

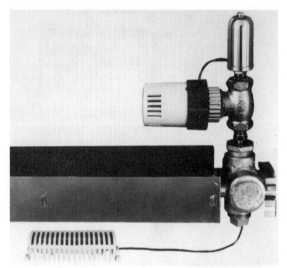

One pipe convector installation of a TLC (Thermostatic Limit Control) valve using the remote sensing operator. TLC valves provide automatic balancing of a heating system giving occupant comfort and eliminating wasteful overheating. This TLC valve is ideal for the private home with an uncontrollable 1 pipe steam system.

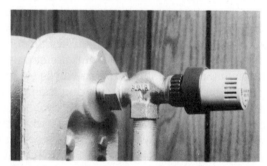

Two pipe radiator installation using remote setting/sensing TLC operator on a NPT union angle standard body. TLC (Thermostatic Limit Control) valves reduce fuel consumption by providing positive control of the flow of steam or hot water into a radiator, convector or baseboard heat exchanger.

FIGURE 13-34 Thermostatic radiator valves. (Courtesy of Flair International, Inc.)

boilers are 15 PSI. Pressure-relief valves for steam systems discharge at this setting. For safety purposes, this pressure is much lower than the 30 PSI of maximum pressure allowed in residential hydronic installations. Commercial steam pressures run much higher, depending on the specific application.

Pressure-control switches: The operating cycle of a steam boiler is initiated by a call for heat from the room thermostat. Whereas hydronic boilers activate the burner mechanism according to the water temperature in the boiler, steam boilers begin the burner sequence based upon the steam pressure in the boiler. Assuming that the pressure in the steam boiler is below the cut-in point and the water is at the specified level, the cut-in switch initiates burner operation. The burner continues to fire until the cut-out pressure has been reached. An examination of the cut-in adjustment on the steam boiler pressure-control switch shown in figure 13-35 illustrates the action of the switch. The cut-in pressure determines the pressure at which the burner mechanism will fire. The cut-out pressure is equal to the cut-in pressure plus a pressure differential, which is set on an adjustment wheel inside the switch housing.

Switch settings for most residential applications feature a cut-in pressure of 3 pounds with a differential of 2 pounds, which would turn off the burner when the boiler reached 5 pounds of steam pressure.

FIGURE 13-35 Steam pressure boiler switch. (Courtesy of Honeywell, Inc.)

Sight (gauge) glass: Safe and proper operation of any steam boiler depends on maintaining the correct water level in the boiler. Although there are specialized controls (to be discussed shortly) that are designed to maintain safe operating conditions in the boiler in the event of low-water conditions, one of the most reliable methods for ascertaining boiler water level is the use of a sight or gauge glass (figure 13-36). The location of the tappings on the boiler and the length of the glass tube are determined by the boiler manufacturer. The glass column is installed into watertight fittings between two specially designed valves. Most glass columns have two inscribed lines that indicate the recommended high and low water limits of the boiler. The valves on the gauge fitting are used to turn off the water supply to the gauge for purposes of cleaning and general maintenance procedures.

Low-water cut-off controls: Through its cyclical evaporation and condensation, some of the water in a steam heating system will be lost over a period of time. However, maintaining a safe water level in the boiler is critical for its safe operation. If the water drops below a certain level while the burner is operating, an overheat condition can cause the boiler either to crack, explode, or both. To eliminate this possibility, control packages are installed in *all* steam boilers that sense the water level in the boiler and deenergize the burner until the water level is brought back up to a safe level. (Low-water controls are sometimes also used in hot-water systems as an additional safeguard.) Several different types of controls are available that provide this feature. Most use a float mechanism in a sealed chamber that is connected to an electrical snap switch. Other types of controls use solid-state sensing devices that determine water level in the steam boiler and send a signal to a microprocessor for interpretation and corrective action, if necessary (see Chapter 10 for a description of these sensors).

The control illustrated in figure 13-37 senses low water levels and through the action of an internal float mechanism. When the water level drops, a set of electrical contacts open, which cut off power to the burner.

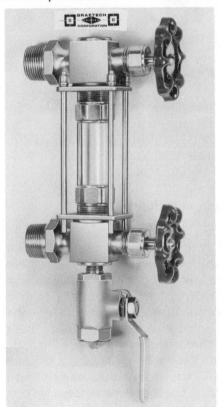

FIGURE 13-36 Gauge glass assembly. (Courtesy of Conbraco Industries)

FIGURE 13-37 Low-water cutoff used in steam boilers.

Note how this switch operates, (see figure 13-38). The float is housed in the switch casting. The height of water in the casting raises and lowers the float that operates the electrical snap switch housed in the end plate of the cut-off assembly.

FIGURE 13-38 Component arrangement of low-water cutoff switch used in steam boilers.

The control in figure 13-37 features only a safety circuit to sense water level and to react to low-water conditions. This control will automatically recycle when the water level within the boiler is raised (some low-water cutoff switches must be reset manually before the burner will reenergize, depending on the make and model of the switch). Some feed-water controls feature both a low-water cutoff and an automatic feed-water mechanism that automatically raises the water level in the boiler to the proper level. The water level is maintained by the use of a float mechanism, similar in operating characteristics to the float mechnanism used to maintain water levels in toilets.

Feedwater cutoff and boiler flushing procedures: All low-water controls must be cleaned and flushed periodically if they are to operate reliably. The constant evaporation and condensation of the water and steam in the piping system create a great deal of scale and sedimentation, which does not happen in other types of heating systems. If not flushed and purged on a regular basis, these sediments will build up in both the boiler and mechanical low-water control switches, causing them to malfunction. To avoid this problem, all low-water controls feature a flushing valve that should be opened once a week during the heating season to prevent a buildup of foreign material inside the valve. This simple procedure is illustrated in figure 13-39.

FIGURE 13-39 Flushing the low-water cutoff.

Steam boilers must also be flushed periodically to remove sedimentary buildup. The boiler in figure 13-40 features a skimmer trough in which sediment is collected and purged during the cleaning procedure.

Gauges and Relief Valves

Although pressure gauges and relief valves designed for use in steam systems appear to be similar to their hydronic counterparts, the pressure ranges and pressure activation points are quite different. As residential steam systems operate at low pressures, pressure-relief valves are set to discharge at only 15 PSI (residential hydronic relief valves discharge at 30 PSI). Steam pressure gauges normally have a range of up to 30 PSI. Commercial steam heating and power-generating plants operate at very high pressures, and will be equipped with relief valves and operating gauges that reflect these conditions.

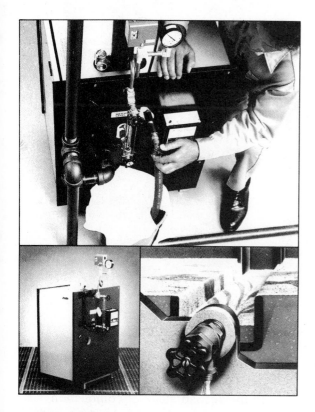

FIGURE 13-40 Flushing a steam boiler.

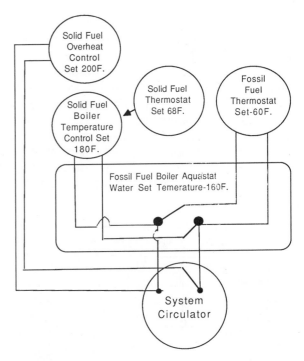

FIGURE 13-41 Operating logic of fossil-fuel and add-on solid-fuel heating system.

SOLID-FUEL ADD-ON CONTROLS

The revitalization of the solid-fuel heating industry in the early 1970s brought with it a great deal of experimentation focusing on the integration of solid-fuel heaters with existing fossil-fuel heating systems. When the solid-fuel boiler is piped into a fossil-fuel heating system, the natural-gas- or oil-fired fossil-fuel boiler functions as a back-up to the solid-fuel boiler. The operating logic of this type of system is illustrated in figure 13-41.

In the majority of installations of this type, the solid- fuel boiler will provide most of the heat to the building; the oil- or gas-fired boiler will operate only if the solid-fuel unit is unable to supply all of the necessary heat. To function in this way, a two-stage thermostat is used to control both the solid-fuel boiler and the existing fossil-fuel boiler, as illustrated in figure 13-41. The first stage of the thermostat activates

the damper control in order to admit air into the solid-fuel unit and build up the intensity of the fire. If the room temperature drops to the second-stage thermostat setting, which is usually 5 to 10 degrees below the first-stage setting, the fossil-fuel boiler is energized to provide additional heat. In order to examine how these systems operate, we must first understand the combustion characteristics of wood and coal and the basic design configurations of these boilers.

Wood Combustion

The amount of heat available from the combustion of wood depends on the type of wood being burned. The categorization of wood into hardwoods and softwoods has less to do with how hard or soft these woods are than with whether the trees lose or keep their leaves during the winter. Trees that keep their leaves all year long are known as coniferous, or cone-bearing, trees—pines, cedars, and other evergreens (softwoods). Trees that lose their leaves during the winter are known as deciduous—oak, maple, and the fruit trees (hardwoods). Hardwoods are generally denser than the softwoods, and since the heating value of wood is a function of density, hardwoods are superior to softwoods for heating purposes. The heating values of selected fuel woods are given in figure 13-42. These values are based on comparing different species of wood by the cord, which is a pile of wood 4 feet high, 4 feet wide, and 8 feet long.

KINDS OF WOOD	Millions of BTUs/Cord in 50% Efficient Stove	Pounds/Cord If Air–Dried
Shagbark Hickory	24.6	4240
White Oak	22.7	3920
American Beech	21.8	3760
Sugar Maple	21.3	3680
Red Oak	21.3	3680
Yellow Birch	21.3	3680
Southern Yellow Pine	20.5	3600
White Ash	20.0	3440
Red Maple	18.6	3200
Douglas Fir	18.0	3100
American Elm	17.2	2900
Eastern White Pine	13.3	2180
Aspen	12.5	2160

FIGURE 13-42 Heating values of selected cord woods.

The combustion of wood takes place in three distinct stages, all of which occur simultaneously within the firebox, as heat from the existing fire is used to ignite fresh charges of wood. For this process to be efficient, the firebox construction must allow for the introduction of primary and secondary combustion air. In the following discussion, consult figures 13-43A and 13-43B, which show an add-on solid-fuel boiler and a typical firebox cross section. The three stages of wood combustion are:

1. Heat provided by kindling (or the existing fire) turns the moisture in the fresh wood into steam and drives it from the logs. During this process, the temperature of the wood rises until it reaches about 550°F, the ignition temperature of the wood.

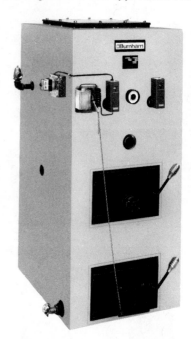

FIGURE 13-43A Add-on wood boiler. (Courtesy of Burnham Corp.)

2. In the second stage of combustion, the actual burning of both the wood and the volatile gases released from the logs takes place. Visible flames are evidence of second-stage combustion. For this process to be efficient, secondary air must be admitted above the firebox in order to burn the volatile gases that are released from the wood during this second stage of the combustion process.

3. The third phase of combustion is the most efficient stage, in which the almost pure carbon remaining after stages one and two is burned at temperatures that exceed 2000°F. This combustion stage is evidenced by the glowing bed of hot coals.

Creosote reduction technology: Inefficiencies in the combustion process can lead to the formation of large creosote deposits in both the boiler combustion chamber and chimney passages. Creosote, a black, gooey, tarlike substance, results from the condensation of unburned

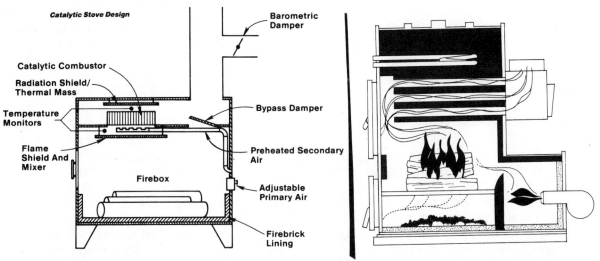

FIGURE 13-43B Firebox cross section for both a wood-burning appliance and a multifuel boiler designed to burn wood, coal, and fuel oil. Note the use of separate combustion chambers for wood/coal and fuel oil. (Wood-only combustion chamber, courtesy of Corning; multifuel boiler, courtesy of the Northland Corp.)

volatile gases. Although some creosote deposition is a normal by-product of combustion, excessive deposits can lead to chimney fires. The ignition temperature of creosote is higher than the normal combustion temperatures in a firebox. As creosote builds up in the heat exchange passages and chimney piping, part of it eventually will ignite, spreading the fire within the chimney. The major technological advance in the control of creosote deposition was the development of the catalytic combustor. Similar in design principle to the catalytic converter used in an automobile, the catalytic combustor is inserted into the firebox to enhance the combustion of volatile gases as they are driven from the logs, prior to their entering the chimney. The design of the catalytic combustor and its use in a typical wood-heating appliance are illustrated in figure 13-44.

In addition to catalytic combustors, there are other, less sophisticated, but nevertheless effective methods to control creosote deposition. Some of these are:

1. Regular cleaning of heat exchange passages and chimneys during the heating season.

2. Using dry hardwoods in lieu of wet woods or softwoods whenever possible.

3. Avoiding the use of a solid-fuel appliance when outside temperatures are above 50–55°F.

4. Maintaining high water temperatures in the boiler. High temperatures minimize condensation on interior boiler surfaces.

Burning Coal

The combustion characteristics of coal are quite different from those of wood—so much so that experienced wood burners often have difficulty in maintaining a fire when they first begin to burn coal. Whereas a wood fire burns primarily from the top layers down, a coal fire does just the opposite, burning from the bottom layer up (figure 13-45).

Since the ignition temperature of coal is much higher than that of wood, a good wood fire or a well-established coal fire is necesary before coal can be added to the firebox. The greatest amount of combustion in this process takes place in the oxidation zone directly above the grates, producing carbon dioxide as the primary by-product. In the reduction zone, carbon dioxide is reduced to carbon monoxide—which is odorless, colorless, and highly toxic. Fresh charges of coal are added above the distillation zone. Heat from the fire drives gases and moisture from the fresh coal, and these collect at the top of the distillation zone. The introduction of secondary air above the distillation zone helps to oxidize the carbon

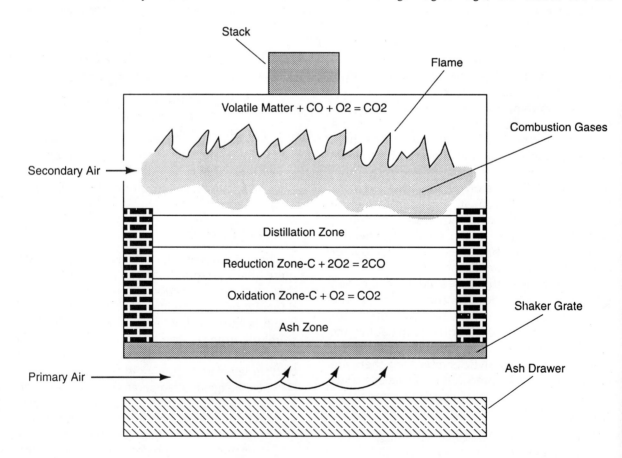

FIGURE 13-44 Catalytic combustor for solid-fuel appliances and installation in typical stove. (Courtesy of Corning, Inc.)

monoxide to carbon dioxide and burn other collected volatiles to complete the combustion process. However, much highly poisonous carbon monoxide still remains, and is vented from the stove through the stove pipe and chimney.

Varieties of coal: Coal is categorized according to variety and size. The varieties of coal are based on geological age; the oldest are the

FIGURE 13-45 Combustion of coal.

densest and best burning, and contain the most heat energy. This ranking of coals by geological age is shown in figure 13-46.

Anthracite coal, sometimes referred to as "hard coal," is the preferred coal for use in residential and commercial fired heating applications. Anthracite yields long burns with little smoke. The soft coals vary widely in physical characteristics and often contain large amounts of sulfur.

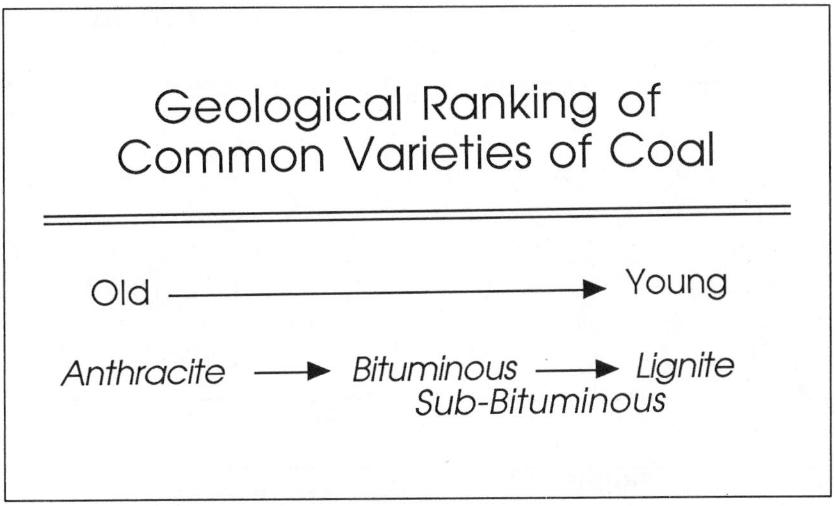

The categorization of coal by size is illustrated in figure 13-47. Coal sizing is determined by the number of openings per inch in different sizes of wire mesh screen that the coal will fit through. Most often, a particular coal-burning appliance is designed to burn one size of coal, based on the construction of the grates and the location of the primary and secondary air inlets.

Operating Controls

The electrical and mechanical controls for regulating draft and for connecting solid- and multifuel boilers into existing heating systems are primarily the same as those used for conventionally fueled systems. Where appropriate, motorized actuators are modified for use in a specific application or appliance.

Draft controls: Three types of draft controls are used in most solid-fuel boilers. In each instance, the control is responsible for the entrance of combustion air into the firebox, either at atmospheric pressure or under forced (motorized blower) conditions.

Geological Ranking of Common Varieties of Coal

Old ⟶ Young

Anthracite ⟶ Bituminous Sub-Bituminous ⟶ Lignite

FIGURE 13-46 Geological ranking of common varieties of coal.

Immersion controls: Immersion aquastat controls employ a bimetallic heat-sensitive element installed in an immersion well. The water in the boiler causes the bimetal element to either expand or contract. This action is converted into an up-and-down motion of a lever that is connected to the primary air intake door by a chain (figure 13-48). When the boiler water cools, the lever moves upward, opening the primary air intake door.

This type of control operates without any electric power. As more air is admitted into the firebox, the fire builds, raising the temperature of the boiler water. As the temperature within the boiler rises, the lever slowly begins to rotate downward, reducing air intake until the set temperature of the aquastat has been reached. In this arrangement, the air intake door does not completely close when the aquastat is satisfied, but rather allows enough air into the firebox to maintain a small fire until the next heating cycle is required from the boiler.

Induced-draft fans: Motorized fans are used in many solid- and multifuel applications to provide combustion air under pressure. This pressurization helps to promote turbulence in the combustion zone and to supply excess air in order to increase the efficiency of the combustion process. Figure 13-49 shows an induced-draft fan that can be used in a variety of forced-draft solid-fuel boilers. Given the length of travel of the air through the various combustion and refractory chambers found in many solid-fuel boilers, a fan-forced arrangement is required to provide the pressure necessary to move the air to where it is needed and still leave the chimney under negative draft conditions.

Motorized dampers: A third type of draft control utilizes a motorized damper to open and close the primary air intake passage. This type of control (figure 13-50) usually responds directly to the thermostat or a thermostatically controlled relay to regulate primary combustion air.

Overheat controls: The biggest potential problem in any solid-fuel heating appliance is that posed by boiler water overheating. In this case, the water in the boiler turns to steam that triggers the pressure/temperature-relief valve to discharge. Although the release of steam from the boiler can be a problem depending on the location of the boiler room, the greatest problem in these instances is a resulting air pocket in either the boiler or associated piping, which will prevent water circulation through the piping circuit. This prevents the delivery of heat to the building, and can create additional overheat conditions as well. Trapped air within the system must always be bled manually before circulation of the water can be reestablished.

To prevent a boiler from overheating, an aquastat is installed either on the output pipe of the boiler or in an unused boiler tapping. An aquastat of this type, which can be used to control a number of different electrical circuits,

Coal Ranking by Size	
Coal Grade	**Mesh Size (Approximate)**
Pea	1/2" to 3/4"
Nut	3/4" to 1 1/2"
Stove	1 1/2" to 2 1/2"
Egg	2 1/2" to 3 1/2"

FIGURE 13-47 Classification of coal by size.

is shown in figure 13-51.

The overheat aquastat is designed to close a set of electrical contacts on temperature rise, and should be set approximately 25 degrees above the boiler water set temperature (and below the temperature setting that triggers the pressure-relief valve on the boiler). A setting between 200°F and 220°F is usually sufficient for adequate overheat protection, given a well-designed and -installed heating system.

One recommended installation procedure is to connect the overheat aquastat contacts in parallel with the thermostat wires that control the largest piping circuit in the heating system; see figure 13-52. In this way, if the boiler water overheats, it will either energize the circulator control in a single-zone system or a zone valve or circulator motor switching relay in a multizone system. The specific heating circuit then acts as a heat dump zone, allowing excess hot water from the boiler to circulate throughout the baseboard system until the boiler water temperature drops to a safe level. This heat dumping process takes place regardless of the thermostat setting in the heat dump zone.

Aquastat and Thermostat Settings for Dual Fuel Systems

Whereas the control settings for single-boiler installations are the same whether the boilers are fueled by solid fuels or fossil fuels, dual

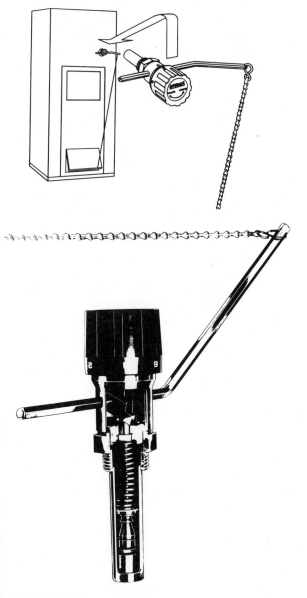

FIGURE 13-49 Induced-draft fan. (Courtesy of Tjernlund, Inc.)

heating system. The two most popular configurations for two boilers are series piping and parallel piping. Each of these methods is illustrated, along with the control features of each.

FIGURE 13-50 Motorized damper used to control primary air intake in a solid-fuel appliance. (Courtesy of Honeywell, Inc.)

Series Piping and Control

Figure 13-53A illustrates an add-on solid-fuel boiler connected in series to an existing single-zone heating system, and figure 13-53B shows the series piping configuration in a multiple-zone/zone-valve system. Note that in the series

FIGURE 13-48 Immersion aquastat used to regulate primary combustion air in solid-fuel appliance. Heating of the aquastat causes the bimetallic spring to expand, controlling a lever to open or close the draft door. (Courtesy of Ammark Controls, Inc.)

boiler controls are somewhat more complicated. The control features in these installations are determined by how the solid- and fossil-fuel boilers are connected to each other and to the

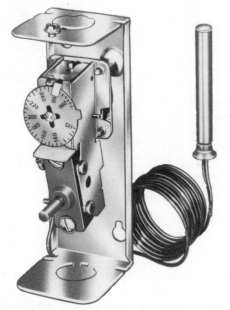

FIGURE 13-51 Aquastat used to control overheating in a solid-fuel boiler. (Courtesy of Honeywell, Inc.)

piping arrangement, water returning from the baseboard circuit is directed through the solid-fuel boiler, where it is preheated, and then

through the backup fossil-fuel boiler. Thus, the water in the heating system first is exposed to any heat available from the solid-fuel heater. If the water in the solid-fuel boiler is sufficiently heated, it travels through the backup boiler and into the heating system without the need for any additional heat from the fossil-fuel boiler.

If, however, the water is not adequately heated by the solid-fuel boiler, the boiler will be energized to provide the additional heat necessary to raise the water temperature to the minimum required for effective space heating. In this arrangement, the boilers act independently of each other. The fossil-fuel heater acts as a backup to the solid-fuel unit. In order to accomplish this type of operation, the boiler aquastats are set according to the control logic in figure 13-54.

Note from figure 13-54 that the add-on solid-fuel boiler is set to maintain a water temperature that is 20 to 40 degrees higher than the backup fossil-fuel unit. As long as the water in the fossil-fuel boiler is above its aquastat high-limit set point, the burner mechanism will not turn on. Should the temperature of the water

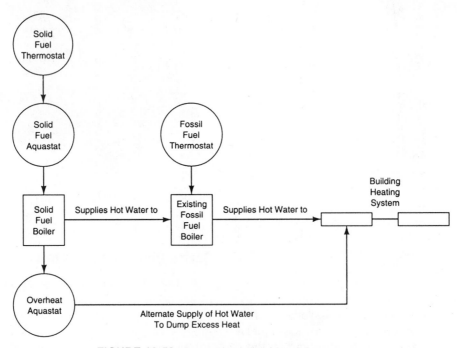

FIGURE 13-52 Operating logic of overheat aquastat.

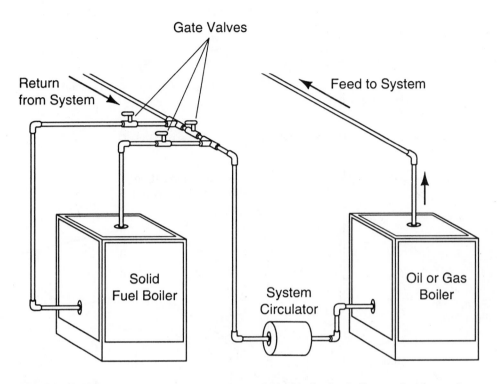

FIGURE 13-53A Series installation of add-on solid-fuel boiler in single-zone heating system.

entering the fossil-fuel unit fall below the high-limit cut-in differential (usually 10 to 15 degrees below the high-limit aquastat setting), the burner will come on to heat the water to the high-limit setting, and then turn off. In this way, the existing fossil-fuel unit acts as a backup to the add-on wood/coal boiler.

An additional variation of this control logic is the use of a two-stage thermostat in place of the existing room thermostat. The first-stage setting, which is the higher temperature, controls the add-on solid-fuel boiler. The second, lower-temperature stage controls the fossil-fuel unit. If the room temperature falls to the second-stage setting, because either the heat loss from the residence exceeds the ability of the solid-fuel unit to heat adequately or the solid-fuel fire goes out, the backup fossil-fuel unit will be activated. Two separate single-stage thermostats can be used in place of one two-stage thermostat, and should be installed next to each other.

Parallel Piping and Control

Parallel piping differs significantly from a series installation. In a parallel arrangement, water is circulated between the solid- and fossil-fuel boilers using a separate circulator. Upon examining the parallel piping arrangement shown in figure 13-55, you will see that the major components of the heating system remain connected to the fossil-fuel boiler. The solid-fuel unit has been added to the system by tapping off the supply and return lines of the fossil-fuel boiler. A flow check valve is installed in the supply pipe between the two boilers to prevent a back-flow of hot water from the existing boiler to the solid-fuel unit. The add-on has been equipped with two aquastats, both of which close on temperature rise. One aquastat triggers the new circulator to turn on when the water temperature in the add-on reaches the set point for water circulation between the two boilers. The other aquastat acts as the overheat control, and can be connected either to an

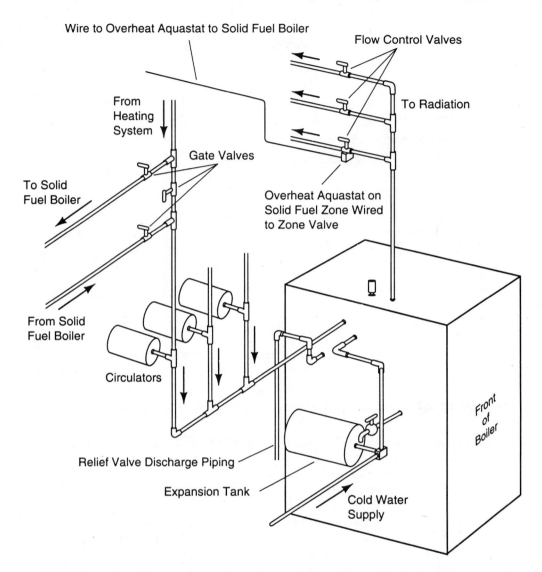

FIGURE 13-53B Series installation of add-on solid-fuel boiler in multiple-zone heating system.

optional bypass valve or to an existing zone thermostat in order to dump extra heat and prevent the boiler water from overheating.

For this installation to function properly, the aquastat that energizes the circulator between the two boilers is set approximately 20 degrees higher than the main operating aquastat on the fossil-fuel boiler. This control logic is illustrated in figure 13-56.

The operating characteristics of a solid-fuel unit in the parallel piping system are different from those of the series installation. In the parallel arrangement, the solid-fuel boiler acts to maintain a constant supply of hot water circulating between itself and the fossil-fuel boiler. On a call for heat from the room thermostat(s), heated water is supplied from either the solid- or fossil-fuel boiler, or both. Normally, water returning from the heating system is drawn into the add-on solid-fuel unit by virtue of the operation of the second circulating pump. Note in figure 13-55 that the

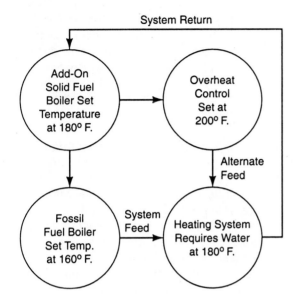

FIGURE 13-54 Operating control logic for series solid-fuel add-on.

arrow in the existing boiler supply pipe points in both directions. As long as the water temperature from the add-on is higher than the high-limit set point (minus the differential) on the fossil-fuel boiler, the fossil-fuel burner will not be energized. If the heat loss in the house is greater than the ability of the solid-fuel boiler to supply heat to the dwelling, then the fuel burner will be turned on, based on a drop in the temperature of the water coming from the add-on unit. The room thermostat(s) operate in a normal manner, energizing the circulator to supply heat throughout the baseboard circuit. The operation of the fossil-fuel burner depends solely on the water temperature in the boilers. As long as the add-on supplies a sufficient quantity of hot water at the high temperature required to keep the fossil-fuel burner from turning on, the heating system will function solely on heat supplied from the add-on solid-fuel unit.

Steam-Heating Solid-Fuel Add-on Arrangements

An arrangement for adding a steam heating boiler to an existing fossil-fuel steam heating

system is shown in figure 13-57.

It can be seen that the steam boilers are connected together in a parallel piping circuit. Each boiler acts independently, with the fossil-fuel boiler again acting as a backup to the solid-fuel unit. The solid-fuel boiler is fitted with a conventional steam control package, including low water cutoff and gauge glass assembly. For this system to work properly, the water levels in both boilers must be exactly the same.

SOLAR HEATING ADD-ON CONTROLS

The integration of solar heating devices with existing fossil- or solid-fuel heating systems produces a reliable hybrid heating configuration. Installations of this type generally place the existing solid- or fossil-fuel unit in a backup role to the solar device, similar to the logic employed with the addition of solid-fuel appliances to existing fossil-fuel heating systems. The term "backup" is used to describe the control features of the system, and not the relative amount of heat produced by either the solar or existing heating system.

Differential Controls

A differential controller is a device that senses two or more separate temperature inputs and initiates some action (usually by closing an electrical relay to turn on a circulating pump or blower) in response to the temperature difference between the two inputs. Figure 13-58 illustrates a solar domestic hot-water heating system that uses a differential controller to regulate the circulation of heat exchange fluid through the solar collector and heat exchanger piping circuit.

Sequence of operation: The system illustrated in figure 13-58 operates in the following way.

1. Sensors located in both the bottom of the solar storage tank and the top of the solar collector array supply temperature values to the differential controller. When the solar collectors are

between 5 and 12 degrees warmer than the water in the bottom of the solar storage tank (depending on the differential threshold of the particular controller), the controller energizes the circulating pump in the collector loop.

2. As the heat transfer fluid (antifreeze) circulates in the collector loop, it picks up heat from the solar collectors and transfers it to the cooler water in the solar collector tank. This heat transfer takes place through an internal heat exchanger located in the bottom of the solar water heater.

3. The solar water heater begins to stratify, with the hot water staying at the top of the water heater and the cooler water,

which is denser, falling to the bottom of the storage tank. Simultaneously, the heat transfer fluid, which has now cooled, exits the heat exchanger and is circulated back up to the solar collectors, where it is reheated to continue the heating process.

4. The differential controller continuously monitors temperatures in both the bottom of the solar storage tank and the collector array. When the water in the solar tank is sufficiently hot that the temperature differential between the collectors and the storage tank falls to a preset cut-out point, the controller deenergizes the circulating pump. Some controllers feature an adjustable shutoff

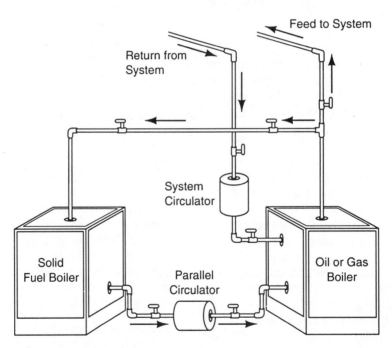

☐ = Gate Valve Locations

Gate valves are used to isolate either the solid fuel
or fossil-fuel boiler if necessary

FIGURE 13-55 Parallel piping of add-on solid-fuel boiler. Note that in a parallel installation, system zoning is not a factor since the add-on shares its heated water with the fossil-fuel boiler. The existing fossil-fuel boiler acts as the system supply for the heating system.

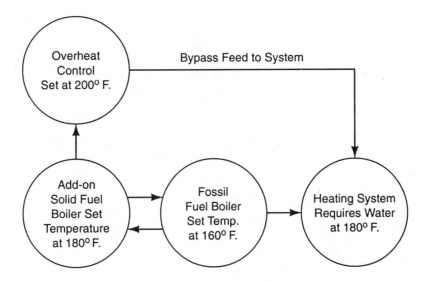

FIGURE 13-56 Operating control logic for parallel solid-fuel add-on.

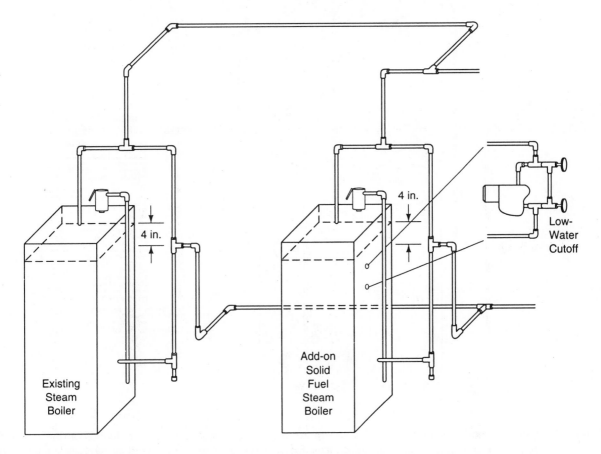

FIGURE 13-57 Solid-fuel steam add-on piping configuration. (Courtesy of the Northland Corp.)

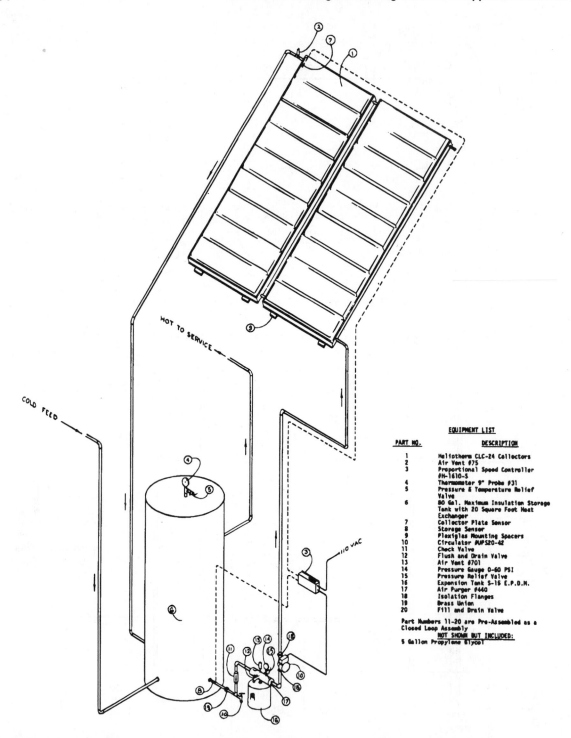

FIGURE 13-58 Solar domestic hot-water heating system. In this closed-loop system, heat is transferred from the collector loop to an internal heat exchanger in the solar storage tank. System sizing approximates one 120-gallon solar storage tank for every three 4-foot by 8-foot solar collectors. (Courtesy of Monitor Data Corp.)

FIGURE 13-59 Solar differential controller. The solar controller compares the temperature in the solar collector array with the bottom of the solar storage tank. When a differential of usually more than 5–8 degrees is reached, the controller activates the collector circulator to move a heat exchange fluid from the collectors through the heat exchanger in the solar storage tank. (Courtesy of Independent Energy, Inc.)

differential. As a general rule, the shutoff differential should be no less than 5 degrees, since heat transfer efficiency at these temperature differentials is very low.

5. As the house calls for hot tap water, preheated water from the solar tank

enters the fossil-fuel water heater. If this water temperature is sufficiently high, the burner mechanism or heating elements in the backup water heater will not be energized. In this way, the solar system provides preheated water to the existing water heater, and only enough the energy to bring the water up to its final temperature is required. Most solar domestic hot-water heating systems operating in this way can supply approximately 70% of the average hot-water requirements of a family of four people.

Differential controllers are available with a variety of options. The controller illustrated in figure 13-59 features a digital temperature display that reads three or more inputs, with one or more controlled AC outlets available.

Many of these controllers enable pumps and blowers to operate at proportional speeds. Proportional speed control is an effective method of matching the flow rate of circulation through a collector loop, for example, with the temperature differential between the collector and storage tank. In this way, as the temperature of the solar collectors rises quickly on a hot day, maximum flow rate of the heat transfer fluid through the collector array helps to transfer as much heat as possible. Proportional control is accomplished by supplying pulses of power rather than varying the level of the voltage to the circulating pumps and air blowers. Thus the pump will operate in timed pulses that increase in frequency as temperature differentials between the collectors and storage tanks increase and higher circulating speeds are required, and, conversely, decrease in frequency as these temperature differentials decrease.

Temperature sensors: The differential controller is activated by changes in the electrical resistance in a thermistor-type temperature sensor caused by changes in temperature. These sensors are installed in the exact locations where the temperature measurements are to be taken, and must be in good thermal contact with the surface through which the temperature will be measured. If the sensor is being used outdoors (for example, in a solar collector array), then it must be protected from the elements. Although these sensors are water resistant, exposure to the weather over time will render them useless. A sensor of this type is illustrated in figure 13-60.

FIGURE 13-60 Temperature sensor of the type used with solar differential controllers.

Problems associated with differential controllers can often be traced to malfunctions in the temperature sensors. Defective sensors can be located by using an ohmmeter to determine the resistance of the sensor. Temperature/resistance charts are available from the manufacturer of the specific sensors to help in analyzing correct sensor operation.

Solar Controls in Hydronic Heating Systems

When a solar system is integrated into an existing fossil-fuel hydronic heating system, the logic of the control circuitry places the solar system in the role of primary heating system and the fossil-fuel system as backup. Again, it is emphasized that these designations refer only to control logic, and not to the amounts of heat delivered to the building by either the solar or fossil-fuel heating system. A system of this type, shown in figure 13-6, uses a two-stage thermostat for operation.

The first stage of the thermostat controls heating contributions from the solar component, while the second stage controls the fossil-fuel boiler. The first stage is set approximately 5 degrees above the second. When the first stage is activated, the solar system is connected

into the baseboard circuit and circulates water from the solar storage tanks throughout the baseboard system. This system works effectively even when the temperature of the water from the solar storage tanks is only moderately above room temperatures (80°F is usually the minimum temperature level required for a space heating contribution from the solar system). If the room temperature drops to the second-stage thermostat setting, the solar system is disconnected from the baseboard circuit and the backup fossil-fuel boiler takes over the heating function until the second-stage thermostat setting is satisfied. At this point, the solar system is again connected to the heating system. Operating in this way, the solar system is continually cycling to contribute to meeting the overall heating requirements of the building.

To prevent the backup boiler from heating water and circulating it through the solar storage tanks, a differential controller compares the water temperature in the hottest part of the solar storage tanks at sensor B with the temperature of the water returning from the baseboard at sensor C. The solar system can only be energized if the water in the solar storage tanks is hotter than the water returning from the baseboard loop. Note the use of a separate expansion tank and pressure-relief valve assembly in both the collector loop circuit and solar water tank circulation piping. Because of the large amount of water contained in the solar storage tanks (each tank in figure 13-61 holds 120 gallons of water), additional expansion capacity is required when integrating the solar system into the existing heating system. The main circulator in the solar storage loop (no. 15) is used to circulate water through the baseboard circuit, as well as solar-heated water through the heat exchanger in the solar domestic hot-water heater, whenever the solar-heated storage water is hotter than the domestic hot water in the heater.

TROUBLESHOOTING LOGIC

For troubleshooting general operating problems, the student is referred to specific operat-

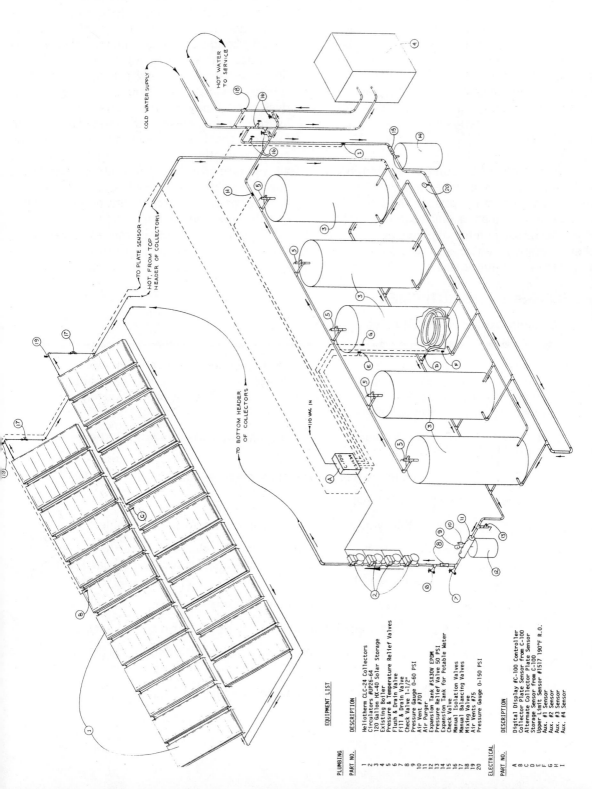

FIGURE 13-61 Integration of a solar heating system with an existing heating system. (Courtesy of Monitor Data Corp.)

EQUIPMENT LIST

PLUMBING

PART NO.	DESCRIPTION
1	Heliotherm CLC-24 Collectors
2	Circulators #UP26-64
3	120 Gallon HX-40 Solar Storage
4	Existing Boiler
5	Pressure & Temperature Relief Valves
6	Flush & Drain Valve
7	Fill & Drain Valve
8	Check Valve 1-1/2"
9	Pressure Gauge 0-60 PSI
10	Air Vent #701
11	Air Purger
12	Expansion Tank #SX30V EPDM
13	Pressure Relief Valve 50 PSI
14	Expansion Tank for Potable Water
15	Check Valve
16	Manual Isolation Valves
17	Manual Balancing Valves
18	Mixing Valve
19	Air Vents #75
20	Pressure Gauge 0-150 PSI

ELECTRICAL

PART NO.	DESCRIPTION
A	Digital Display #C-100 Controller
B	Collector Plate Sensor from C-100
C	Alternate Collector Plate Sensor
D	Storage Sensor from C-100
E	Upper Limit Sensor #1517 190°F R.O.
F	Aux. #1 Sensor
G	Aux. #2 Sensor
H	Aux. #3 Sensor
I	Aux. #4 Sensor

ing sequences described earlier in this chapter. Figures 13-13 through 13-16 illustrate the logic and operating sequences that can be used to trace through most problems that will be encountered in most hydronic installations. Figures 13-41 and 13-58 refer to the sequencing on add-on solid-fuel and solar-assisted systems that should be helpful in this regard as well.

SUMMARY

The technology of hydronic and steam heating systems covers a wide range of system configurations, control devices, and auxiliary heating options. In an examination of hydronic system designs, a number of piping options and boiler operating sequences were highlighted. Each of these variations was shown to offer the installer and technician operating sequences and controls that are individualized for specific installation applications. The use of vent dampers and solid-state programmable thermostats maximizes the efficiency of any system design. Steam systems, suffering a loss in popularity with the availability of other heating alternatives, use many of the same controls that hydronic systems use, with modifications incorporated for lower operating pressures and higher operating temperatures (they can, after all, operate from the same boiler with some minor modifications). Hydronic systems lend themselves to interfacing with either solid-fuel or solar-powered heating systems.

Chapter 14 continues the examination of heating system controls and applications, with a focus on forced convection warm-air heating.

PROBLEM-SOLVING ACTIVITIES

Troubleshooting Case Studies

A. A service technician is called to a home that is heated by a three-zone/zone-valve hydronic heating system. One of the zones in the system is cold even though the room thermostat has been turned all the way up. The following conditions have been observed by the serviceperson.

1. The thermostat in the cold zone is operational and shows 24 volts in the operating circuit.

2. The other two zones in the heating system seem to be functioning normally.

3. The system circulator is functional.

Outline the procedures that you would use to isolate and solve the problem.

B. A two-zone/zone-valve hydronic system does not allow the gas burner to fire, even though there is a call for heat by both room thermostats. Upon closer inspection, the following conditions are found to exist.

1. There is 110-volt power present in the main aquastat.

2. Both thermostat circuits are working, and 24-volt power is present in each circuit.

3. Both zone valves open and close in response to a call for heat by the room thermostats.

4. The circulator functions when the zone valves are fully opened.

Indicate the steps you would take to isolate and remedy the problem.

Review Questions

1. Distinguish between two-pipe direct return and reverse return circuits.

2. Describe or list the operating sequence of a multicirculator gas-fired boiler. For each sequence, be specific as to the operation that takes place and the control responsible for the operation.

3. Distinguish between a cold-start relay and a conventional high/low-limit controller.

4. Highlight the reasons why proportional indoor/outdoor controllers can save a building owner money over conventional single-set-point controls.

5. Explain the control relationship between the sight glass and the low-water cutoff on a steam boiler.

6. Identify at least three major differences between the combustion of wood and of coal in an add-on boiler.

7. What are the main operating differences between a hydronic aquastat that allows for heating domestic hot water in a boiler tankless coil and an aquastat without the domestic hot-water control function?

CHAPTER **14**

Forced Warm-Air Heating

Heating systems that operate by circulating warm air, via either gravity feed or forced fan convection, predate all other types of heating system design. Fueled primarily by wood until the end of the 19th century and by coal in the early 20th century, warm-air heating continues to be a viable option in terms of installation costs and operating efficiency, as compared with other types of heating systems used in residential and commercial applications.

Hot-air heating systems have two major components—the furnace and the air distribution system. Air distribution is handled by a series of supply ducts that channel heated air from the furnace to the individual rooms in the building and return ducts that draw cold air from within the building and feed it into the heat exchanger in the furnace for reheating and distribution back to the building. Hot-air furnaces come in a variety of component designs and fuel preferences to meet almost all application requirements.

BASIC SYSTEM COMPONENTS

The following are the major components of the warm-air system, with their basic functions.

Furnace: The furnace consists of several subcomponents, notably, the burner and the heat exchanger. The burner, firing in a combustion chamber, sends the heat of combustion up through a heat exchanger. The heat exchanger is located in the airstream of the system. Air

flowing over the heat exchanger surfaces picks up the heat of combustion and transfers this heat throughout the building. Depending on the type of fuel to be used and the design of the specific furnace, heat exchangers are usually made of either stainless steel or cast iron.

Blower/fan: The blower/fan is located in a return air box adjacent to the furnace heat exchanger. Some blowers are direct drive, whereas others use separate motors coupled to the blower with a drive belt and pulley arrangement. In direct drive units, the fan speed usually can be changed by utilizing various wiring options for the motor. On belt drive units, the speed of the blower is adjusted by changing the drive belt and pulley on the motor. In older, gravity-fed warm-air systems, there is no blower to move the heated air. Rather, the system relies upon convection currents and thermosiphon effects to move the air from the furnace to throughout the building.

Plenum: The plenum, a sheet metal box built onto the top of the furnace, is the location at which all distribution ductwork is attached.

Fan/limit switch: The fan/limit switch turns the blower on and off at preset limits. Usually, the fan is set to turn on at 150 degrees and shut off when the plenum temperature drops to 100 degrees. This switch also contains a high-limit control to turn off the burner if the plenum temperature goes above a preset high limit, usually 200 degrees.

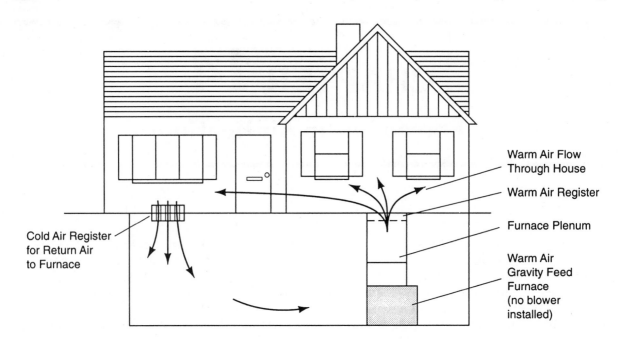

Cold Air Register
for Return Air
to Furnace

Warm Air Flow
Through House

Warm Air Register

Furnace Plenum

Warm Air
Gravity Feed
Furnace
(no blower
installed)

FIGURE 14-1A Gravity warm-air heating system. Gravity systems use oversize ducts to minimize frictional losses in heat distribution. These systems are usually found in older homes or in those where electrical supply is either limited or not available.

Duct distribution system: The ductwork is usually fabricated from galvanized sheet metal and consists of rectangular and/or circular ducts that act both to distribute hot air to all of the rooms in the building and to channel cold return air back to the furnace for reheating and redistribution.

WARM-AIR SYSTEM DESIGNS

Gravity and Forced-Convection Heating Systems

The earliest warm-air systems used little or no ductwork. Rather, they relied on the fact that warm air becomes less dense (lighter) as it is heated, and it therefore rises. This constant supply of heated air sets up convection currents to move heated air around the building. In ducted systems, the furnace was usually located in a basement and a series of ducts attached to the plenum of the furnace ran directly to the rooms above. Cold-air return to the furnace was accomplished through open grates in the floor that allowed the cooler air

to drop back into the furnace area for reheating. This ducting arrangement was sometimes referred to as an octopus because of its appearance.

If a building had no basement in which to put the furnace and ductwork, the unit could be installed below the floor in a centrally located compartment. This type of installation gave rise to the labeling of these systems as floor furnaces. No ducts were used in this arrangement; instead, it relied on convective air currents to circulate warm air from the furnace throughout the building. Although gravity-fed systems are easy to install, the resulting distribution of heat is not very efficient. With the availability of high-efficiency forced-convection heating sytems, the gravity systems have all but disappeared. At present, they are used primarily in applications where either no electricity is available, or a forced-convection system is impractical to install.

Forced-convection warm-air heating systems rely on a fan, or motor-driven blower, to circulate air from the furnace through the

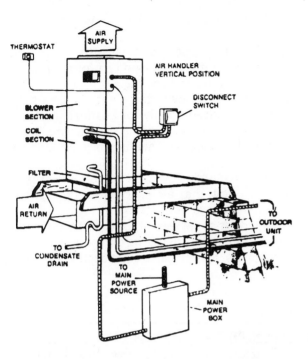

FIGURE 14-1B Forced convection warm-air heating system. The unit shown also incorporates a coil for summer air conditioning. (Courtesy of York International)

supply ducts to the individual rooms and then back to the furnace through a system of return ducts. Before we discuss specific ducting installations, the principles relating to the air volume delivered and to the static pressure, or resistance to airflow, in the heating system need to be understood. Gravity and forced-convection systems are illustrated in figure 14-1A and 14-1B.

Determining Proper Air Volume

Temperature rise: In order for any furnace to heat a building properly, the amount of hot air that it can deliver must fall within certain limits. Although the hot-air output of most furnaces can be made to vary over their operating Btu range, the most advantageous air delivery of a particular unit is based on the temperature rise that it is rated to deliver. Temperature rise is the degree difference between the supply and return air. These tempera-

ture measurements should be taken in the supply and return ducts located about 6 feet away from the furnace. Made in this way, the temperature readings will not be affected by any radiant heat produced by the furnace. In general, furnaces that are used strictly for space-heating should operate with a temperature rise of approximately 80~85°F. Furnaces that are used for both heating and cooling should produce a temperature rise of between 70° and 75°F. This difference in temperature rise arises from the fact that the air ducts in combination heating/cooling systems are larger than those designed for heating-only systems. Combination systems also use higher air velocities, which accounts for the lower temperature rise.

Air quantity: Once temperature rise has been measured, air volume can be calculated by using the following formula, which is applicable to most indoor oil or gas furnaces.

$$\text{Volume of delivered air} = \frac{\text{Btu input of furnace}}{\text{Temperature rise (°F)} \times 1.40}$$

If, for example, a 100,000-Btu/hr furnace is operating with a temperature rise of 75 degrees, then the volume of the air delivered from the furnace is:

$$\text{Volume} = \frac{100,000 \text{ Btu/hr}}{75 \times 1.40} = \frac{100,000}{105}$$

Volume = 952.3 cubic feet per minute (CFM)

Note that this calculation assumes a properly installed and operating furnace, including a Btu input adjusted to factory specifications.

Adjusting the temperature rise: Temperature rise should be adjusted on the basis of the nameplate rating on the specific furnace. To make this adjustment, the blower speed is increased to decrease temperature rise, or decreased to increase temperature rise. Many furnaces have two-speed or multispeed direct drive fans that can be adjusted by changing the electrical connections to the blower. If there are no multispeed electrical taps, the blower speed must be changed by switching the fan pulley and belts on belt-driven units. Opening the space between the pulleys (smaller drive motor pulleys) decreases blower speed. Closing the pulleys (larger drive motor pulleys) increases blower speed. For each 1-inch diameter increase in pulley size, the belt length must be increased by 2 inches and vice versa. Some furnaces use step pulleys on both the drive motor and fan. In these installations, pulley adjustment automatically compensates for belt length, since a switch to a smaller drive motor pulley is accompanied by a larger blower pulley,

which is matched to keep the drive belt distance the same.

Different fan speed settings also depend on whether the furnace is used just for heating or for both heating and cooling. Furnaces used for both operate at lower temperature rises than do furnaces used strictly for heating, since the duct sizes in air-conditioning systems are larger than in corresponding heating-only systems.

Static pressure: Static pressure is the resistance to flow that air encounters throughout the heating system. All components of the system offer some resistance to airflow, including duct elbows, grilles, air diffusers, and heat exchangers, as well as straight sections of duct

BLOWER PERFORMANCE

MODEL	SPEED TAP	EXTERNAL STATIC PRESSURE, INCHES WC									
		0.1	0.2	0.3	0.4	0.5	0.6	0.7	0.8	0.9	1.0
P2UDD06†03801	HI	795	760	720	665	625	550	505	435	---	---
	MED	485	465	455	440	435	395	350	295	---	---
	LOW	---	---	---	---	---	---	---	---	---	---
P2UDD10†05701	HI	1150	1105	1050	1000	960	905	850	790	720	655
	MED	900	885	850	815	780	760	705	655	605	515
	LOW	710	700	690	660	615	585	530	475	---	---
P2UDD12†07601	HI	1400	1360	1315	1255	1210	1135	1080	1035	935	820
	MED	1140	1110	1090	1055	1025	960	895	820	775	745
	LOW	840	825	815	795	765	710	690	635	585	560
P2UDD12†09501	HI	1490	1440	1375	1320	1255	1175	1115	985	945	845
	MED	1155	1110	1065	1010	955	850	830	765	---	---
	LOW	---	---	---	---	---	---	---	---	---	---
P2UDD20†09501	HI	2152	2089	2029	1955	1884	1810	1720	1639	1538	1438
	MED	1749	1723	1689	1656	1599	1556	1481	1426	1343	1242
	LOW	---	1403	1400	1387	1376	1362	1332	1266	1187	1104
P2UDD16†11401	HI	1830	1790	1710	1645	1550	1455	1375	1210	1155	1030
	MED	1690	1615	1550	1480	1420	1340	1240	1150	1040	950
	LOW	1380	1365	1325	1300	1235	1200	1125	1035	965	870
P2UDD20†13301	HI	2190	2130	2060	1980	1930	1780	1685	1635	1535	1390
	MED	1755	1735	1705	1665	1615	1545	1500	1425	1330	1240
	LOW	---	---	---	---	---	---	---	---	---	---

† May be P = Propane (LP), or N = Natural Gas
    Airflow expressed in standard cubic feet per minute.
NOTES: 1. Return air is through side opposite motor (left side).
    2. Air filter installed.
    3. Motor voltage at 115 V.

FIGURE 14-2 Static pressure ratings for a variety of warm-air furnaces. Note the ratings for three different blower speeds based on the electrical hook-up to the blower motor. (Courtesy of York International)

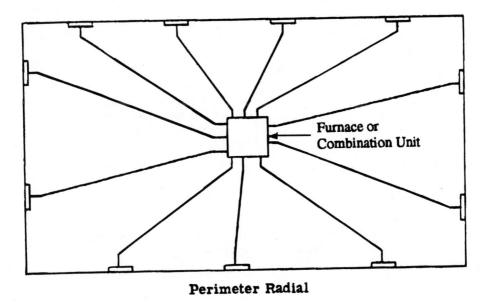

Perimeter Radial

HEATING

Limitations:

a. 1000 to 1200 square feet floor area.

b. Distance between any 2 diffusers should not exceed approx. 15 ft.

c. Furnace should be centrally located.

d. No warm air radial substantially longer than 20 ft.

e. Each diffuser –
   maximum 7,000 BTU's

Calculations:

Use NESCA Manual K., or ACCA Manual D.

SUMMER AIR CONDITIONING

This system is not recommended

FIGURE 14-3 Perimeter radial duct design. (Courtesy of the Williamson Corp.)

pipe. Total static pressure is made up of external static pressure consisting of components external to the furnace, as well as internal or furnace-component–related pressure. Furnace fans are rated to supply a given quantity of air at a specified maximum static pressure. Therefore, if a furnace is designed to deliver 1000 CFM of air at a total static pressure of .30 inch, it means that the total static resistance of all system components must be below .30 inch of w.c. measured in a manometer. Figure 14-2 illustrates manufacturers' ratings for a series of furnaces, giving the CFM air delivery at different static pressures using any one of three speed taps for running the fan motor. It should be remembered that the term *static* means with the

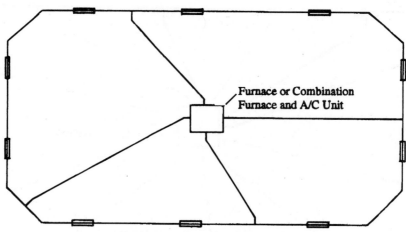

Perimeter Loop (Method "A")

HEATING

Limitations:

    a. Heat loss - maximum 60,000 BTU's.

    b. Perimeter of structure - 140 ft.

    c. Furnace centrally located.

    d. Maximum length of a feeder -- 30 ft.

Specifications:

    a. 6,000 to 8,000 BTU's per diffuser.

    b. 15,000 BTU's per feeder.

    c. Maximum distance between feeders on loop -- 35 ft.

    d. 3 diffusers in loop, between any 2 feeders.

    e. 18 inches minimum distance between diffuser and feeder on loop.

    f. 15 ft. maximum distance between diffuser and feeder on loop.

    g. 20 ft. maximum distance between 2 adjacent diffusers on loop.

    h. Never install a diffuser on a feeder duct.

    i. Largest feeder diameter determines perimeter loop pipe diameter.

    j. First feeder location - coldest corner of room with the largest heat loss.

Calculations - feeders and loop sizes:

    Use NESCA Manual K., or ACCA Manual D.

SUMMER AIR CONDITIONING

Limitations:

    a. Heat gain - maximum 60,000 BTU's.

    b. Perimeter of structure 140 ft.

    c. Evaporator coil - centrally located.

    d. Maximum length of a feeder -- 30 ft.

Specifications:

    a. Allow 2400 BTU's total heat per diffuser (10 x 6).

    b. Total heat per feeder - use NESCA Manual K., or ACCA Manual D.

    c. 18 inches minimum distance between diffuser and feeder on loop.

    d. Never install a diffuser on a feeder duct.

    e. Loop pipe diameter same as feeder diameter.

Calculations - Feeder & loop sizes:

    Use NESCA Manual K.

FIGURE 14-4 Perimeter loop duct design. (Courtesy of the Williamson Corp.)

furnace and blower operating; it does not mean with the blower off.

Larger fan motors are generally installed in dual-purpose heating/air-conditioning furnaces since they operate at higher static pressures than do heating-only furnaces. Therefore, a small 100,000-Btu/hr hot-air heating-only furnace might use a 1/3-horsepower fan motor, whereas the same unit used for both heating and cooling would require a 1/2-horsepower fan motor to overcome the additional static resistance imposed by the larger ducts and evaporator coils found in the air-conditioning system.

Supply and Return Ducting

Supply ducts from the furnace are designed and installed to provide efficient heating distribution in a manner similar to that of their hydronic and steam heating system counterparts. For example, the perimeter radial distribution system illustrated in figure 14-3 is similar to a one-pipe steam system. A central plenum is used to collect heat from the furnace, which is then channeled out through individual ducts for delivery to each room in the building.

Perimeter radial duct systems are usually limited to installations of less than 1200 square feet, and are not recommended for use in air-conditioning applications.

The perimeter loop system shown in figure 14-4 is similar in design to the supply loop hydronic arrangement. The perimeter loop uses feeder and loop ducts of the same diameter. This type of arrangement is employed in applications where the heat loss from the building is less than 60,000 Btu, along with a maximum perimeter distance of 140 feet.

If the heat loss of the building is greater than 60,000 Btu, a second type of perimeter loop design, illustrated in figure 14-5, can be installed. This type of loop utilizes a central feeder trunk while maintaining the perimeter design.

Another common type of supply system incorporates a main supply trunk with smaller branch circuits feeding the individual rooms, as shown in figure 14-6. The main supply trunk

acts as an extension of the furnace plenum and can be brought out from either side of the furnace for convenient connection to individual feeder circuits. These trunk arrangements can be used in either supply or return ducting systems.

Duct installation practices: Given the variety of duct systems available for installation in any given system, some general practices that are applicable to all ducted systems still apply.

1. The technician should be aware of the CFM that the system blower can deliver. This information should be available in the product literature. Note from figure 14-2 that most units can overcome a static pressure of .2 inch and deliver between 1000 to 2000 CFM.

2. The air capacity of each size of duct should be known. Listed below are the air capacities for both round and trunk ducts at .1 inch static pressure.

Round Pipe		Trunk Ducts	
SIZE, INCHES	CFM	SIZE, INCHES	CFM
6	100	10 x 8	400
7	150	14 x 8	500
8	200	18 x 8	800
9	300	20 x 8	1000
10	400	24 x 8	1200
12	650		
14	1000		

3. The longest duct runs should be kept under 100 feet of equivalent length. Equivalent length in this setting is defined as the actual distance from the plenum plus the equivalent foot totals from all fittings, such as elbows and transition pieces, on a particular run.

4. Air volume dampers should be used at each branch takeoff to enable fine-tuning of the air supply.

5. The use of quality components is recommended in order to maximize system performance.

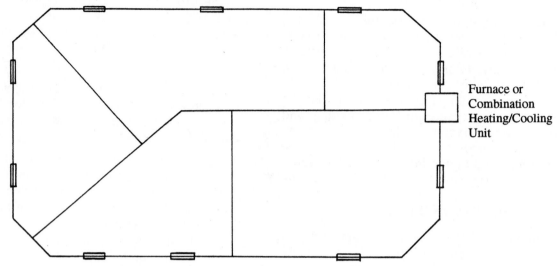

Perimeter Loop (Method "B") - with trunk duct

HEATING

Limitations:

   a. Heat loss maximum 100,000 BTU's -- if more than 100,000 use 2 furnaces.

   b. Perimeter of structure - 210 ft.

   c. Maximum length of feeder 30 ft.

   d. Maximum length of duct 50 ft.

Specifications:

   Same as Method "A" - Perimeter Loop

Calculations - Feeders & Ducts:

   a. Use NESCA Manual K., or ACCA Manual D.

   b. Return air - Same as Method A -- Perimeter Loop.

SUMMER AIR CONDITIONING

Limitations:

   a. Heat gain - maximum 60,000 BTU's.

   b. Perimeter of structure - 210 ft.

   c. Maximum length of feeder - 30 ft.

   d. Maximum length of duct - 50 ft.

Specifications

   a. Allow 2400 BTU's total heat per diffuser (10 x 6).

   b. Total heat per feeder - See NESCA Manual K., or ACCA Manual D.

   c. 18 inches minimum distance between diffuser and feeder on loop.

   d. Never install a diffuser on a feeder duct.

   e. Loop pipe diameter same as feeder diameter.

Calculations - Feeder, loop and duct:

   Use NESCA Manual K., or ACCA Manual D.

FIGURE 14-5 Perimeter loop with trunk duct. (Courtesy of the Williamson Corp.)

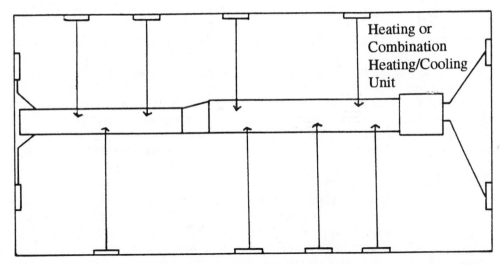

Extended Plenum System

HEATING

Limitations:

a. Maximum **7000 BTU's per diffuser.**

b. Any one extended plenum should not exceed 35 ft. in length.

c. 1200 CFM maximum per extended plenum duct.

d. Whenever any one extended plenum exceeds 20 ft. in length, make one reduction midway in order to maintain velocity.

Calculations - Pipe & Ducts:

Use NESCA Manual K., or ACCA Manual D.

SUMMER AIR CONDITIONING

Limitations:

a. Refer to NESCA Manual K.

b. For return air - Use NESCA Manual K., or ACCA Manual D.

FIGURE 14-6 Extended plenum system. (Courtesy of the Williamson Corp.)

Return air systems: The total area of the return air system must equal that of the supply system if the heating system is to be balanced. One easy method of determining whether the return air system contains adequate duct area is to turn on the furnace fan with the door to the basement or furnace room slightly ajar, and then remove the cover over the fan compart- ment. If the furnace room door is drawn shut, then the return air ducting to the furnace is inadequate. Return air ducts should be located a distance from supply ducts in order to prevent a short cycle of air from moving from the hot-air supply system directly into the cold-air returns. Because of the contaminants in the air, return ducts are not installed in garages, kitchens, or bathrooms. These rooms utilize adjacent ducts for return air purposes.

Although it is common practice to locate return air ducts along the lower parts of walls near the floor, this location usually creates floor-level drafts. To avoid this problem, return air ducts can be located higher up on inside walls, as long as they are a sufficient distance from the supply air registers.

Duct sizing: Sizing the ductwork properly in any heating or combination heating/air-conditioning system is a necessary prerequisite for efficient system operation. Many companies that sell heating and air-conditioning equipment also supply worksheets and specially designed

sizing slide rules that can be used to determine the proper duct sizes based on the Btu capacity and physical size of the heating system and building. Heating and cooling engineers will usually consult original architects' drawings when designing any HVAC system.

Improperly sized ductwork is often at the root of many complaints of improper heating and cooling system function. For best results, duct dimensions should be as square as possible, with the straightest runs practical.

For further reference, the reader is directed to the Association of Heating, Refrigeration and Air-Conditioning Engineers (ASHRAE) handbook, as well as publications from the Air Conditioning Contractors of America (ACCA).

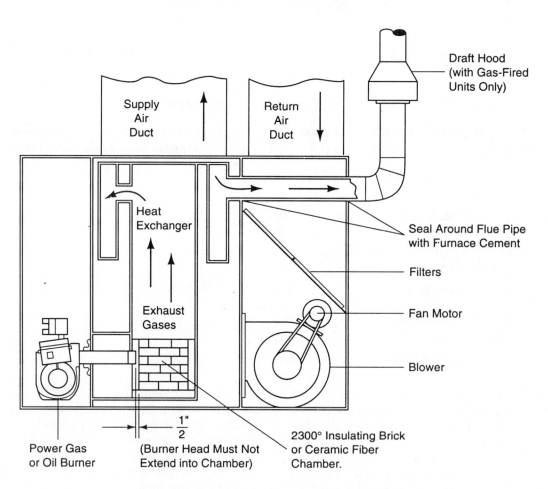

FIGURE 14-7 Typical warm-air furnace components arrangement.

FURNACE OPERATING SEQUENCES

Introduction to Common Components

Most of the characteristics and components of hot-air furnaces are similar regardless of the fuel used to supply the heat. Figure 14-7 illustrates a typical warm-air furnace, along with the location of common components.

In this component arrangement, return air from the building enters the blower cabinet through the air filter assembly. As air is forced over the external surfaces of the heat exchanger, it is heated, and then exits the warm-air outlet. Traveling through the ductwork, the warm air enters the rooms through wall or floor registers. Colder room air enters the return ductwork in order to continue the heating cycle as long as the thermostat is calling for heat.

Furnace Design Categories

Furnace designs fall into one of four categories, based on the arrangement of the components, as shown in figure 14-8. Note the low-boy

TYPICAL APPLICATIONS

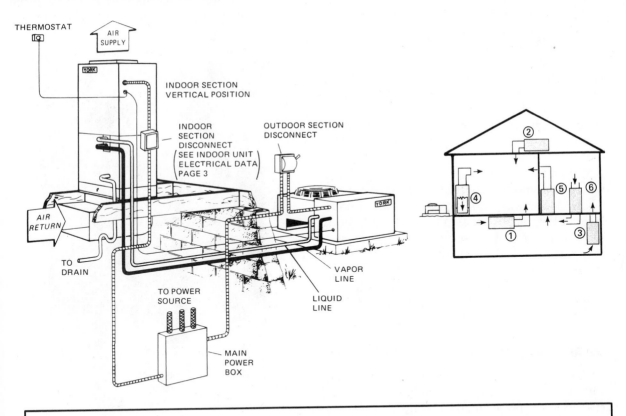

① Horizontal Suspended   ③ Upflow wall-mounted   ⑤ Upflow-basement or crawl space return

② Horizontal (Attic or Crawl)   ④ Upflow Plenum Return   ⑥ Counterflow

FIGURE 14-8 Furnace design categories. The designs are based on the direction of airflow through the furnace. The use of any particular furnace design is based on the space available and the type of construction in the dwelling (slab, crawl space, etc.). (Courtesy of York International)

component arrangement. This design features the blower and heat exchanger compartments side-by-side, with a low-height profile that is accomplished by increasing the floor space of the unit. The low profile makes this the unit of choice for most basement installations where low ceilings and limited headroom require maximum clearances above the furnace for hot and return air plenum installations.

In contrast to the low-boy arrangement, high-boy configurations are sometimes referred to as upflow furnaces, since return air enters the blower cabinet located below the heat exchanger and is forced upward through the warm-air plenum. High-boy furnaces are most commonly installed where floor space is limited, such as in utility rooms and small basement areas that have sufficient headroom.

The horizontal furnace is commonly installed in attics and crawl spaces that present height restrictions. In the horizontal design, air is drawn into one end of the unit and forced by the blower horizontally over the heat exchanger and out of the warm-air plenum on the other end of the unit.

Counterflow furnaces are used in buildings where the supply ductwork is located under the floor in an area too small for the furnace installation, such as small crawl and work spaces. Return air enters the furnace blower compartment at the top of the unit. The fan forces the air down over the heat exchanger, which is located below the fan compartment, and out of the bottom of the furnace into the warm-air distribution network.

Oil and Gas Furnace Sequences

The basic principles and controls associated with gas- and oil-fired technology were discussed in Chapters 10 and 11. These principles are relevant to most warm-air furnaces. This section focuses on the operating cycles and controls of warm-air furnaces that are applicable to both oil and gas-fired combustion systems. Operating distinctions that are fuel dependent will be highlighted.

The operating cycle of a gas- or oil-fired furnace is started by a call for heat from the room thermostat. (See Chapter 12 for a complete discussion of thermostat controls.) The operating logic of a typical warm-air furnace is illustrated by the following sequence of events.

Thermostat → Low-voltage relay → High limit → Gas valve or oil burner primary → Fan/limit blower

This operating sequence is the same regardless of the component design of the furnace or the fuel being burned (with the exception of solid-fuel and solar applications, which will be discussed later in this chapter). A drop in room temperature results in a call for heat from the

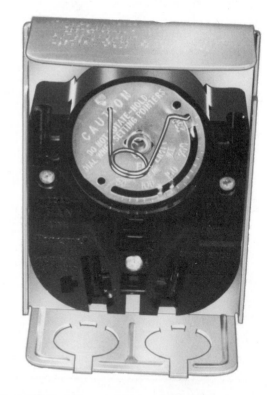

FIGURE 14-9 Fan limit switch. These switches are available with different probe lengths, which are selected based on the physical size of the warm-air plenum. The heat-sensing probe extend far enough into the plenum accurately to sense the temperature in the airstream as it emerges from the heat exchanger. (Courtesy of Honeywell, Inc.)

room thermostat, which closes the electrical circuit to either the gas valve or oil burner primary control. On gas-fired furnaces that incorporate a standby pilot system, the gas valve is energized to fire the main burners. Intermittent spark ignition systems generate a pilot light that lights the main gas burner ports. In oil-fired systems, the oil burner primary initiates burner operation.

With continued fuel combustion, the temperature in the heat exchanger rises to the fan-on setting point of the fan/limit switch. When this setting has been reached, the control closes the electrical circuit to the blower to begin moving air throughout the heating system. Note that the blower depends on the temperature of the heat exchanger for operation, and not on the room thermostat. A typical combination control is illustrated in figure 14-9.

Note from figure 14-9 that the combination fan/limit switch contains three different set points: the fan-off set point, the fan-on setpoint; and the high-limit setting. Most of these controls also incorporate a switch for turning on the fan manually. The manual switch is generally used for running the fan during the summer to cool the structure by circulating basement air throughout. Recommended settings for these switch points are as follows:

Fan off: 100°F
Fan on: 150°F
High limit: 200°F

The combination control is mounted directly onto the furnace plenum. Fan operation is controlled by the rotation of the bimetallic coil that protrudes into the plenum from the back of the switch (figure 14-10).

During heating system operation, the air in the plenum is heated by fuel combustion and heat transfer from the heat exchanger. The fan/limit switch dial rotates until it hits the fan-on set point and the electrical contacts close. When the fan is turned on, air circulates over

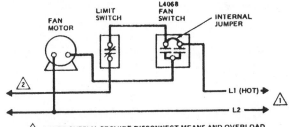

Typical wiring **used in forced** **air heating system.**

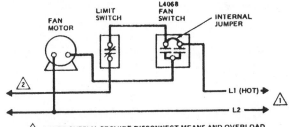

FIGURE 14-11 Electrical circuit in forced hot-air heating system. Note from this schematic that in the event that the high-limit switch opens, power to the fan is still maintained, where power to the burner is shut off.

the heat exchanger surfaces, which causes the temperature in the plenum gradually to drop to the fan-off setting as air is circulated throughout the heating system. At the fan-off switch point, the fan circuit is deenergized, allowing the plenum temperature to rise gradually to the fan-on set-point. This on-off cycle continues until the room thermostat has been satisfied and the burner has cycled off.

Gas or oil burner operation is controlled by a combination of the thermostat circuit and high-limit setting of the fan/limit switch (sometimes called a combination control). During the heating cycle, the burner will remain on until

FIGURE 14-10 Bimetallic coil on fan limit switch. The bimetallic coil expands or contracts depending on the temperature of the air. The expansion and contraction are changed to rotary motion, which opens and closes the contacts in the switch.

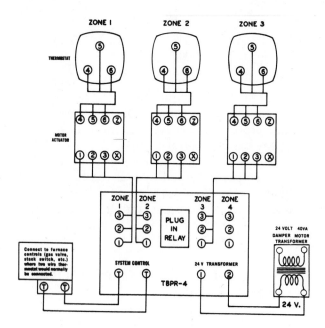

For systems which supply heating only, each thermostat terminal is wired to the correspondingly numbered terminal of the damper motor actuator. Motor terminals 1, 2 and 3 are wired to corresponding terminals of TBPR-4 Panel. 24 volt power to operate up to four zone motors is provided by 40VA transformer.

A call from zone thermostat causes the damper motor to move to open position. As damper opens the built-in end switch in the motor actuator makes and causes the plug in relay to pull in and make the TT circuit. The TT circuit connects to the furnace controls, (or cooling controls) wherever a thermostat would normally be connected.

FOR SYSTEMS WHICH SUPPLY COOLING ONLY.

To make thermostats open actuators on temperature rise instead of temperature drop, reverse wires 4 and 6 on thermostat. Connect thermostat terminal 6 to motor actuator terminals 4, and connect thermostat terminal 4 to motor actuator 6.

This cooling system wiring arrangement will not provide for Continuous Air Circulation as all dampers are closed when thermostats are satisfied. For CAC operation on heating or cooling, use Mastertrol Panel.

The use of a TBPR-4 Panel simplifies wiring and provides a relay between the gas valve or other heating or cooling control contactors or relays. The renewable plug in relay of the panel lengthens and protects the life of the actuator end switch.

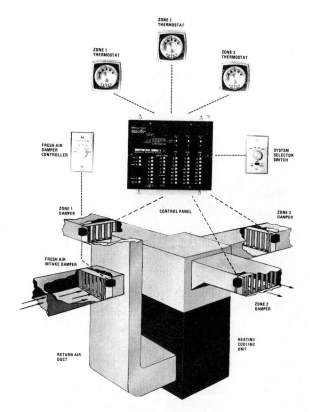

FIGURE 14-12 Three-zone warm-air/air-conditioning control with fresh-air intake. (Courtesy of Trol-A-Temp)

the limit setting on the fan/limit switch is reached. Most combination controls feature a fixed differential of 25 degrees. Thus, if the high-limit is set at 200 degrees, the burner will turn off when the plenum temperature reaches 200 degrees, and will reenergize when the plenum temperature drops to 175 degrees (assuming there is still a call for heat from the room thermostat). The high-limit setting on the limit switch serves to interrupt the flow of electric power to either the oil burner or gas valve, and functions as the main overheat safety switch in the system. Figure 14-11 illustrates a hot-air furnace electrical circuit that contains a fan switch with a separate high-limit controller.

Most high-limit switches come factory preset at 200 degrees, with a built-in fixed differential of approximately 25 degrees. Whether a separate high-limit or combination fan/limit control is used, the operating action of all switches should be checked periodically to determine both the accuracy of the switching points and the proper activation of all switch controls.

The operating sequence just described is true of single-zone hot-air systems only. When the heating system is designed for multizone heating, operating sequences are somewhat different (just as they are in multizone hydronic systems).

MULTIZONE HOT-AIR SYSTEM OPERATION

Duct systems for hot-air heating (and air conditioning) are sized to provide heating and cooling for all rooms at the same time, regardless of the needs in the individual zones. Heat gain from the sun as it tracks across the sky and stratified air in a home, caused by warm air rising to the second story and cooler air dropping to the first floor, create varied heating demands in different parts of a building. Zoning applications follow the same logic in warm-air installations as in hydronic systems— to block the flow of heat in the fluid (or air) carrier that goes to specific zones. In hot-air systems, this is accomplished through the use of motorized dampers that function in a manner similar to that of the motorized zone valves utilized in hydronic applications.

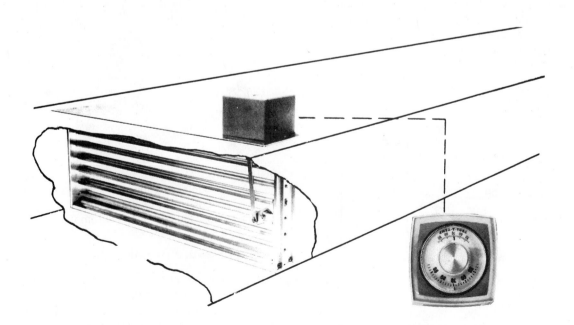

FIGURE 14-13 Motorized damper for zoning warm-air system. (Courtesy of Trol-A-Temp)

System Logic

Figure 14-12 illustrates a three-zone warm-air/air-conditioning control system equipped with a fresh-air intake vent. This type of system uses a motorized blade damper that is inserted into the duct (figure 14-13).

The electrical circuits for controlling the zoning functions are similar to those of their hydronic counterparts, and are shown in figure 14-14. The sequence of system operation is as follows:

1. On a call for heat from the thermostat, the low-voltage circuit closes, activating the motor actuator to open the damper.

2. The 24-volt transformer provides power to the damper motor to open the damper blades in the ducts feeding the area, or zone, that is calling for heat.

3. When the damper is completely open, the end switch contacts close in the motor actuator, which closes the thermostat terminals in the furnace, beginning burner operation.

4. When the thermostat has been satisfied, the thermostat circuit opens, which causes the damper to close. Switch contacts open the circuit to the burner, turning it off when the damper has closed.

In applications where the ductwork is not accessible for the installation of internal motorized dampers, motorized registers can be used in place of conventional wall or floor registers for zoning regulation. Low-voltage wiring to these registers is usually snaked through the existing ductwork.

To correct problems in rooms that are either too warm or too cool, motorized registers can be installed that open and close in response to a room thermostat. These registers control only the flow of heated or cooled air to that room without affecting the operation of the central heating or cooling unit. Therefore, if the room is heated above the thermostat set temperature, the motorized register will close. If the temperature in the room falls below the thermostat set temperature, the motorized register will open. However, the room will only be heated when the main furnace thermostat again calls for heat. A motorized wall register of this type is shown in figure 14-15.

FIGURE 14-15 Motorized wall register. This type of register is used to control the heat in an individual room and does not control the operation of the burner control on the furnace. (Courtesy of Trol-A-Temp)

Static Pressure Regulation

Regulation of static pressure is required on all forced-convection heating and cooling systems. This need arises as a result of the automatic closing of various zone dampers, which increases air pressure and velocity through

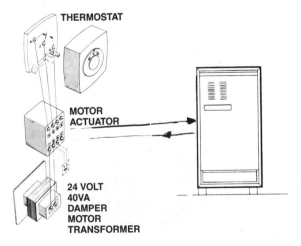

FIGURE 14-14 Damper control logic. (Courtesy of Trol-A-Temp)

the duct system. If enough zones are closed, the increase in air pressure may, in some instances, overcome the static pressure rating of the blower motor, resulting in a reduction of air volume throughout the system. To remedy this situation, a static pressure regulating damper of the type in figure 14-16 can be installed in the duct system.

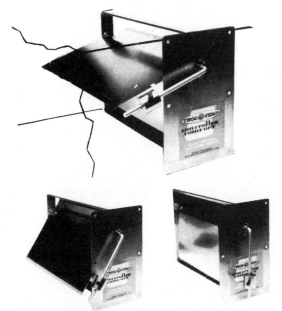

FIGURE 14-16 Static pressure damper. These dampers are used to control static pressure in the duct system. (Courtesy of Trol-A-Temp)

The static pressure damper has a weighted arm that controls a single-blade barometric-type damper similar to that used in oil-fired systems to control draft. The weight is adjusted to open the damper blade when the static pressure reaches a specified level. Ordinarily, this bypass air is directed into a hallway, basement, or other area of the building where temperature levels are not critical.

Fresh-Air Differential Economizers

Differential temperature controllers are useful for controlling interior comfort levels as outdoor temperatures vary with the seasons. Admitting varying amounts of fresh air into a

building allows for better control of humidity levels. It enables the circulation of fresh air when indoor and outdoor temperatures are about the same, and also permits the proper mixing of indoor and outdoor air based on varying temperture differences. We have seen how this logic is handled in hydronic systems by varying the boiler water temperature with changes in outdoor temperatures. In forced-convection systems, this same control logic applies by admitting varying amounts of outdoor air to temper the air that is delivered from the furnace to the building interior. These systems can either be simple, as in a small residential installation, or complex, as required in large commercial installations.

A simple fresh-air economizer system suitable for small residential installations is shown in figure 14-17.

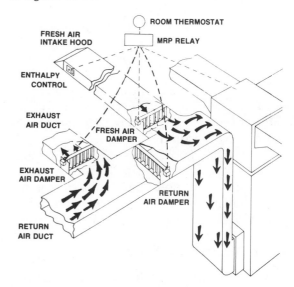

FIGURE 14-17 Residential and light commercial fresh-air economizer system. (Courtesy of Trol-A-Temp)

Note that the fresh-air intake and ducted return air registers are controlled from a single source. The amount of fresh air to be admitted is selected and inputted to a controller, which opens the fresh-air damper shutters to any one of a number of preset positions. This controller simultaneously closes the return air damper to

COMMERCIAL ROOF TOP

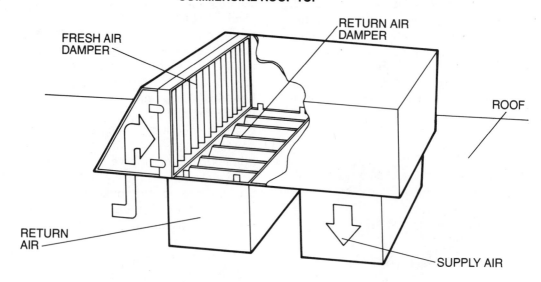

FIGURE 14-18 Commercial fresh-air economizer system. (Courtesy of Trol-A-Temp)

a position that restricts return air from entering the furnace in an amount proportional to the fresh air entering the system. Installing a differential temperature control allows this system to function as a true economizer by regulating fresh-air intake based on varying outdoor temperatures during both the heating and cooling seasons.

Commercial applications generally require not only regulation of fresh and return air, but also control of both the furnace and air-conditioning compressor controls. A commercial system of this type is illustrated in figure 14-18.

Air-to-Air Heat Exchangers

With the advent of thick insulation, energy-saving doors and windows, and newer building materials, less energy is necessary to provide for heating and cooling. However, a buildup of indoor air pollutants and other materials that would be vented under normal air exchanges associated with conventional construction practices takes place. To solve this problem, air-recovery systems are available that exchange stale indoor air for fresh outdoor air. In this process, heat is transferred from the indoor air to the fresh outdoor air through an air-to-air

plate-type heat exchanger. This type of device, which is shown in figure 14-19, uses counterflow airstreams that are brought into close proximity to one another, separated by the heat exchanger surface.

Air Flow Diagram

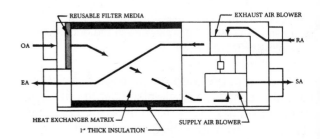

SA = SUPPLY AIR
RA = RETURN AIR
OA = OUTSIDE AIR
EA = EXHAUST AIR

FIGURE 14-19 Air-to-air heat exchanger. Heat transfer from outgoing warm air into incoming cooler air allows for efficient fresh-air makeup, minimizing additional heating requirements by maximizing heat reclamation. (Courtesy of Des Champs Laboratories, Inc.)

In this design, two separate air passages are created. As both streams of air flow through the heat exchanger in this counterflow air pattern, heat is transferred from the warmer outgoing air to the cooler incoming air. During the summer, the opposite action takes place. Cooler, outgoing indoor air drops the temperature of the warmer, incoming outdoor air. This transfer of heat during both the summer and winter can account for substantial savings over ventilation systems that do not utilize heat exchanger technology. The effectiveness of a specific unit depends on airflow and humidity and temeprature levels, although efficiencies of the process should range between 70% and 90%.

AIR CLEANERS

The use of electronic air cleaners has grown in popularity as more attention has been directed to the issue of indoor air quality. As building codes continue to require tighter structural envelopes to mimize air infiltration, the buildup of contaminants in a building becomes more of a problem. The efficiency of standard air filter elements to remove contaminants depends on the size of the offending particles and the filter's ability to remove them. To put this in perspective, figure 14-20 shows the sizes of various common airborne particles, along with the ability of three common types of filter media to remove them from the airstream. Particle sizes are given in microns (1 micron is equivalent to 1/1000 of a millimeter!).

Note from figure 14-20 that standard furnace filter elements are generally effective only in removing particles that are visible. As particle size is reduced, common furnace filters become relatively ineffective. Media air filters are more effective than standard furnace filters, and utilize a specially constructed filtration element that can remove particles as small as .5 micron. A media filter consists of a specially constructed cabinet containing the filter element, which is installed in the return air duct of the warm-air heating system; see figure 14-21.

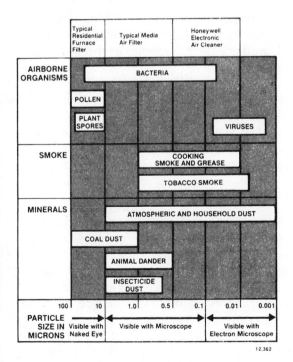

FIGURE 14-20 Particle sizes of selected air contaminants.

FIGURE 14-21 Media air filter element designed to filter out a variety of contaminating particles that are too small to be effectively removed by standard air filter elements. (Courtesy of Honeywell, Inc.)

Media air filters do not require any electrical connections, although some units can be modified for electronic use.

Electronic air cleaners come in a variety of designs for use in residential and commercial applications that have different installation requirements. Figure 14-22 shows two electronic

FIGURE 14-22A Electronic air filter installed in return air duct. (Courtesy of Honeywell, Inc.)

FIGURE 14-22B Electronic air filter installed as grille-mounted unit. (Courtesy of Honeywell, Inc.)

air cleaners; a unit designed to be installed in the return air duct (A), and a grille-mounted unit that replaces a standard wall or ceiling grille for central forced-air applications (B).

In small office buildings and large commercial establishments, air cleaning is usually accomplished by the use of separate units installed throughout the building. These air cleaners are available for wall or ceiling mounting, and come equipped with their own filter and cell assemblies. Each separate unit also requires its own connection to electric power.

Many electronic air cleaners are equipped with a prefilter element for removing larger particulate matter and an electronic cell to remove the small airborne particles. Most of these units place an electrostatic charge on the particles in the airstream passing through the filter. Stationary plates in the cell maintain a charge opposite to that of the airborne particles, drawing the particles onto the plates and out of the passing air. A simplified arrangement of these components is illustrated in figure 14-23.

An inexpensive alternative to the electrostatic air cleaner is to use electrostatic filters in place of ordinary filter elements in the furnace; see figure 14-24.

The type of electrostatic filter shown is washable and permanent. Some manufacturers offer charcoal inserts to filter out the smell of tobacco smoke and other odor contaminants.

HUMIDIFICATION

Operating Environment

The purpose of humidification in any heating system is to maximize comfort levels. As pointed out in Chapter 2, relative humidity is a key factor in determining human comfort in either a heated or a cooled environment. The cooler the air is, the less moisture it will hold, and vice versa. Therefore, when outdoor temperatures are low, the air contains relatively little moisture. When this air is heated in the furnace, its relative humidity drops unless moisture can be added to it.

F52E Electronic Air Cleaner

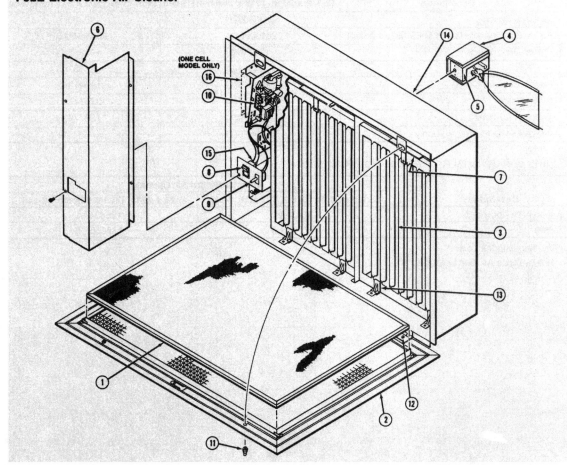

FIGURE 14-23A Parts of a typical electronic air cleaner. (Courtesy of Honeywell, Inc.)

FIGURE 14-23B

Maintaining relative humidity levels of between 35% and 55% is a desirable goal during the winter heating months. When humidity levels are too low, not only do a building's occupants feel cold, but

FIGURE 14-24 Electrostatic air cleaner element. (Courtesy of Newtron Products, Inc.)

RELATIVE HUMIDITY TABLE (In Percent) +30 +120 °F.

Difference Between Wet and Dry Thermometers

Dry Bulb Temp. °F.	0	1	2	3	4	5	6	7	8	9	10	11	12	13	14	15	16	17	18	19	20	21	22	23	24	25
30	100	89	78	67	56	46	36	26	16	6																
32	100	89	79	69	59	49	39	30	20	11	2															
34	100	90	81	71	62	52	43	34	25	16	8															
36	100	91	82	73	64	55	46	38	29	21	13	5														
38	100	91	83	75	66	58	50	42	33	25	17	10	2													
40	100	92	83	75	68	60	52	45	37	29	22	15	7	0												
42	100	92	85	77	69	62	55	47	40	33	26	19	12	5												
44	100	93	85	78	71	63	56	49	43	36	30	23	16	10	4											
46	100	93	86	79	72	65	58	52	45	39	32	26	20	14	8	2										
48	100	93	86	79	73	66	60	54	47	41	35	29	23	18	12	7	1									
50	100	93	87	80	74	67	61	55	49	43	38	32	27	21	15	10	5	0								
52	100	94	87	81	75	69	63	57	51	46	40	35	29	24	19	14	9	4								
54	100	94	88	82	76	70	64	59	53	48	42	37	32	27	22	17	12	8	3							
56	100	94	88	82	76	71	65	60	55	50	44	39	34	30	25	20	16	11	7	2						
58	100	94	88	83	77	72	66	61	56	51	46	41	37	32	27	23	18	14	10	6	1					
60	100	94	89	83	78	73	68	63	58	53	48	43	39	34	30	26	21	17	13	9	5	1				
62	100	94	89	84	79	74	69	64	59	54	50	45	41	36	32	28	24	20	16	12	8	4	1			
64	100	95	90	84	79	74	70	65	60	56	51	47	43	38	34	30	26	22	18	15	11	7	4	0		
66	100	95	90	85	80	75	71	66	61	57	53	48	44	40	36	32	29	25	21	17	14	10	7	3	0	
68	100	95	90	85	80	76	71	67	62	58	54	50	46	42	38	34	31	27	23	20	16	13	10	6	3	
70	100	95	90	86	81	77	72	68	64	59	55	51	48	44	40	36	33	29	25	22	19	15	12	9	6	3
72	100	95	91	86	82	77	73	69	65	61	57	53	49	45	42	38	34	31	28	24	21	18	15	12	9	6
74	100	95	91	86	82	78	74	69	65	61	58	54	50	47	43	39	36	33	29	26	23	20	17	14	11	8
76	100	96	91	87	82	78	74	70	66	62	59	55	51	48	44	41	38	34	31	28	25	22	19	16	13	11
78	100	96	91	87	83	79	75	71	67	63	60	56	53	49	46	43	39	36	33	30	27	24	21	18	16	13
80	100	96	91	87	83	79	75	72	68	64	61	57	54	50	47	44	41	38	35	32	29	26	23	20	18	15
82	100	96	92	88	84	80	76	72	69	65	61	58	55	51	48	45	42	39	36	33	30	28	25	22	20	17
84	100	96	92	88	84	80	76	73	69	66	62	59	56	52	49	46	43	40	37	35	32	29	26	24	21	19
86	100	96	92	88	84	81	77	73	70	66	63	60	57	53	50	47	44	42	39	36	33	31	28	26	23	21
88	100	96	92	88	85	81	77	74	70	67	64	61	58	54	51	48	46	43	40	37	35	32	30	27	25	22
90	100	96	92	89	85	81	78	74	71	68	65	61	58	55	52	49	47	44	41	39	36	34	31	29	26	24
92	100	96	92	89	85	82	78	75	72	68	65	62	59	56	53	50	48	45	42	40	37	35	32	30	28	25
94	100	96	93	89	85	82	79	75	72	69	66	63	60	57	54	51	49	46	43	41	38	36	33	31	29	27
96	100	96	93	89	86	82	79	76	73	69	66	63	61	58	55	52	50	47	44	42	39	37	35	32	30	28
98	100	96	93	89	86	83	79	76	73	70	67	64	61	58	56	53	50	48	45	43	40	38	36	34	32	29
100	100	96	93	89	86	83	80	77	73	70	68	65	62	59	56	54	51	49	46	44	41	39	37	35	33	30
102	100	96	93	90	86	83	80	77	74	71	68	65	62	60	57	55	52	49	47	45	42	40	38	36	34	32
104	100	97	93	90	87	83	80	77	74	71	69	66	63	60	58	55	53	50	48	46	43	41	39	37	35	33
106	100	97	93	90	87	84	81	78	75	72	69	66	64	61	58	56	53	51	49	46	44	42	40	38	36	34
108	100	97	93	90	87	84	81	78	75	72	70	67	64	62	59	57	54	52	49	47	45	43	42	39	37	35
110	100	97	93	90	87	84	81	78	75	73	70	67	65	62	60	57	55	52	50	48	46	44	42	40	38	36
112	100	97	94	90	87	84	81	79	76	73	70	68	65	63	60	58	55	53	51	49	47	44	42	40	38	36
114	100	97	94	91	88	85	82	79	76	74	71	68	66	63	61	58	56	54	52	49	47	45	43	41	39	37
116	100	97	94	91	88	85	82	79	76	74	71	69	66	64	61	59	57	54	52	50	48	46	44	42	40	38
118	100	97	94	91	88	85	82	79	77	74	72	69	67	64	62	59	57	55	53	51	49	47	45	43	41	39
120	100	97	94	91	88	85	82	80	77	74	72	69	67	65	62	60	58	55	53	51	49	47	45	43	41	40

HOW TO USE THE HYGROMETER TABLE

Suppose the bulb temperature reads 80° and the wet bulb temperature is 68°. The difference between the wet and dry thermometers is therefore 12°. Under the column head "Dry Bulb Temp." locate the "80" line; then follow over to the right until you reach the vertical column headed "12". This gives the figure "54", indicating a relative humidity of 54%.

FIGURE 14-25 Relative humidity table.

wood and plaster can shrink, and waking up with a dry throat is a common experience. Walking across a carpeted surface and getting a shock when you touch a light switch or pet a dog or cat is also a sign of humidity levels that are too low. Conversely, if humidity levels are too high, condensation problems can occur, since the dew point will be reached on many surfaces, such as the inside surfaces of windows, building sheathing, and insulaton.

The relative humidity of the environment can be determined by taking a wet- and dry-bulb temperature reading of the area, and then consulting the relative humidity chart shown in figure 14-25.

Humidistats

Humidistats sense relative humidity in the air and regulate a power humidifier to cycle on or off in response to the humidity level that has been set on the humidistat. A moisture-sensitive nylon ribbon or similar material is used to sense the humidity and provide the switching action that opens or closes the low-voltage circuit to the humidifier. Humidistats can be mounted directly in the return air duct of the heating system, or on the wall like a room thermostat (figure 14-26).

FIGURE 14-26 Humidistat for power humidifier installed on warm-air furnace. (Courtesy of Herrmidifier Co., Inc.)

Types of Humidifiers

Humidifiers are available in a variety of design and operating configurations, including evaporative, rotating-drum/pad, atomizer, and steam-assisted types. The humidifer is usually installed so that the evaporator surfaces of the pads and drums are in contact with the passing airstream in the plenum. Some humidifiers are equipped with a duct that connects the cold-air return plenum to the body of the humidifier in order to siphon a portion of the returning cold air, as shown in figure 14-27.

Evaporation humidifers: Most evaporation humidifiers are of the plate type, in which a float valve controls the level of water in a small reservoir. This reservoir feeds water into a shallow tray in which a number of absorbent humidifier plates, or pads, are suspended. These pads have a large surface area and humidify the air as it passes over and through the moisture-laden pads. Relative humidity is controlled by either increasing or decreasing the number of humidifier plates installed in the evaporation tray. Figure 14-28 illustrates a plate-type humidifier, and includes the moisture output capacity of the humidifier based on the number of plates in the unit. Evaporative-type humidifiers are easily installed and have no electrical connections or humidistat controls associated with their operation. Whenever the furnace fan is operating, the humidifier operates by virtue of the location of the pads in the passing airstream.

Keeping the float valve assembly free of scaling and impurities is critical to the proper operation of these units. These impurities, if allowed to collect, can build up and jam the float mechanism, causing the water reservoir to overflow and leak onto the heat exchanger assembly of the furnace. These units are equipped with an overfill drain tube that is incorporated into the reservoir of the humidifier to help prevent such a condition. It should be noted that continual overfilling of the reservoir can, over time, cause the heat exchanger to corrode and eventually fail. For this reason, the drain should be checked periodically during the heating season to make sure that it is free of the buildup of mineral deposits.

In general, evaporative pad-type humidifiers are generally limited as to the size of buildings

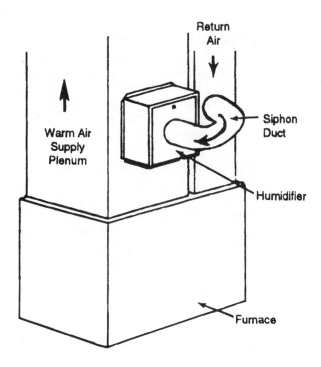

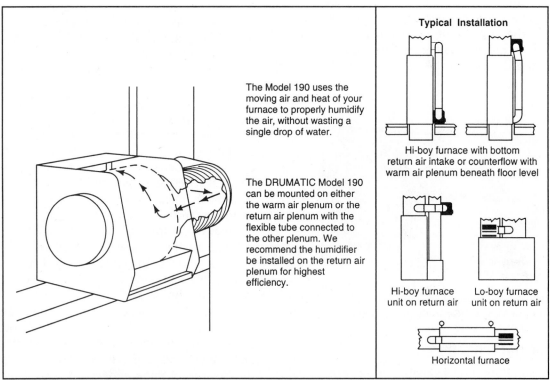

The Model 190 uses the moving air and heat of your furnace to properly humidify the air, without wasting a single drop of water.

The DRUMATIC Model 190 can be mounted on either the warm air plenum or the return air plenum with the flexible tube connected to the other plenum. We recommend the humidifier be installed on the return air plenum for highest efficiency.

Typical Installation

Hi-boy furnace with bottom return air intake or counterflow with warm air plenum beneath floor level

Hi-boy furnace unit on return air

Lo-boy furnace unit on return air

Horizontal furnace

FIGURE 14-27 Installation of humidifier equipped with cold-air siphon.

that they can effectively serve. For most small pad units, 1000 to 1500 square feet is considered the maximum surface area that can be effectively humidified. In larger buildings, central power humidification systems capable of large-capacity water delivery are employed.

FIGURE 14-28 Plate-type humidifier. (Courtesy of Skuttle Manufacturing Co.)

Rotating drum/pad humidifiers: Several different types of rotating and pad-type power humidifiers are available to suit a variety of installation requirements. A humidifier of the

FIGURE 14-29 Rotating drum humidifier. (Courtesy of Skuttle Manufacturing Co.)

type shown in figure 14-29 incorporates a rotating drum that houses the humidifier pad. The motorized pad sits in a water-filled tray (figure 14-30).

The humidifier is energized only when the furnace fan is operating. Air from the return air ducts passes over the rotating pad, which picks up fresh supplies of water as it rotates through

FIGURE 14-30 Motorized pad humidifier and exposed water-filled tray exposing pad for service and/or replacement. (Courtesy of Skuttle Manufacturing Co.)

the water reservoir. This moisture is transferred to the passing airstream, increasing its relative humidity during furnace operation. A humidistat supplies power to the motorized pad and energizes the unit when the relative humidity of the building air is below the set point on the humidistat.

Note from figure 14-30 that the float valve assembly is exposed for cleaning and servicing when the bottom water tray is removed. The pads can be either cleaned or replaced, when required, based on mineral buildup and individual service requirements of the specific system.

Atomizing humidifiers: Spray mist technology is used in the design of atomizing humidifiers that employ a spray nozzle to break up the water into fine droplets that evaporate quickly within the airstream of the furnace; see figure 14-31. While all atomizing humidifiers are controlled by the operation of a humidistat, some units also incorporate a built-in thermostat that prevents the atomizer from operating until the air temperature in the heating system has reached a predetermined level. This allows the humidifier to function efficiently and prevents dew-point condensation in the ductwork.

FIGURE 14-31 Atomizing humidifier. (Courtesy of Herrmidifier Co., Inc.)

Atomizing humidifiers use less energy than do motorized damper units. The only power draw on the atomizer is the electrical solenoid valve thay controls the flow of water to the nozzle. For effective atomization, water pressure should be a minimum of 40 PSI. For this reason, atomizing humidifiers are not generally used in homes and buildings that supply water from their own wells. Rather, they require the higher water pressures generally found in centrally supplied water systems.

Steam-assisted humidifiers: Steam-assisted humidifiers incorporate their own heating source in the humidifier assembly, which allows for effective humidification, regardless of the temperature of the air in the furnace. Also, since the unit does not depend on air temperature for efficient operation, it can be installed in

either the warm-air or cold-air ducts. These units are capable of high-capacity humidification, and should be sized carefully to meet the building requirements, following the specific manufacturer's recommendations. A steam-assisted humidifier is illustrated in figure 14-32. Although low-mineral-content water is an important prerequisite for the proper operation of any humidifier, it is more critical with steam units. Periodic cleaning of the heating element is important in preventing mineral coating. In-line water conditioners are available for use with all types of humidifiers in order to minimize mineral buildup.

FIGURE 14-32 Steam humidifier. (Courtesy of Skuttle Manufacturing Co.)

Tips on Operating Humidifiers

The following operating considerations apply to all types of humidifiers installed in both residential and commercial environments.

1. The action of evaporation produces mineral deposits that were previously suspended in the water supply. Where possible, the installation of a water softener should be considered in order to minimize these deposits.

2. In operating conditions where the water is hard, mineral deposits may occur on the interior surfaces of air ducts.

3. When first installed, a new humidifier may operate almost continuously in order to raise the relative humidity of the indoor air. This happens because all

of the objects inside of the buiding will absorb moisture until normal operating levels have been reached.

4. If moisture condensation occurs on interior walls or windows, the moisture setting on the humidistat should be lowered. In the case of plate-type humidifiers, some of the plates should be removed.

5. The humidifier power and water supply should be turned off at the end of each heating season. The float valves should be cleaned, and all pads either cleaned or replaced. In the case of atomizing humidifiers, a new nozzle should be installed at the beginning of each heating season.

6. In-line water filters should be removed from the system and washed.

7. The seal and operation of all saddle supply valves should be inspected and replaced if necessary.

8. The interior of the ductwork and heat exchanger surfaces of the furnace surrounding the humidifier should be inspected to make sure that no local corrosion or significant buildup of mineral deposits has occurred during the operating season. If possible, an inspection plate should be installed in the plenum that enables the heat exchanger to be inspected visually.

It should be remembered that humidifiers, if not properly inspected and maintained, can present numerous problems to the building owner and technician. For example, standing water in the humidifier water tray is an ideal breeding place for bacteria. The tray should be inspected, emptied if necessary, and cleaned. Following the above guidelines should minimize operating and maintenance problems.

SOLID- AND MULTIFUEL HOT-AIR HEATING ALTERNATIVES

Operating Environment

The economics of heating with solid- and/or multifuel hot-air furnaces is simliar to heating with hydronic systems. These furnaces are available in configurations that burn either wood, wood and coal, or wood/coal and a conventional fossil fuel such as oil or gas. Depending on the combustion characteristics, these furnaces can be either used as the primary heating source, or integrated into an existing fossil-fuel heating system to carry a percentage, or all, of the heating load of the building, based on system design.

For specifics concerning the combustion characteristics of wood and coal, the reader is directed to Chapter 12.

Add-on Hot-Air Solid-Fuel Furnaces

When solid-fuel furnaces are integrated into an existing heating system, the unit of choice is sometimes an add-on furnace of the type shown in figure 14-33.

FIGURE 14-33 Solid-fuel (wood and coal) add-on furnance. These units can be ducted into existing gas- or oil-fired heating systems for increased heating flexibility. (Courtesy of Riteway Manufacturing Co.)

SERIES INSTALLATION

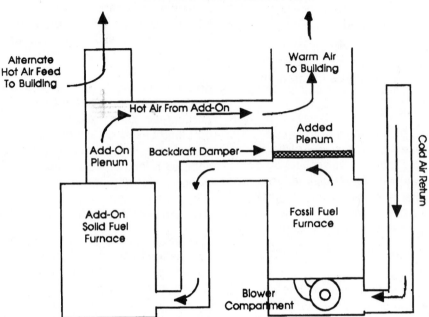

PARALLEL INSTALLATION

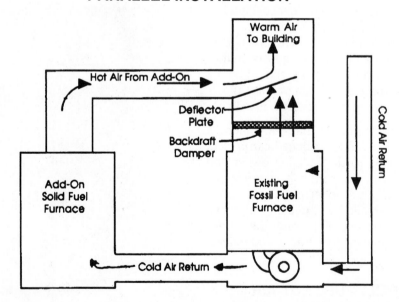

FIGURE 14-34 Series and parallel installation of add-on solid-fuel furnaces. In the series configuration, air is channeled first through the existing fossil-fuel furnace, and then into the solid-fuel add-on. Note the use of the added plenum with a diverter that directs airflow through the two furnaces. In contrast, the parallel configuration allows the existing fossil-fuel furnace to blow warm air directly into the duct system without channeling it through the add-on. The choice of installation is usually determined by available sjpace, plenum sizes, and duct configuration.

This furnace is equipped with a plenum jacket that allows interfacing into the existing ductwork. An optional blower is available that provides additional hot air to the integrated system or allows the furnace to be used as a stand-alone heater.

The installer is cautioned that the use of any solid-fuel device, whether as a stand-alone or an add-on to an existing heating system, requires a separate chimney or an unused flue in an existing chimney to vent the solid-fuel appliance. Venting the solid-fuel furnace into a chimney flue used for any other appliance can cause the chimney to overheat, with potentially lethal consequences. Both homeowners and technicians should consult local codes that govern these installations to ensure a safe, trouble-free installation.

When installing the solid-fuel furnace as an add-on to an existing heating system, one of two installation methods is generally used—either a series or a parallel hookup, as shown in figure 14-34.

In a series installation (figure 14-34A), the original furnace blower is used to move all the air in the heating system through both furnaces. Note that this has a major effect on static pressure in the system. In this type of installation, adjustment of the fan speed to maintain the proper temperature rise may be required. Static pressure measurements should be taken after the intallation is complete to assure proper blower operation.

A separate fan/limit switch is installed in the solid-fuel plenum, and wired to energize the blower when the air in the plenum reaches the desired cut-in temperature (approximately 150 degrees). Cold air returning from the building is channeled through the solid-fuel furnace, where it is preheated, and then into the conventionally fueled furnace. If the air has been sufficiently heated by the solid-fuel fire and can maintain the building's thermostat setting, no additional heating is required from the fossil-fuel unit. The air continues on its way through the supply ductwork and into the building. If the solid-fuel unit is unable to heat the air sufficiently, the fossil-fuel backup furnace is energized to pro-

vide whatever additional heat is required. Operating in this way, the fossil-fuel furnace acts as a backup to the solid-fuel unit.

In contrast to the series installation, consider the operating characteristics of the parallel installation in figure 14-34B. In the parallel mode, the solid-fuel unit either can pull return air from the fossil-fuel furnace, or can be separately ducted to draw and supply conditioned air, relying on its own, separately powered blower. Note the use of a back-draft damper to prevent a short cycling of air through the fossil-fuel furnace.

Both the series and parallel types of interconnection will operate effectively in most situations. The type of installation chosen is usually determined by the size of the existing duct system and its availability for modifications to allow insertion of the solid-fuel furnace. Note

FIGURE 14-35 Multifuel furnace. These units typically feature two separate combustion chambers—one for the wood/coal and the other for the conventional fuel (oil or gas). The use of a combined or separate stack outlet is a function of the design of the particular furnace. (1) Heat exchanger. (2) Firebrick combustion chamber lining. (3) Separate fossil-fuel combustion chamber. (4) Afterburner combustion air system. (5) Cast iron shaker grate system. (6) Oil burner or gas power burner. (7) Stack pipe cleanout access door. (Courtesy of Yukon Enterprises, Inc.)

that two furnaces increase the possibility of potentially dangerous contamination of room air as the result of combustion by-products, especially if the solid-fuel unit is burning coal. The heat exchangers in both furnaces should be inspected on a regular basis to ensure proper and safe operating conditions.

Stand-Alone Multifuel Hot-Air Furnaces

Multifuel hot-air furnaces usually feature two separate combustion chambers, one for burning wood and coal and the other for the combustion of either fuel oil, natural gas or propane gas. The two combustion chambers are manifolded together so that one flue outlet can be used to vent the appliance. This type of component arrangement is illustrated in figure 14-35.

The sequence of operation for the typical multifuel furnace follows the two-stage heating logic introduced previously for hydronic applications. A two-stage thermostat is used to set room temperatures for both the wood/coal and backup oil or gas system. The first stage, the higher thermostat setting, controls the operation of the air inlet control to the solid-fuel combustion chamber, allowing combustion air to enter the firebox when there is a call for heat. The second stage, the lower thermostat setting, operates the control circuit to the oil or gas burner, energizing the backup burner if and when the room temperature falls to the lower temperature setting. In most instances, the two thermostat settings should be separated by at least 5 to 8 degrees in order to prevent an overlap in operating modes of the thermostat's switching from one heat source to the other, and to take into account residual heat that will be supplied by the furnace between staging. Most of these furnaces also incorporate automatic ignition of the wood or coal charge, using the fossil-fuel burner to ignite fresh fuel charges.

Both the add-on and multifuel units have installation and operating advantages, depending on the requirements of the application. A major advantage of the multifuel furnace is that only one appliance is needed to accomplish the space heating, with the flexibility of using two

or three fuel sources. However, the efficiency of the single multifuel unit will be lower than that of the two-unit add-on system, since design compromises must be made to incorporate two combustion designs into one furnace.

SOLAR HOT-AIR HEATING SYSTEMS

The integration of solar heating with conventional hot-air heating systems is accomplished through the use of a storage medium, which is usually water. This water, which is heated by the solar collectors, transfers heat into the conventional heating system through a water-to-air heat exchanger.

In the early days of solar space heating system design, attempts were made to heat materials that contained large thermal masses, such as rock bins and large water-storage tanks. This stored heat would then be transferred over an extended period into a building. It soon became obvious, however, that large thermal mass storage facilities were unworkable. After the initial charge of heat was withdrawn, it took an excessively long time to recharge the storage medium to the level where it was able to provide meaningful contributions to meeting the space heating needs of the building. The technology then began to focus on heating relatively small storage tanks of water quickly, transferring this heat into a building, and then reheating the storage medium to continue the heating cycle. In this way, the solar heating system was able to cycle quickly from a discharged to a charged state, and to utilize almost all available incoming solar heat with a high degree of efficiency.

Solar space heating systems are classified as either active or passive. With passive solar heating systems, construction features are incorporated into the home or building that are designed to maximize the infiltration of solar energy during the winter and to minimize solar heat gain during the summer. This is accomplished through the generous use of south-facing windows, properly designed roof overhangs that provide adequate shade during the summer, and open interior spaces to allow

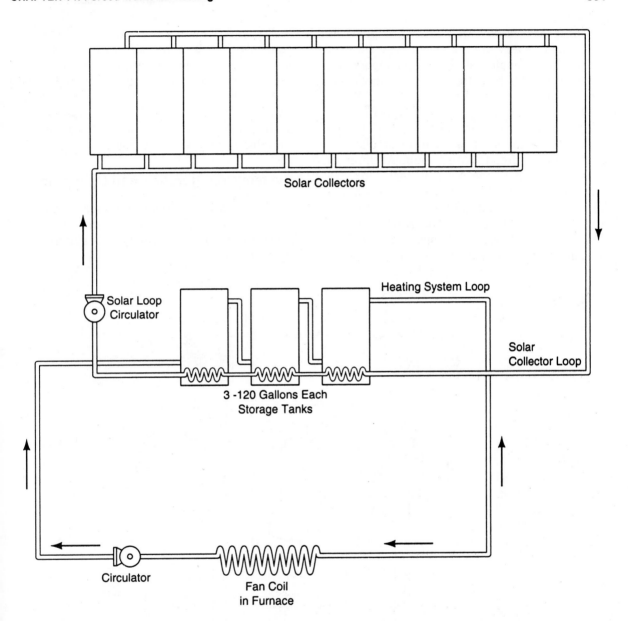

Solar Collectors

Heating System Loop

Solar Loop
Circulator

Solar
Collector Loop

3 -120 Gallons Each
Storage Tanks

Circulator

Fan Coil
in Furnace

FIGURE 14-36 Heating logic of a solar warm-air heating system. The collector loop maintains temperature levels in a series of water storage tanks. On a call for heat, a separate circulator is energized, along with the furnace fan. Hot water from the solar storage tanks circulates through the furnace fan coil, providing warm, heated air to the building. (Courtesy of Monitor Data Corp.)

interior heated air to circulate easily. Passive solar buildings are also superinsulated to minimize either heat loss or gain from the interior to the exterior of the building. The primary focus of this section is on design features relating to active solar systems involving the use of a heat transfer fluid that is circulated between the solar collectors and a storage medium. The operating logic of a solar space heating system is illustrated in figure 14-36.

The space heating logic shown emphasizes efficiency at every point at which heat transfer takes place. Single-wall heat exchangers are used to transfer heat from the solar collector loop to the storage tanks. The system in figure 14-36 uses an external heat exchanger between the solar collector loop and the solar storage tanks and operates with a separate circulator in each heating circuit in this loop. An internal heat exchanger in the solar storage tank can also be used for heat transfer between the solar collectors and storage tanks. In this arrangement, only one circulator is required, and double-wall heat exchangers can replace the single-wall heat exchangers. This is done in order to isolate the storage water from the collector fluid in anticipation of the failure of one of the walls or surfaces in the heat exchanger. Note that the logic illustrated accommodates the production of both space heating and domestic hot water. An external heat source, such as a wood stove or boiler, can also be used with this system operating logic.

System Configuration and Operation

Solar heating systems interface easily with existing warm-air heating systems. A typical system integration of this type is illustrated in figure 14-37.

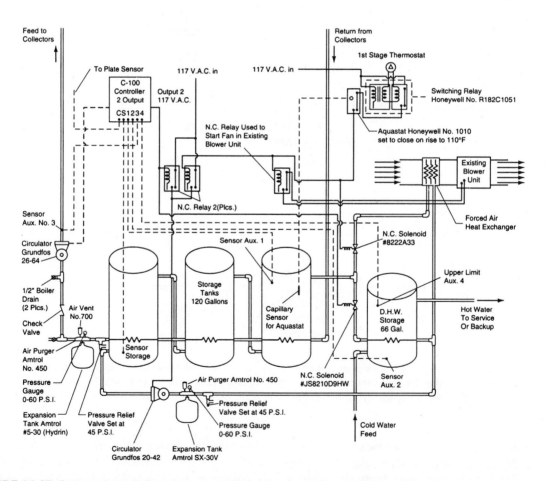

FIGURE 14-37 Piping schematic for solar system with existing hot-air heating system. This schematic is an expansion of the logic diagram illustrated in figure 14-36. This system also features a separate loop for heating domestic hot water, in addition to providing for space heating. (Courtesy of Monitor Data Corp.)

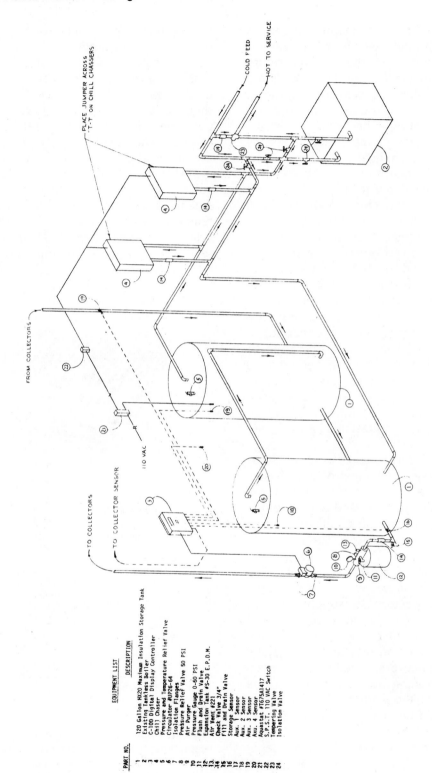

FIGURE 14-38 Solar system with fan convectors. In situations where there is no existing warm-air furnace, the use of separate fan-coil convectors to provide warm-air heating to separate rooms or areas is recommended. These fan-coil convectors resemble automobile heaters. Each unit contains a separate circulating pump and blower/fan located behind a fin tube coil. On a call for heat from a separate thermostat, the pump circulates warm water from the solar storage tanks through the coil; at the same time, the convector blower is enrgized. Air is heated as it is forced through the coils by the fan to heat the room. (Courtesy of Monitor Data Corp.)

EQUIPMENT LIST

PART NO.	DESCRIPTION
1	120 Gallon HX20 Maximum Insulation Storage Tank
2	Existing Tankless Boiler
3	C-100 Digital Display Controller
4	Chill Chaser
5	Pressure and Temperature Relief Valve
6	Circulator #UP26-64
7	Isolation Flanges
8	Pressure Relief Valve 50 PSI
9	Air Purger
10	Pressure Gauge 0-60 PSI
11	Flush and Drain Valve
12	Expansion Tank #S-30 E.P.D.M.
13	Air Vent #221
14	Check Valve 3/4"
15	Fill and Drain Valve
16	Storage Sensor
17	Aux. 2 Sensor
18	Aux. 3 Sensor
19	Aux. 4 Sensor
20	Aquastat #T675A1417
21	S.P.S.T. 110 VAC Switch
22	Tempering Valve
23	Isolation Valve
24	

Combination of the two heating systems is accomplished through the use of a heat exchange coil installed in the cold-air return plenum of the hot-air furnace. These coils are available in a number of design configurations, and are sized on the Btu output that the coil is expected to deliver. The system operates through the use of a piping circuit that runs between the heat exchanger coil and the solar storage tanks. The operation of both the solar system and conventional fossil-fuel furnace is controlled by a two-stage thermostat. On a call for heat at the higher, stage-one setting of the thermostat, the furnace fan and circulating pump are energized to pump water from the solar storage tanks through the heat exchanger coil and back to the storage tanks. Heat from the exchanger coil is transferred to the passing airstream in the plenum. If the building temperature drops to the second-stage thermostat setting, the fossil-fuel burner is energized to provide any additional heating required to satisfy that setting. An aquastat provides a low-limit cutoff if the temperature of the water in the storage tanks becomes too low for space heating purposes (this aquastat is generally set at 85 degrees).

The number of storage tanks used in a solar system depends on the number of solar collectors on the roof. As a general sizing rule, three to four solar collectors are used to charge one 120-gallon solar storage tank (this sizing rule assumes a minimum collector surface of 24 square feet).

A second type of solar warm-air heating application, shown in figure 14-38, uses separate fan-coil convectors to supply heat to individual rooms or areas in a building and operates independently of any existing heating system. This installation may affect the static pressure in the ductwork, based on the characteristics of the solar-heated coil placed in the cold-air return of the warm-air furnace. Measurements should be taken with a manometer to ensure proper blower operation.

The use of fan convectors allows for separate zoned warm-air heat. Each convector is powered by its own thermostat and an internal circulating pump coupled to a fan motor. When the system is operating, the internal pump in the convector draws hot water from the solar storage tanks and circulates it through the internal fan coil in the convector and back to the solar tank. As the fan blows air across the coils, heat in the coils is transferred to the room air. The use of fan-coil convectors is often specified when either no heating system is available in the building to interface with the solar system, or the existing heating system is steam based. Because of the high operating temperatures encountered in a steam heating system, the integration of a solar system (which relies on low-design temperatures for most efficient operation) with the steam system is usually impractical. Both of the solar systems

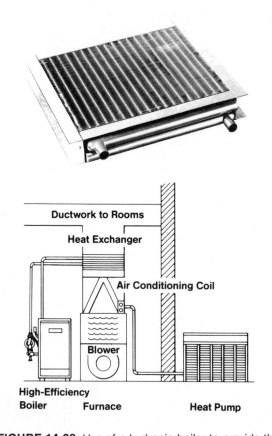

FIGURE 14-39 Use of a hydronic boiler to provide the heat source for a warm-air heating system fueled by either a conventional furnace or heat pump. (Courtesy of Monitor Data Corp.)

discussed in these warm-air applications are operated by the differential controllers discussed in Chapter 13.

HYDRONIC HOT-AIR CONVERSIONS

Existing hot-air heating systems, electric furnaces, or heat pump installations can be modified so that they can be converted to lower-cost gas or oil heat powered by a conventional boiler. This conversion is accomplished by the use of a fan-coil heat exchanger installed in the plenum, above the furnace. The coil is then connected to a high-efficiency gas- or oil-fired boiler. The original furnace or heat pump blower is still used to circulate warm air throughout the house, but the air is heated by the boiler instead of by the electric furnace or electric resistance coils in the heat pump. The fan-coil heat exchanger and a simplified system diagram for this conversion are illustrated in figures 14-39A and 14-39B.

GENERAL MAINTENANCE AND TROUBLESHOOTING PROCEDURES

Most troubleshooting procedures performed on warm-air furnaces will relate to either the gas- or oil-fired combustion systems that power them. The student is referred to Chapters 11 and 12 for information on these troubleshooting procedures. There are, however, certain basic maintenance and troubleshooting operations that can be used for all warm-air furnaces, regardless of the fuel source, to maximize system safety and performance. Remember that properly maintained furnaces minimize the possibility that potentially lethal exhaust gases will enter airflow to the interior of the building and home. Some of these general maintenance and troubleshooting procedures are:

1. Check burner start-up. On gas furnaces, inspect standing pilot and main burner adjustments. Adjust for normal starting. On oil-fired units check for delayed ignition. On both oil and gas systems, check for afterburning.

2. Adjust combustion. On gas-fired furnaces, adjust pilot and main burner flames according to the procedures outlined in Chapter 11. On oil-fired furnaces, make ignition adjustments according to the instructions in Chapter 10. Consult the manufacturer's specifications for final adjustments.

3. Check for fuel leaks. Check all joints on gas feed lines. Make sure that the safety rollout switch is operating properly. To perform this check, the exhaust gases must be blocked for a short period. Consult the manufacturer's recommendations for specifics concerning the proper operation of this swtich. Check oil lines for both suction and pressure leaks.

4. Check operation of fan and limit control. When the furnace reaches operating temperature, block all return air ducts and allow the furnace to shut down on high limit. In this situation, the fan must continue to run. Remove restrictions on return air and the burner should ignite within a short time.

5. Replace all filters. Air filters should be replaced yearly whether or not they appear to be dirty. Consult the manufacturer's recommendations regarding water line and conditioning filters, electrostatic air cleaner filters, and fuel filters.

6. Check for proper operation of humidifier and electrostatic air cleaners (if installed).

7. Visually inspect the integrity of all heat exchanger areas. This inspection might require removal of vent pipe or furnace access panels. It should be undertaken in strict accordance with the manufacturer's recommendations.

Following these simple procedures, along with the manufacturer's recommendations, will help keep operating problems to a minimum.

SUMMARY

In this chapter, we examined a variety of warm-air heating system designs and installation options. The use of air as the heat transfer medium is similar in many respects to the use of water in hydronic systems. The proper sizing of supply and return ducts as a function of the quantity and temperature of the air moving through the system is critical to efficient system performance. Hot-air systems allow for zoning with the use of motor-controlled dampers that are similar in their control function to hydronic zone valves.

The ease with which warm-air heating systems allow for the addition of air-quality control devices, such as electrostatic air cleaners and humidifiers, is a significant advantage of these systems over their hydronic counterparts. Also, the incorporation of air conditioning into a warm-air heating system, given adequate sizing of the existing ductwork, increases the system flexibility. Interfacing both solid-fuel and solar-based heating systems is more easily accomplished in warm-air heating than with hydronic systems, since only minor modifications of the existing ductwork are required. In most instances, this involves far less work than with the piping modifications that must be made to either hot-water or steam heating systems for this purpose.

PROBLEM-SOLVING ACTIVITIES

Troubleshooting Case Study

A. A service technician responds to a complaint of no heat at a building with a gas-fired hot-air heating system. Upon examination, the technician observes the following.

1. The gas burners operate for a short period of time and then shut down.
2. The furnace fan does not turn on.
3. The thermostat circuit appears to function properly, and 24 volts is present in the circuit.
4. The furnace is equipped with a drum-type power humidifier, which appears to be inoperable.

Identify the procedures that you would use to isolate the problem and return the furnace and humidifier to a working condition.

Review Questions

1. Warm-air heating systems should be balanced for equal amounts of supply and return air to the furnace. Indicate a simple procedure that can be used to determine if there is a proper amount of return air reaching the furnace.

2. Indicate the difference between a gravity feed and a forced-convection warm-air heating system. How do these two systems vary in control components and their function?

3. A 150,000-Btu/hr furnace is operating with a temperature rise of 73 degrees. What is the volume of air delivered from the furnace?

4. Highlight the differences between a perimeter loop and extended plenum supply trunk duct system.

5. Highlight the differences among low-boy, high-boy, horizontal, and counterflow furnaces. In this discussion, indicate the preferred installation application for each type of furnace.

6. Discuss the operating sequence of a typical warm-air furnace. In your discussion, include a description of the control sequence of the combination fan/limit switch.

7. Indicate the operating characteristics of four different types of furnace humidifiers. Highlight the methods of operation of each of the humidifiers, along with one unique feature or advantage of each type.

8. Indicate by a simple schematic drawing the differences in installing a solid-fuel add-on furnace in a series and in a parallel ducting configuration. For each illustration, discuss the sequence of operating controls and characteristics of airflow.

CHAPTER 15

Air-Conditioning and Heat-Pump Technology

THE REFRIGERATION CYCLE

To understand the technology of refrigeration cycle devices such as air conditioners and heat pumps, we will be drawing upon information in earlier chapters that dealt with the various heat categories (sensible and latent heat, for example). New information will focus on pressure and temperature relationships that exist in all closed-circuit refrigeration systems.

Heat flows "downhill," that is, from the hotter to the cooler object. There is no need for any external energy to move the heat. However, air conditioning involves moving heat "uphill," from a cooler area (the inside of a home at 75 degrees) to a warmer one (outdoor air during the summer). To accomplish the uphill movement of heat, external power in the form of a refrigeration compressor is required.

Heat pumps and air conditioners are almost identical devices in their basic system and component design. Air conditioners remove heat from indoor air and deposit it outdoors. The heat pump is, in effect, a reversed air conditioner. In addition to cooling a building during the summer, it can be reversed during the winter to extract heat from the outdoor air and deposit it in the building. The reader is referred to Chapter 2 for a review of some basic terminology that refers to Btu's, heat, and calories.

Change of State

The temperature of a particular substance often determines the form or state of the substance. Many substances can exist in either the solid, liquid, or gaseous state at different temperatures. Water is perhaps the best example to illustrate this point. Figure 15-1 shows the state of water based on its temperature, along with the Btu's involved in the various state changes.

One Ton of Refrigeration

Note from figure 15-1 that to melt 1 pound of ice, 144 Btu of latent heat is required (remember, latent heat changes the substance from a solid to a liquid or a liquid to a vapor without a temperature change in the substance). During this melting process, the heat required to melt the ice may come from surrounding air. The air must be warmer than the ice. This heat flows downhill, from the warmer air to the ice, and cools the air.

If we were to melt a ton of ice to water, 288,000 Btu of heat would be absorbed (2000 X 144 = 288,000). In extending this melting process over the 24-hour period of the day, we find that 12,000 Btu/hr is required to melt a ton of ice in 24 hours. One ton of refrigeration can also be expressed as the cooling rate of 12,000 Btu/hr. When sizing air conditioners for a

CHANGE OF STATE　　Temperatures in °F

Substance	Amt	From Temp	To Temp		Temp Rise		Spec Heat		Amt of Heat	Type of Heat	Name of Change
Ice (solid)	1 lb	0°	32°	=	32°	at	1/2 BTU	=	16 BTU's	Sensible	(Heating)
Ice (solid)	1 lb	32°	0°	=	-32°	at	1/2 BTU	=	-16 BTU's	Sensible	(Cooling)
Ice to Water	1 lb	32°	32°	=	0°	at	144 BTU	=	144 BTU's	Latent	(Melting)
Water to Ice	1 lb	32°	32°	=	0°	at	-144 BTU	=	-144 BTU's	Latent	(Freezing)
Water (liquid)	1 lb	32°	212°	=	180°	at	1 BTU	=	180 BTU's	Sensible	(Heating)
Water (liquid)	1 lb	212°	32°	=	-180°	at	1 BTU	=	-180 BTU's	Sensible	(Cooling)
Water to Vapor	1 lb	212°	212°	=	0°	at	970 BTU	=	970 BTU's	Latent	(Vaporization)
Vapor to Water	1 lb	212°	212°	=	0°	at	-970 BTU	=	-970 BTU's	Latent	(Condensation)
Vapor (gas)	1 lb	212°	222°	=	10°	at	1/2 BTU	=	5 BTU's	Sensible	(Heating)
Vapor (gas)	1 lb	222°	212°	=	-10°	at	1/2 BTU	=	-5 BTU's	Sensible	(Cooling)

FIGURE 15-1 Change of state of water.

particular location, the heat gain from all sources (from solar infiltration, electrical appliances, stoves, as well as the number of rooms in the building) is calculated to determine the number of Btu's per hour that must be removed by the air conditioner in order to maintain a specified indoor design temperature.

Pressure and Temperature Relationships

The relationship between the pressure and temperature of a substance and its resulting boiling point is an important concept in understanding air conditioning and heat-pump technology. We know from our previous readings that pressure has a definite effect on the boiling point of a substance. For example, water boils at 212°F at normal atmospheric pressure (sea level). When the pressure is increased, the boiling point of the water is increased, and vice

versa. Due to pressure in the cooling system, the water in an automobile radiator boils at a temperature well above 212 degrees under pressure in the cooling system. However, if the radiator cap is removed from a hot radiator, the pressure in the radiator will drop to zero, and the boiling point of the water drops to 212 degrees (this rapid boiling results in a dangerous situation, and the reader is cautioned not

Temp. ° F.	Inches Mercury Vacuum
212°	0
200°	6.45
150°	22.3
100°	28
70°	29.18
40°	29.67

FIGURE 15-2 Boiling point of water at different atmospheric pressures.

Refrigerant Characteristics

(Standard Ton Conditions)	(All Figures Approximates)	
	R-22 (CHClF$_2$)	R-12 (CCl$_2$F$_2$)
Boiling Point (one atm)	-41.4° F.	-21.6° F.
Refrigerating effect/lb.	69 BTU	50 BTU
Refrigerant circulated lbs./min.	3 lbs.	4 lbs.
Freezing Point	-256° F.	-252° F.
Flammable or Explosive	No	No

FIGURE 15-3 Characteristics of R-12 and R-22 refrigerant. (Courtesy of the Williamson Co., Division of Metzger Machine Corp.)

to try this experiment). At 1000 feet above sea level, the boiling point of water is only 193 degrees since the pressure at 1000 feet of elevation is less than the pressure at sea level. Figure 15-2 shows the boiling point of water at different pressures.

REFRIGERANT CHARACTERISTICS

A common characteristic of all fluids used as refrigerants is their low boiling point, which enables them to absorb heat under pressure. The basic characteristics of two common refrigerants are listed in figure 15-3.

Determining Proper Charge

All refrigeration devices must be properly filled (charged) with the correct weight of refrigerant if they are to work efficiently. Note from figure 15-3 that R-22 refrigerant will absorb 69 Btu of heat. To obtain the number of pounds of refrigerant per ton that must be circulated for proper cooling, divide 12,000 by 69, which gives 173 pounds per hour. To reduce this figure to minutes, divide 173 by 60, which equals approximately 3 pounds/minute/ton of refrigerant that must be circulated in this example. The normal charge for a standard air conditioner can usually be estimated by using these calculations and adding a small amount

of extra charge for the tubing length. Using R-12 and R-22 as the refrigerants in a 2-ton air conditioner, we can calculate the charges as follows:

R-12 (50 Btu/lb), 2-ton unit, 4 lbs x 2 = 8 lbs + tubing

R-22 (69 Btu/lb), 2 ton unit, 3 lbs x 2 = 6 lbs + tubing

This procedure can be used only for units with receivers for storing excess liquified refrigerant. In all instances, the manufacturer's recommendations must be consulted for the specified charge for each unit. Figure 15-4 is a temperature–pressure chart that shows the pressure required to achieve a variety of temperatures using both R-12 and R-22 refrigerants.

Determining Cooling Load

Many factors affect the performance and efficiency of air conditioners and heat pumps, one of the most important of which is the sizing of the unit relative to the amount of heat that must be removed from a particular building. The heat gain in a building due to lighting and machinery; external heat gain through windows, walls, and ceilings; and the number of people in the building all must be taken into account

TEMPERATURE PRESSURE CHART

°F	R-12	R-22		°F	R-12	R-22
-20	0.5	10.3		26	25.4	50.2
-18	1.3	11.5		27	26.1	51.5
-16	2.0	12.7		28	26.9	52.7
-14	2.8	13.9		29	27.7	54.0
-12	3.6	15.2		30	28.5	55.2
-10	4.4	16.6		31	29.3	56.5
- 8	5.3	18.0		32	30.1	57.8
- 6	6.2	19.4		33	30.9	59.2
- 4	7.1	20.9		34	31.7	60.5
- 2	8.1	22.5		35	32.6	61.9
0	9.2	24.1		40	37.0	69.0
1	9.7	24.9		45	41.7	76.6
2	10.2	25.7		50	46.7	84.7
3	10.7	26.6		55	52.0	93.3
4	11.2	27.4		60	57.7	102.5
5	11.8	28.3		65	63.7	112.2
6	12.3	29.2		70	70.1	122.5
7	12.9	30.1		75	76.9	133.4
8	13.5	31.0		80	84.1	145.0
9	14.0	32.0		85	91.7	157.2
10	14.6	32.9		90	99.7	170.1
11	15.2	33.9		95	108.2	183.7
12	15.8	34.9		100	117.1	197.9
13	16.5	35.9		105	126.5	212.9
14	17.1	36.9		110	136.4	228.7
15	17.7	37.9		115	146.7	245.3
16	18.4	38.9		120	157.6	262.6
17	19.0	40.0		125	169.0	280.7
18	19.7	41.1		130	181.0	299.3
19	20.4	42.2		135	193.5	319.6
20	21.0	43.3		140	206.6	341.3
21	21.7	44.4		145	220.3	364.0
22	22.4	45.5		150	234.6	387.2
23	23.2	46.7		155	249.5	410.8
24	23.9	47.8		160	261.1	434.6
25	24.6	49.0				

Explanation of Pressure and Temperature Chart:

A. Use of -
     a - To determine condensing temperature
     b - To determine proper charge
     c - To determine non-condensibles
     d - To determine evaporator temperature
     e - To use as a thermometer

B. Results -
     a - How to approximate head pressure on units with receivers
        1. Outside ambient temperature plus 25° equals condensing temperature,
          when unit is running.

              Example -   85° Outside   110° - 228.7 P.S.I. ⎱
                     25° Plus   (-5°)105° - 212.9 P.S.I. ⎬ R-22
                110° Total   (+5°)115° - 245.3 P.S.I. ⎰

FIGURE 15-4 Temperature and pressure chart for R-12 and R-22 refrigerants. (Courtesy of the Williamson Co., Division of Metzger Machine Corp.)

when determining what the cooling load of a particular building will be.

Heat gain, like its heat-loss counterpart, is rated in Btu's per hour. Heat gain calculations are performed using assumptions about the outdoor and indoor temperature levels that are to be found in a particular geographical region. Chapter 2 offered a brief description of some procedures that are designed to place the heat loss/gain of a building in perspective in order to get an idea of how these systems will be sized.

There are several guides that can be consulted in performing detailed heat-loss and heat-gain analyses; one of the most widely used is the *Handbook* of the American Society of Heating, Ventilating and Air-Conditioning Engineers (ASHRAE). Virtually all of the factors affecting heat gain, from the surface area of the building roof to the number and types of lighting fixtures in the building, are taken into account in order to arrive at an accurate heat gain figure. This figure can then be used when selecting an air conditioner or heat pump of the proper capacity for a specific application. In new construction, heat loss and gain calculations are usually performed by the building's design architects, and are kept on file with the building plans. In some older homes, these figures are sometimes attached to the original architect's drawings. For further information regarding the process of performing heat loss and heat gain calculations, the student is referred to the ASHRAE handbook or other comprehensive calculating guides.

DEVELOPMENT OF THE REFRIGERATION CYCLE

The refrigeration cycle is based on the ability of a fluid with a low boiling point, the refrigerant, to absorb heat from an enclosed area, such as a building, as it circulates through a coil of tubing known as an evaporator coil. As heat is absorbed by the refrigerant in the coil, the refrigerant boils and vaporizes. The vaporized refrigerant travels through a compressor, where its pressure and temperature are raised.

It is then circulated to a coil called the condenser, where it releases its absorbed heat to the atmosphere and condenses back to a liquid in the process. The development of this cycle is illustrated in figures 15-5 through 15-8.

In the simple cooling system of figure 15-5, a closed vessel, or evaporator, that contains R-22 refrigerant is placed in a box that is to be cooled. The box is insulated to prevent heat penetration and to maintain even temperature levels inside the box.

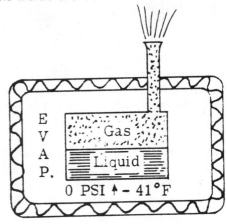

FIGURE 15-5 Step 1 in the development of the basic refrigeration cycle. (Courtesy of the Williamson Co., Division of Metzger Machine Corp.)

Note that the pipe from the evaporator extends into the air so that the pressure in the evaporator is 0 pounds per square inch (PSI). Assuming a sufficient quantity of refrigerant, the box would be cooled to −41°F (see figure 15-3).

Since −41°F is too cold for most refrigeration applications, let us raise the temperature inside the insulated box to 20°F. To do this, we find, by looking at the temperature–pressure chart in figure 15-4, that the pressure of the refrigerant must be raised to 43 PSIG. To raise the pressure to this point, we install a valve in the evaporator outlet pipe that allows this pressure to develop. This setup is illustrated in figure 15-6.

With regard to another modification of the process in figure 15-6, we see that, in order to maintain a pressure of 43 PSIG, we must regulate the flow of liquid refrigerant into the

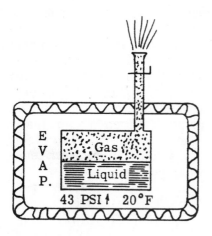

FIGURE 15-6 Step 2 in the development of the basic refrigeration cycle. Raising the pressure of the refrigerant. (Courtesy of the Williamson Co., Division of Metzger Machine Corp.)

evaporator. This modification is shown in figure 15-7. But although all of these designs will work in maintaining the required pressure and evaporator temperatures, they are wasteful in that the refrigerant vapor is vented into the air.

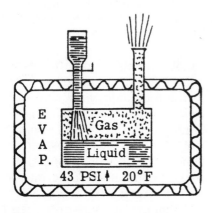

FIGURE 15-7 Step 3 in the development of the basic refrigeration cycle. Regulation of the flow of refrigerant into the evaporator. (Courtesy of the Williamson Co., Division of Metzger Machine Corp.)

Figure 15-8 represents the final stage in refrigeration cycle development. The refrigerant loop is closed and a condenser is added to the system.

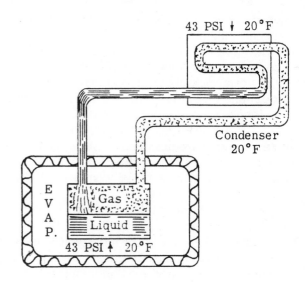

FIGURE 15-8 Step 4 in the development of the basic refrigeration cycle. Closing the refrigerant loop to establish the refrigeration cycle. (Courtesy of the Williamson Co., Division of Metzger Machine Corp.)

Refrigeration in this design is accomplished by absorption of heat in the evaporator, in which the refrigerant absorbs heat and boils, changing from a liquid to a gas. The temperature and pressure of the refrigerant are raised by the system compressor. The liquid refrigerant is heated to a gas, which is then sent to the condenser coils where the absorbed heat is discharged to the surrounding air. With the heat removed, the gas travels through an expansion valve as it enters the evaporator coils. In the evaporator, the liquid refrigerant absorbs heat once again, and the refrigeration cycle continues.

BASIC AIR-CONDITIONING SYSTEM OPERATION

Open and Closed Systems

The most popular type of air-conditioning system operates in a closed-loop environment, totally isolated from the atmosphere. This unit is referred to as a hermetically sealed compression-type system, and relies on a mechanical compressor to raise the temperature and pressure of the refrigerant to ensure proper heat transfer before the refrigerant enters the condensing coils. Many of the compressors are similar in design to the piston-and-valve arrangement in a conventional automobile engine.

The refrigeration compressor operates in a sealed environment in which the circulating refrigerant, along with some lubricating oil, serves as the compressor lubricating agent. Many compressors also utilize a separate oil lubrication system, in addition to the lubricating effect of the system refrigerant. Hermetically sealed compressors are not easily field reparable, since opening the compressor dome breaks the seal, and it is difficult to reseal. However, hermetically sealed units eliminate both belt service and crankshaft seal service procedures, which account for much of the work on open-type systems.

Open systems are typical of older refrigeration units and use a separate motor to drive the compressor through a connecting belt. Open compressors use a separate oil crankcase for lubrication. Refrigerant lines run directly into and out of the compressor head. These units can be field repaired more easily than can their sealed counterparts.

Within the system, the compressor maintains a low refrigerant pressure on the evaporator, or suction side of the compressor, and a high refrigerant pressure on the condenser, or output side of the compressor. In the majority of open systems, the flow of refrigerant into the evaporator coil is controlled by either a capillary tube, an automatic expansion valve, or a thermostatic expansion valve. Compressor design is examined in more detail later in this chapter. The basic refrigeration system configurations are presented in the following illustrations to show their basic operation and differences in refrigerant metering practices.

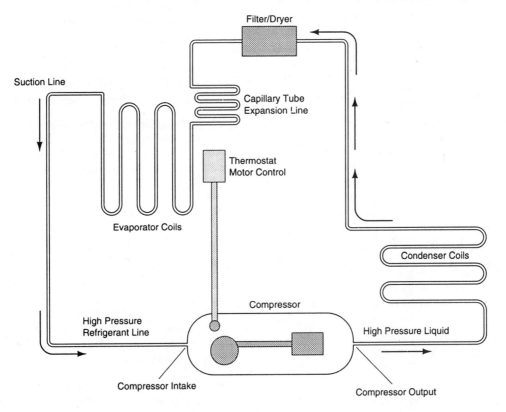

FIGURE 15-9 Mechanical compression refrigeration system using a capillary tube expansion valve. In this system, a thermostat in the cooled space controls the operation of the compressor.

Mechanical Compression Systems

With capillary tubes: Figure 15-9 illustrates a compression system that uses a capillary tube as a refrigerant expansion control.

The complete operating cycle, using this system as a reference, is as follows:

1. Liquid refrigerant at low pressure in the evaporator coils absorbs heat from the indoors, causing the refrigerant to boil. Low pressure in the evaporator coils helps to ensure proper heat absorption from the indoor air (note that the evaporator coil picks up about 10 degrees of superheat between the inlet and outlet of the coil).

2. Low-pressure refrigerant vapor from the evaporator coil travels through the suction line to the compressor, which raises the temperature and pressure of the refrigerant. The compressor pumps the refrigerant to the condenser coils. In this process, the compressor oil, acts to lubricate and cool the compressor.

3. The refrigerant travels through the condensing coils, giving off heat to the surrounding air. As the temperature of the refrigerant drops, it condenses back into a high-pressure fluid. A filter and/or drier is usually installed in the line between the condenser and the evaporator coils to remove any trapped moisture or contaminants in the refrigerant gases.

4. The refrigerant travels through the capillary tube metering device. The capillary tube is a long small-diameter tube (about 1/8 inch in outside diameter). The capillary tube expansion device reduces the high pressure of the refrigerant from the condenser coils to the low operating pressure required by the evaporator. This pressure reduction is accomplished in part by the resistance to the flow of the refrigerant along the length of the tubing. The capillary tube is engineered for the specific appliance in which it will be installed. It is designed to permit just enough liquid refrigerant through the capillary tube to make up for the amount of refrigerant that is vaporized in the evaporator coils.

5. In the evaporator, the low-pressure refrigerant absorbs heat from the surroundings, continuing the cooling process.

This cycle of compression and expansion of the refrigerant through the system will continue until the room thermostat has been satisfied. The compressor will cycle on and off, based on the pressure and temperature of the refrigerant.

With automatic expansion valves: A compression system that uses an automatic expansion valve to meter the flow of refrigerant into the evaporator coil is shown in figure 15-10.

1. Refrigerant from the compressor flows through the condenser coils, where it condenses into a high-temperature liquid. From the condenser, the liquid refrigerant flows into a receiving tank, and then through the filter/drier and into the automatic expansion valve.

2. The expansion valve is so designed that no refrigerant will flow through it unless the compressor is operating, which reduces the pressure in the evaporator coils.

3. Liquid refrigerant is sprayed from the expansion valve into the evaporator, where, due to the low evaporator pressure, it absorbs heat from the surroundings and quickly boils.

4. The boiling refrigerant is drawn by suction into the system compressor, where it is heated and pumped back to the condenser coils to continue the cooling cycle.

With thermostatic expansion valves: Automatic expansion valve systems are typical of older refrigerators, as well as of current small commercial air-conditioning systems. Newer systems

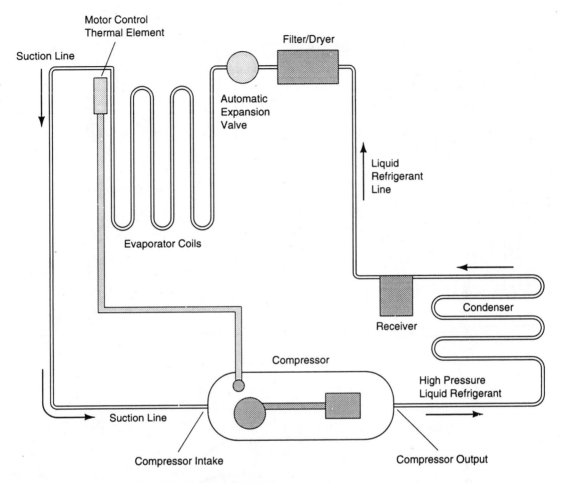

FIGURE 15-10 Mechanical compression refrigeration system using an automatic expansion valve for refrigerant control. A thermostat controlling compressor operation senses the temperature of the refrigerant at the output of the evaporator coil, at the beginning of the suction/refrigerant return line.

commonly use thermostatic expansion valves for metering refrigerant to the evaporator coils; see figure 15-11.

1. Refrigerant in the system flows from the liquid receiver, through a filter/drier, and into the thermostatic expansion valve. A typical thermostatic expansion valve is illustrated in figure 15-12.

2. Operation of the valve is controlled by three pressures:

   a. Pressure from the temperature sensing bulb.

   b. Evaporator pressure.

   c. Spring pressure in the valve.

3. The sensing bulb contains both liquid and vapor refrigerant. Note that the sensing bulb is located at the exit of the evaporator coils. If the valve is not feeding enough refrigerant, bulb temperature is increased by the warmer vapor leaving the evaporator coils. This increases the bulb pressure and causes the expansion valve to open, admitting more refrigerant into the evaporator coil.

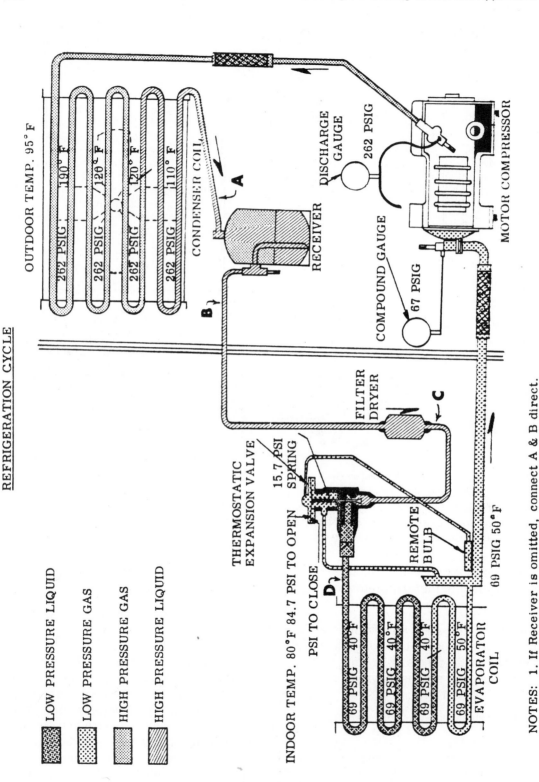

FIGURE 15-11 Mechanical compression refrigeration system using a thermostatic expansion valve for refrigerant control. Compressors in these systems must be capable of starting under load, since system refrigerant pressures do not balance during the off-cycle.

4. If too much refrigerant is admitted to the coils, the bulb temperature will drop and/or the evaporator pressure will increase, causing the valve to close.

Under normal operating conditions, all three pressures—bulb pressure, evaporator pressure, and valve spring pressure—are almost in equilibrium, with the expansion valve working to balance these three pressures.

Absorption Cycle Cooling

Absorption cycle cooling has been most widely used in small refrigerators, primarily in rural areas where utility-supplied electricity is not widely available. It is also in use in large commercial cooling applications. A simple absorption system with separate generating and cooling cycles is found in a small residential refrigerator shown in figure 15-13. In the generating cycle, figure 15-13A, a kerosene or gas-fired burner heats the generator, which is charged with an ammonia–water solution. Heat from the generator drives the ammonia up a tube to the condenser. The ammonia gas is cooled, condenses, and collects in the receiver tank. The generating process usually takes place once a day. When the generator runs out of fuel, the generating process is completed.

Little cooling takes place during the generating cycle. The cooling cycle begins after the burner stops operating (figure 15-13B). At this point, pressure in the system begins to drop. Liquid ammonia flows into the evaporator coil, where it evaporates, cooling the coil. The ammonia vapor then flows down the recovery tube and back into the generator, where it and the heat are absorbed by the water. Note that the system contains no moving parts, thus simplifying maintenance.

HEAT PUMP OPERATION

Heat pumps, like air conditioners, work against the natural flow of heat, extracting available heat from one area and transferring it to another. Unlike conventional air conditioners, however, they are reversible, in that they can provide cooling during the warm months and heating during the cooler months.

Even though the air outside may feel cool, it still contains some heat that can be transferred indoors. As a general rule, as long as the outdoor air is above 40°F, it contains sufficient heat for efficient heat pump operation. Below this tempeature, the heat pump must rely on electric strip heaters to supply warm air to the building. The operating efficiency of the heat pump comes from using electric energy to move existing heat from the outside of the house to within the house, and is measured by the number of units of heat energy output that are obtained for each unit of energy input into the heat pump.

The efficiency figure used to rate heat pumps is called the coefficient of performance (COP), and is expressed as a number (2.0, for example). The 2.0 figure means that the heat pump delivers two units of heat energy for each equivalent input unit of energy. These efficiency figures are interesting when one considers that a COP of 2.0 indicates that the machine is moving heat with an efficiency of 200%. It is

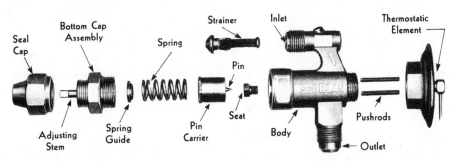

FIGURE 15-12 Component arrangement of thermostatic expansion valve. (Courtesy of Sporland Valve Co.)

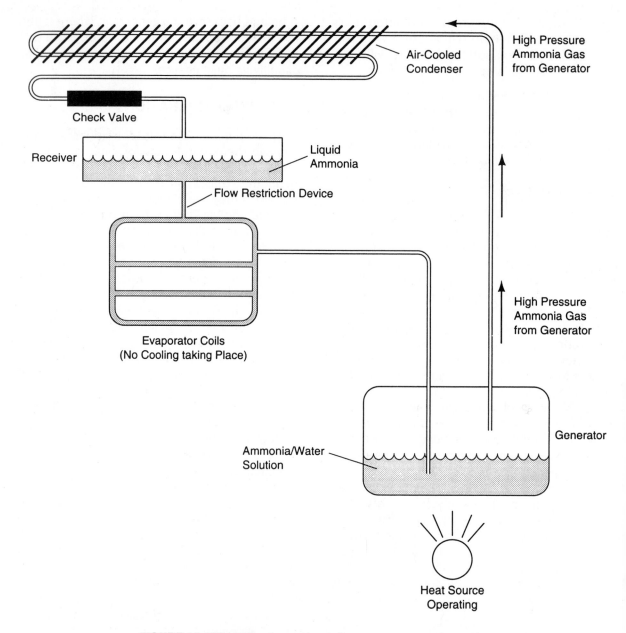

FIGURE 15-13A Absorption cycle cooling system—generating cycle.

important to keep in mind that these efficiency figures apply to the heat pump as an isolated device that is neither creating nor transforming heat energy, but is merely moving the heat from one location to another. The reversible operation of the heat pump is shown in figure 15-14, which details the heat pump operating in the cooling cycle during the summer months and the cycle reversed to provide space heating during the winter.

Note the refrigerant flow of the cooling cycle of a typical heat pump. A four-way valve is used to change the direction of refrigerant flow, depending on whether the unit is heating or cooling the

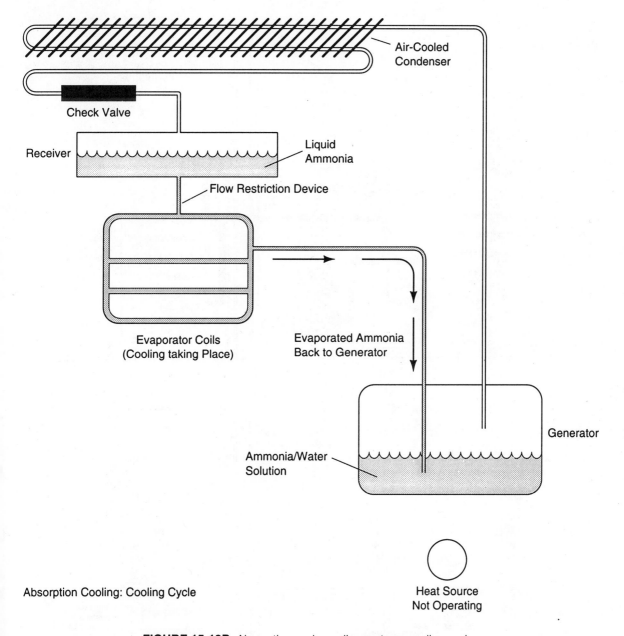

Air-Cooled
Condenser

Check Valve

Receiver

Liquid
Ammonia

Flow Restriction Device

Evaporator Coils
(Cooling taking Place)

Evaporated Ammonia
Back to Generator

Generator

Ammonia/Water
Solution

Absorption Cooling: Cooling Cycle

Heat Source
Not Operating

FIGURE 15-13B Absorption cycle cooling system—cooling cycle.

building. In the cooling mode, refrigerant from the compressor moves through the four-way valve into the condenser (outdoor coil), where heat from the refrigerant is released to the outdoor air. From the condenser, the refrigerant moves through the expansion valve and into the evaporator (indoor coil), where heat is absorbed from indoor air,

moving through the compressor and back again to the condenser. This operation and refrigerant flow is identical to that of a conventional air conditioner.

In the heating mode, note that the position of the four-way valve has been changed, so that it reverses the direction of refrigerant flow to

Heating Cycle

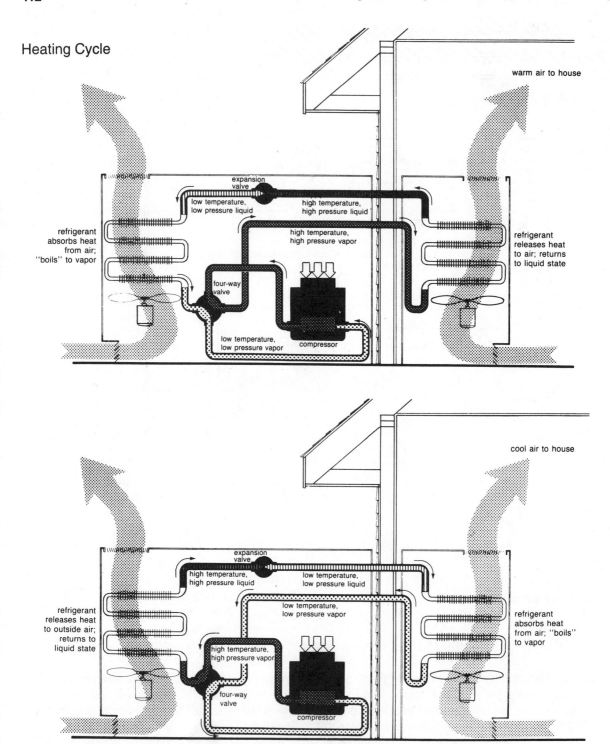

FIGURE 15-14 Heating and cooling cycle of a typical residential heat pump. (Source: U.S. Dept of Energy, 1991)

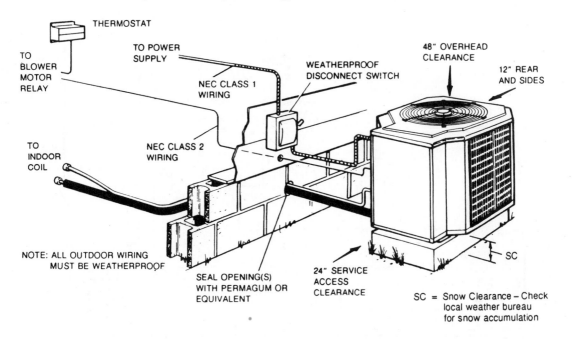

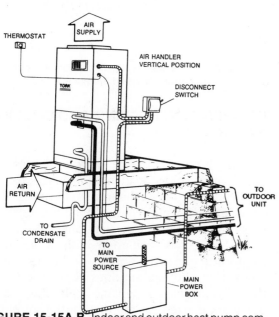

FIGURE 15-15A,B Indoor and outdoor heat pump component arrangement. (Courtesy of York International, Inc.)

the direction of that in the cooling mode. In the heating cycle, refrigerant flows through the outdoor coil, which now serves as the evaporator. Heat absorbed from the outdoor air or some other source, such as groundwater, causes the refrigerant to boil. This low-pressure/low-temperature vapor then travels through the compressor, exiting as high-pressure/high-temperature vapor. The refrigerant then moves on through the indoor coil, which now acts as the condenser coil in the heating mode. Here, the refrigerant releases its heat to the indoor air and condenses back to a liquid state. The refrigerant then travels back through the expansion valve and to the outdoor evaporator coil, where the process continues.

Supplemental electric strip heating elements are located in the indoor coil to provide space heating when outdoor temperatues are too low for efficient refrigeration cycle operation. The lowest temperature at which the heat pump can heat the building without relying on supplemental strip heaters is called the balance point of the heat pump. This point depends on the outdoor design temperature, the building heat loss, and the heating capacity of the heat pump at the outdoor design temperature.

Figure 15-15 illustrates a typical residential heat pump installation. Figure 15-15A shows the indoor component arrangement, and figure 15-15B, the installation of the outdoor components.

413

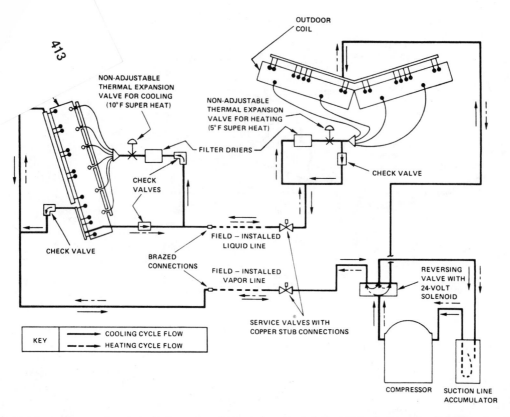

FIGURE 15-16 Refrigerant flow in typical commercial heat pump installation. Note the use of a distributor for distribution of refrigerant to the indoor evaporator and outdoor condenser coil circuits. (Courtesy of York International, Inc.)

In commercial installations, several indoor coils are manifolded together through the use of distribution valves that feed each of the individual indoor coils. A typical refrigerant flow diagram for this type of commercial system is highlighted in figure 15-16.

Although the majority of heat pumps are used in new residential construction, many commercial units are installed to recover waste heat, thus increasing the operating efficiency of these installations. The source of waste heat is found in the air-conditioning condenser water of large air-conditioning and refrigeration systems that operate for most of the year. Also, water-processing equipment and industrial processing machinery, such as welders and extruders, reject large amounts of waste heat that can be efficiently reclaimed through the use of heat-pump waste-heat reclaimers.

COMBINATION HEATING AND COOLING APPLIANCES

In addition to separate air conditioning and heat pumps, many residential installations feature the use of dual-purpose furnaces that provide both heating and cooling functions. In these applications, a conventional oil- or gas-fired furnace is equipped with a separate evaporator coil in the bonnet, or the plenum, of the furnace. Refrigerant lines connect this coil to an outdoor condensing unit. The same ductwork is used for the distribution of either the heated or cooled air, and must be sized properly to accommodate the cooling function of the unit. Since cool air is denser than warm air, cooling systems require a larger duct area and fan capacity than do those used in heat-only applications. If a homeowner wishes to add air conditioning to an existing heating system and the ductwork is not sized properly, either the existing ductwork must be replaced, a separate duct system for the air conditioning must be installed, or a ductless air-conditioning system must be used.

In either combination or retrofit applications, an A coil, named for the shape of the evaporator tube assembly, is installed in the furnace plenum or bonnet adapter as a complete unit. Figure 15-17 shows a typical A-coil assembly, and figure 15-18 the use of an A-coil assembly in conventional upflow, horizontal, and counterflow furnaces.

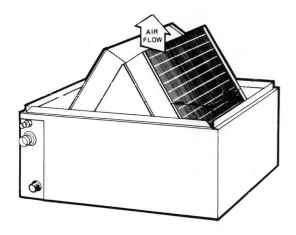

FIGURE 15-17 Typical A-coil assembly. These units either are part of initial heating and cooling installations, or can be retrofitted into existing hot-air furnaces for add-on cooling. (Courtesy of York International, Inc.)

DUCTLESS AIR-CONDITIONING SYSTEMS

An additional option available for both residential and commercial installation is the ductless air-conditioning unit, shown in figure 15-19. These systems incorporate one condenser sized to serve the needs of several remote evaporator units.

The lack of existing ductwork in homes and buildings with either hydronic or electric resistance heating may preclude the possibility of installing a ducted air-conditioning or heat-pump system. Ductless units feature separate air-handler/evaporator units that are similar in appearance to window air conditioners. However, they are surface-mounted units that do not completely penetrate the wall when they are installed. Each evaporator, equipped with its own power supply and thermostat, is connected by refrigeration tubing to a central condenser located outside of the building. Several evaporators can be served by one central condenser, depending on the capacity of both the evaporator and the condenser coils. These units are also available in heat pump configurations in which each indoor air handler can activate the heat-pump reversing valve, enabling the system to provide either heating or cooling. Most ductless systems feature either wall- or ceiling-mounted indoor air handlers, depending on the requirements of the installation.

TEMPERATURE AND PRESSURE RELATIONSHIPS IN CONVENTIONAL AIR-CONDITIONING SYSTEMS

The relationship of temperature and pressure in the various components of the air-conditioning or refrigeration system must be understood in order to ensure proper system operation, as well as to troubleshoot and isolate faulty components when problems occur. Figure 15-20 illustrates the arrangement of a typical central air-conditioning system, detailing the placement and location of critical system components, including the routing of electrical wiring and refrigeration lines.

For proper operation, the condenser, or outdoor coil, must have an adequate supply of air or water (in the case of some heat pumps) to cool the refrigerant after it leaves the compressor. In this operating scheme, the compressor raises the temperature of the refrigerant above the outdoor ambient temperature to make sure that there is an adequate temperature differential for efficient heat transfer to take place in the condenser coils.

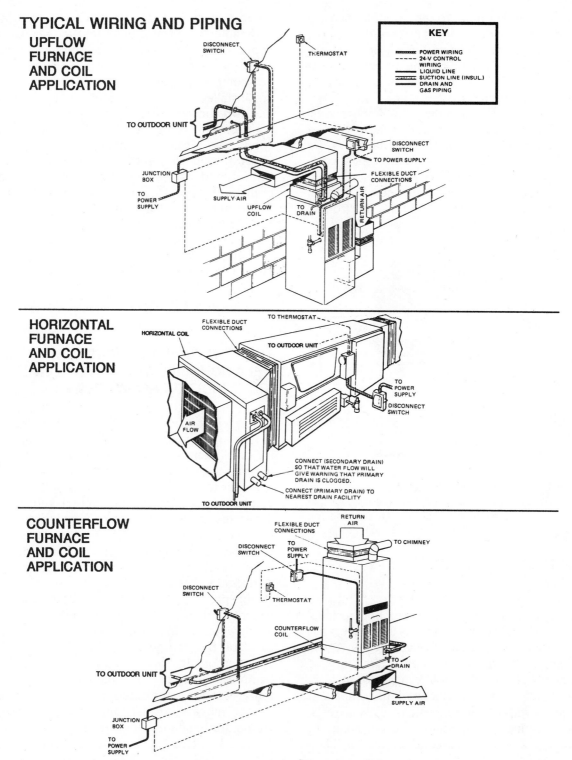

FIGURE 15-18 A-coil installations in upflow, horizontal, and counterflow furnaces. (Courtesy of York International, Inc.)

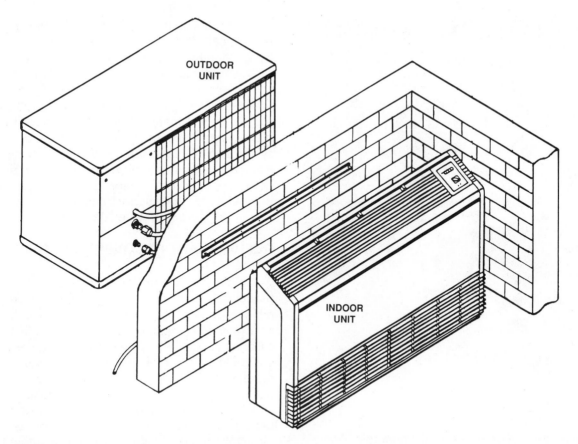

FIGURE 15-19A Ductless air-conditioning system: Separate indoor units are served by one central outdoor condenser. The number of separate indoor evaporators is a function of condenser capacity and distances from the evaporators to the condenser. (Courtesy of the Burnham Corp.)

Most residential air conditioners are sized for an airflow through the evaporator of 400 CFM per ton of capacity, based on an outdoor temperature of 95°F. Airflow through both the condenser and evaporator coils must be unrestricted for proper operation. The use of air filters in the evaporator coils usually prevents the excessive buildup of dirt and dust, as long as the filters are inspected periodically and changed when they become dirty. Outdoor condensing units should be cleaned each year with a vacuum cleaner and/or garden hose to remove accumulated debris. Putting a layer of automobile wax on the cabinet each year is recommended to maintain the painted or anodized exterior finish. (See the end of this chapter

for details regarding evaporator and condenser coil maintenance procedures.)

REFRIGERANT READINGS USING THE MANIFOLD GAUGE ASSEMBLY

Pressure and compound (vacuum and pressure) gauges are diagnostic instruments used in performing a variety of service and maintenance procedures on all air conditioners, heat pumps, and refrigeration systems. Proper use of the gauge set is essential if the technician is to perform system diagnosis and undertake corrective procedures when and where required. Figure 15-21 shows a typical gauge set with a connecting manifold and hoses. The gauges and

FIGURE 15-19B

hoses are all connected to the manifold. One of the gauges, usually the red-cased gauge, is connected to the high-pressure side of the system, while the blue-cased gauge is connected to the low-pressure, or suction, side of the system.

Three hoses connect to the bottom of the manifold. The red hose goes to the high-pressure side of the system, the blue hose goes to the low-pressure side of the system, and the yellow hose is used when evacuating the system (it is connected to the vacuum pump) or when charging the system (it is connected to the charging or refrigerant cylinder). The handwheels on the manifold control refrigerant flow from each hose to the other hoses through internal passages in the manifold. The right handwheel (high-side valve) opens or closes the flow through the right hose flare fitting to the other two fittings, as long as the left handwheel is open. The left handwheel and fitting function

in a manner similar to that of the right handwheel. The internal connections of the gauge manifold are illustrated in figure 15-22.

The manifold gauges are used in a variety of ways, depending on the service or maintenance procedure being performed. Figure 15-23 depicts a typical manifold gauge setup used for recharging hermetically sealed air-conditioning systems. Note that purging the lines is done from the central gauge valve. When purging, both gauge valves are closed. When charging, the valve to the suction line is opened to permit refrigerant to enter the system.

REFRIGERANT RECOVERY SYSTEMS

Almost all commercial refrigerants at present, particularly those used in automotive and small residential and commercial applications, belong to a family of compounds called chlorofluorocarbons (CFCs). When released into the atmosphere because of leaks in the refrigeration system, as well as in normal servicing conditions, these refrigerants attack the ozone layer above the earth. The ozone layer is a gaseous barrier that filters out harmful amounts of ultraviolet (UV) radiation. In excessive amounts, UV radiation has been shown to cause skin cancers and eye damage. Other effects of UV radiation are not fully known at this time, but it is clear that the ozone layer is critical in maintaining a safe and healthy environment. Growing concern over increasing amounts of refrigerants such as R-12, R-22, R-500, and R-502 released into the atmosphere has resulted in the development of refrigerant recovery and recycling units that are designed to eliminate and/or minimize refrigerant lost to the atmosphere during normal service procedures.

In addition to the development of recycling equipment, there are also a growing number of businesses that focus on recovering discarded refrigeration appliances in order to remove the potentially dangerous refrigerants, refurbish those appliances that are in reasonable condition, and safely scrap the remainder for recycling purposes.

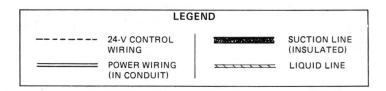

LEGEND		
– – – – – – 24-V CONTROL WIRING	▓▓▓▓▓▓ SUCTION LINE (INSULATED)	
═══════ POWER WIRING (IN CONDUIT)	✕✕✕✕✕ LIQUID LINE	

Evaporator Coil	Rows Deep x Rows Wide	3 x 32
	Finned Length - inches	46
	Face Area - square feet	10.2
	Tube (Copper) OD - inches	3/8
	Fins (Aluminum) per inch	13
Centrifugal Blower (Forward Curve)	Diameter x Width - inches	15 x 15
Filters (Throwaway)	Quantity Per Unit (16" x 25" x 1")	4
	Face Area - square feet	11.1
Distributor	Two Per Unit	5-3-6 per circuit
Operating Charge	Refrigerant-22, Lbs - Oz.	2-0 per circuit
Weight, Lbs.	Shipping (Volume - 53 Cu Ft)	395
	Operating	355

Blower Motor HP	Power Supply Voltage[1]	FLA	Fuse Size,[2] Amps.	Wire Length,[3] Feet
2	208	7.5	10	145
	230	6.8	10	178

[1] All voltages are 3-phase, 60-hertz.
[2] Dual element, time delay fuses.
[3] Based on three 60°C, 14 AWG, insulated copper conductors in steel conduit, a 3% voltage drop and a blower motor power factor of 0.85.

UNIT DIMENSIONS & CLEARANCES

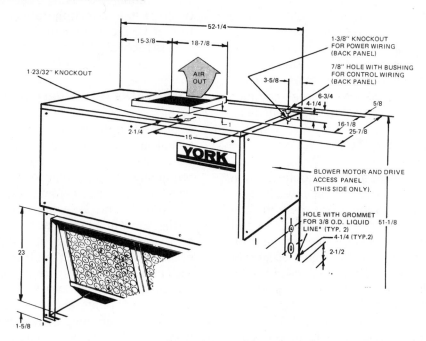

FIGURE 15-20 Component arrangement of typical residential air-conditioning system. (Courtesy of York International, Inc.)

FIGURE 15-21 Gauge manifold with compound gauge set. (Courtesy of Robinair Division, SPX Corp.)

Recovery and recycling units should be used every time a system undergoes diagnostic or repair procedures. The recovery and recycling unit in figure 15-24 is used for automotive air-conditioning systems. Similar units are available for small residential, as well as large commercial, air-conditioning and refrigeration servicing.

EVACUATING A SYSTEM

Prior to charging any air-conditioning system, it is necessary to remove or evacuate all of the old refrigerant. Evacuation allows the technician to remove any moisture, air, or other contaminants that may be present in the refrigeration system in addition to the old refrigerant. Contaminants that are left will reduce operating efficiency and can lead to compressor and/or component failure. During evacuation, the gauges are installed as shown in figure 15-25.

Note that the service line from the compound gauge is connected to the service valve

on the suction line of the compressor. If no service valve is present, a valve adapter needs to be installed in some part of the low side of the system. When attaching the gauge manifold, all gauge valves and the service valve should be closed. After all attachments have been made, the vacuum pump in the recovery unit should be started and the valves opened to evacuate the system. When using a separate vacuum pump for system evacuation, test for system leaks by closing the gauge manifold valve when the gauge registers 29 inches of vacuum. If the gauge reading remains steady for 2 minutes, continue system evacuation. After approximately 20 minutes, a final check for complete system evacuation can be made by checking the vacuum reading for 3 minutes. If the vacuum reading is maintained, the system is ready for recharging. On recovery units with built-in controls, such as that in figure 15-26, the evacuation system will shut off automatically.

The recovery unit shown features an inlet drier for moisture removal and an oil separator that removes any compressor oil, acids, and/or particulates that may be present. The advantages of using a recovery system are twofold: the atmosphere is protected from further CFC contaminants, and the refrigerant that is recovered from the unit can be filtered and reused.

CHARGING A HERMETIC SYSTEM

Several options are available to service technicians when recharging a sealed refrigeration system. One preferred method is to use a refrigerant management system that employs one unit for refrigerant recovery/recharging operations, and a second unit to recycle the refrigerant while removing moisture, acids, and solid particulates after the system has been evacuated. Figure 15-27A illustrates a recovery/recharging unit, and figure 15-27B a companion recycling system. Most of these refrigerant management units give an accurate measurement of the amount of refrigerant recovered from a system. They automatically purge lines to reduce refrigerant losses when switching

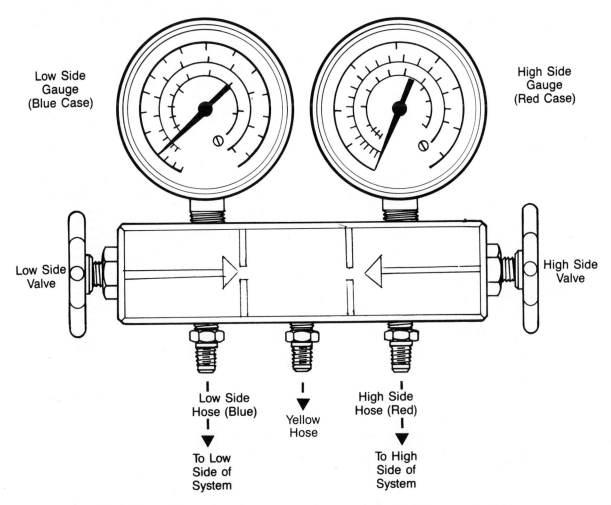

FIGURE 15-22 Internal passageways and connections of typical gauge manifold.

refrigerants, and employ microcprocessor controls for accurate system recharging.

When using conventional equipment to recharge a refrigerant system, either a refrigerant cylinder or a charging cylinder is used to add the selected refrigerant to the unit. Figure 15-28 shows the use of a charging cylinder, along with a gauge manifold to recharge the cooling system.

The charging cylinder is filled with refrigerant from a bulk tank. The cylinder features a calibrated plastic shroud the makes it easy to read the desired amount of refrigerant, by weight, that is being charged into the system. These cylinders also have a heating unit that

makes the charging process faster. The heating unit raises the temperature and pressure of the refrigerant in the cylinder, helping to overcome the equalization of pressure between the air-conditioning or heat-pump cooling system and the charging cylinder, speeding up and increasing charging process efficiency. A charging cylinder of this type is shown in figure 15-29.

When starting the charging process, the service connection lines must be clean and free of air. This is accomplished by purging the line between the refrigerant or charging cylinder and the manifold gauge at the gauge port. With both gauge valves closed, the charging line is left loose for a moment at the gauge port

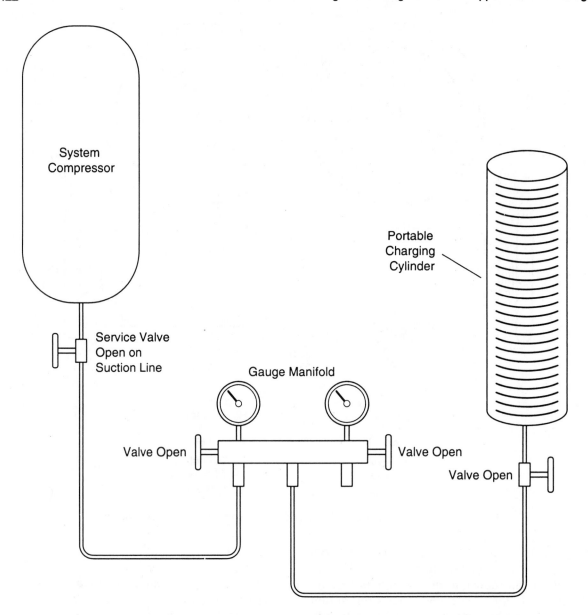

FIGURE 15-23 Gauge setup for charging a typical hermetic refrigeration system.

connection, and then quickly closed. The system is started and the refrigerant container or charging cylinder is opened. When using charging cylinders, the amount of refrigerant supplied to the system is read directly from the calibration scale. When using a refrigerant cylinder, the low-pressure gauge should read a maximum of 5 PSI, adjusted by the valve on the refrigerant cylinder. Allow the system to charge for 3 to 5 minutes and then check the operating pressures supplied by the manufacturer to determine if the system is operating within the pressure limits specified (see figure 15-31). Refrigerant must then be either purged (and recovered) or added to the system to bring the refrigeration charge to within factory specifications.

FIGURE 15-24 Refrigerant recovery and recycling system. The unit illustrated is designed for use with automotive air-conditioning systems. Similar units are available for commercial and residential air-conditioning systems. (Courtesy of Robinair Division, SPX Corp.)

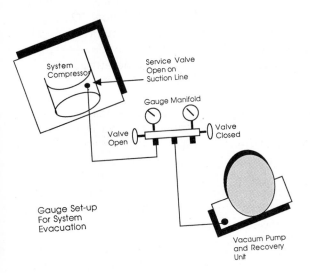

FIGURE 15-25 Gauge setup for system evacuation.

LEAK DETECTION

Leak detection procedures are particularly important after any service or maintenance procedures have been performed on the refrigeration system. There are several techniques that are effective in pinpointing even small refrigerant leaks. Electronic leak detectors, such as that in figure 15-30, are the easiest to use, and are capable of detecting leaks as small as 1/2 ounce of refrigerant per year.

The leak detector in figure 15-31 responds with an audible signal that changes in tone when a leak is found. A halogen gas detector, figure 15-31, detects the presence of halogen gas formed by the reaction of the refrigerant with oxygen in the air.

FIGURE 15-26 Refrigeration recovery system. (Courtesy of Robinair Division, SPX Corp.)

Other time-tested methods, such as the use of gas-fired halide torches that change flame color in the presence of refrigerant, or a soapsuds solution that bubbles when placed over a fitting with escaping gases, are also effective in troubleshooting suspected leaks in the refrigeration system.

If a leak is found at a flare fitting, it can often be stopped by tightening the fitting. Any or all leaks should be located and repaired at one time. Leaks involving soldered connections can only be repaired by releasing the system pressure, repairing the connection, and recharging and testing the system.

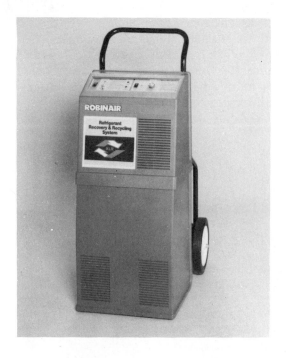

FIGURE 15-27A,B Companion refrigeration recovery and recycling systems. (Courtesy of Robinair Division, SPX Corp.)

When using electronic leak detectors, it should be kept in mind that since the refrigerant is heavier than air, the detector probe should be run along the bottom of lines and fittings, and so on. Also, leak detection should start at the top of the unit, with the technician working downward. Because of the possible accumulation of refrigerant at the bottom of the unit, the leak detector can be activated before the sensing proble is in the direct area of the leak itself.

MEASURING SUPERHEAT IN THE EVAPORATOR COIL

Note in figure 15-32 that the temperature across the evaporator coils ranges between 40 and 50 degrees. The temperature of the evaporator coil should not fall below 32 degrees. Under normal conditions, about 10 degrees of superheat is required for proper operation of an evaporator coil. The temperatures used here are for example only. For specific applications, the manufacturer's specifications should be consulted. The temperature–pressure chart in figure 15-32 can be used for a variety of service calculations, including the determination of superheat across evaporator coils.

To determine the superheat in the evaporator coil using the system chart in figure 15-32 as a reference, this procedure can be followed.

1. Measure the suction line temperature at the point where the remote bulb is fastened to the suction line, or at the suction line exit of the evaporator coil.

2. Determine the suction line pressure at the bulb location. Take a reading of the suction gauge at the condenser service port, and add the estimated pressure drop through the suction line.

3. Note that the estimated pressure drop for this unit is 2 PSIG.

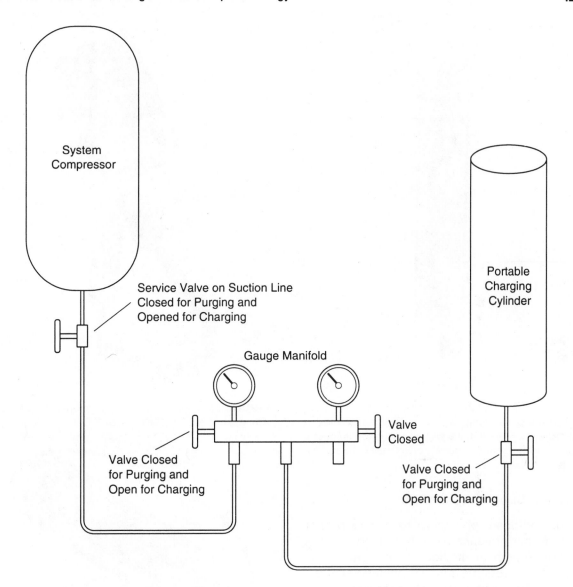

FIGURE 15-28 Recharging a refrigeration system using a portable charging cylinder through a gauge manifold assembly. (Courtesy of Robinair Division, SPX Corp.)

4. Perform superheat calculations as follows:

   Refrigerant in system: R-22
   Suction line temperature = 50°F
   Suction at compressor = 67 PSIG
   Pressure drop estimation = 2 PSIG
      Total pressure = 69 PSIG =

   40°F evaporator temperature (from figure 15-12)

5. Normal superheat should be 6°F to 15°F, depending on specific applications. Consult manufacturers' specifications for air-conditioner and heat-pump recommended superheat specifications.

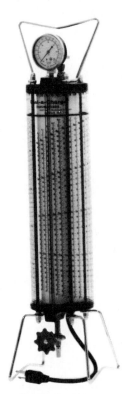

FIGURE 15-29 Portable charging cylinder. The cylinder is filled from a bulk refrigeration tank. Calibrated scales on the cylinder ensure that the proper amount of refrigerant is added to the system. (Courtesy of Robinair Division, SPX Corp.)

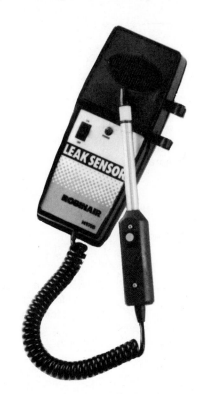

FIGURE 15-30 Electronic refrigeration leak detector. (Courtesy of Robinair Division, SPX Corp.)

Suction Superheat Method

An additional procedure to ensure proper system performance is referred to as the suction superheat method. In this procedure, the suction pressure, outdoor ambient temperature, and suction temperature are measured and compared with a chart supplied by the manufacturer to ensure that the system is operating within acceptable limits. For these calculations, the chart in figure 15-33 should be consulted after making the following system measurements.

FIGURE 15-31 Halogen gas leak detector. (Courtesy of Robinair Division, SPX Corp.)

1. Operate the unit for 15 minutes prior to taking any measurements.

2. Take meausurements of the suction pressure, outdoor ambient temperature, and suction temperature.

3. Enter measured suction pressure and outdoor temperature on the chart to determine recommended superheat.

4. Calculate suction superheat by subtracting saturated suction temperature

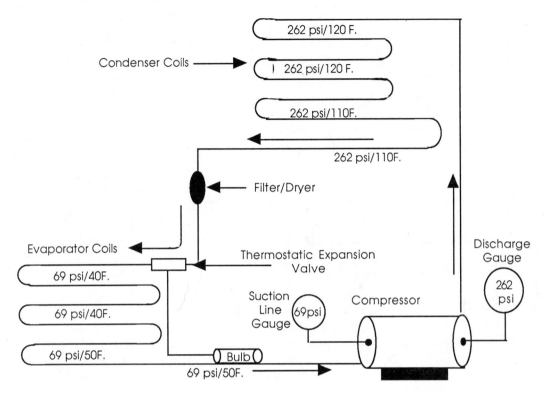

FIGURE 15-32 Measuring superheat in the evaporator coils. Under normal conditions, superheat in both the evaporator and condenser should range between 10 and 20 degrees. System pressures shown are typical.

(shown under suction pressure) from the measured suction temperature.

5. If the actual suction superheat is greater than the recommended value, add refrigerant charge; if the actual superheat is less than recommended, refrigerant charge must be bled from system.

Individual components of residential and commercial refrigeration systems can now be examined with an eye toward how they function in the total cooling system.

BASIC REFRIGERATION COMPONENTS AND CONTROL SYSTEMS

Compressors

Many different types of compressors are used in air-conditioning applications. Design differences are based on the operating condi-

tions and cooling loads that the compressor must be able to supply.

Reciprocating hermetic and semihermetic: I n residential and small commercial installations, the reciprocating type of hermetically sealed compressor is used most frequently. In a hermetic design, both the compressor and the motor are contained in a sealed housing. Operating in a direct drive fashion, the compressor rotates at the speed of the drive motor. Belt drive mechanisms that link the drive motor to the compressor are used only in open systems.

Semihermetic units, in which the motor is bolted to the compressor housing, offer more flexibility than do fully hermetic units, in that they can be partially disassembled for many field service procedures. Hermetic and semihermetic compressors are shown in figure 15-34.

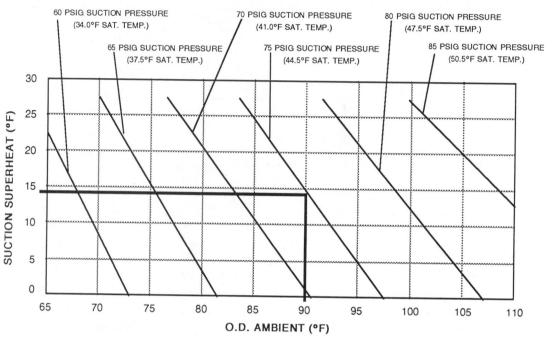

EXAMPLE: 75 PSIG Measured Suction Pressure at 90°F O.D. Ambient shows Superheat to be 14°F
(44.5 Saturated Suction Temperature + 14° Suction Superheat = A Suction Line Temp. of 58.5°)

FIGURE 15-33 Suction/superheat charging chart. (Courtesy of York International, Inc.)

Various methods are available for cooling the compressor motor windings. One method is to place the motor windings against the hermetic dome housing, using direct heat transfer from the motor windings through the casing of the dome. Another method is to have the vapors

(HERMETIC)   (SEMI-HERMETIC)

FIGURE 15-34 Hermetic and semihermetic compressors. (Courtesy of the Williamson Co., Divioision of Metzger Machine Corp.)

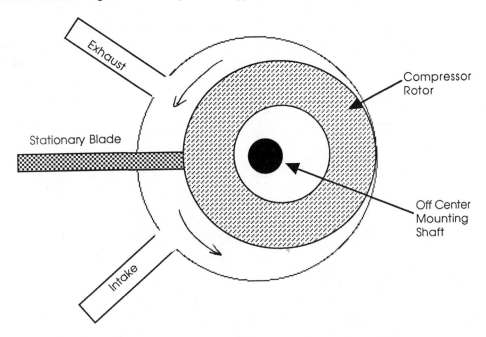

FIGURE 15-35 Stationery blade rotary compressor design.

returning from the evaporator pass through the motor windings before they reach the compressor, cooling the windings in the process. In some instances, cooling fins are incorporated into the compressor dome, which increases the dome's surface area, similar in principle to the cooling fins found on small air-cooled gasoline-engine cylinders.

Lubrication for compressor components is provided by either an oil splash system or oil injection system, depending on compressor design. The splash system is the less expensive of the two alternatives, and uses an oil sling connected to the crank throw, which dips into an oil reservoir at the base of the compressor and flings oil onto the cylinder walls and piston pin bearings. During the lubricating process, the oil drains back into the reservoir at the bottom of the crankcase. An oil sling system is common in most small compressors, and is, in fact, the system of choice for most small gasoline engines.

Large residential and commercial cooling units use an oil injection system. These systems utilize an oil pump that injects oil under pressure to all critical bearing surfaces. The oil pump is usually connected to the cylinder crankshaft, and the system incorporates some type of oil pressure cutoff switch to stop the unit if the oil pressure drops below a preset level.

Most hermetic compressors use mufflers on either the intake or exhaust line (or both) to reduce the pumping noise of the compressor. These mufflers are similar in operating principle to those used on automobiles. Gases expand as they enter the muffler head, slowing down the gas and reducing the noise coming from the compressor.

Rotary compressors: The rotary compressor design is used widely in hermetic compressors. Rotary designs are of either the stationary or rotary blade type; the former is illustrated in figure 15-35.

In the stationary blade design, a roller is mounted on an off-center shaft. As the roller rotates within the cylinder, the space between the roller and cylinder wall changes, either squeezing and compressing the refrigerant as it leaves through the exhaust valve, or creating a partial vacuum to draw fresh refrigerant into the compressor.

In a rotating blade design, two or more blades are mounted in an off-center rotating cylinder. In the two-blade design in figure 15-36, note how the vapor is drawn into the cylinder by the suction action created as the roller passes the suction port, and how it is compressed to discharge into the condenser lines. Rotary compressors use reed-type valves similar to those found on two-cycle engines in the discharge ports to prevent exhaust vapors from backing up into the compressor.

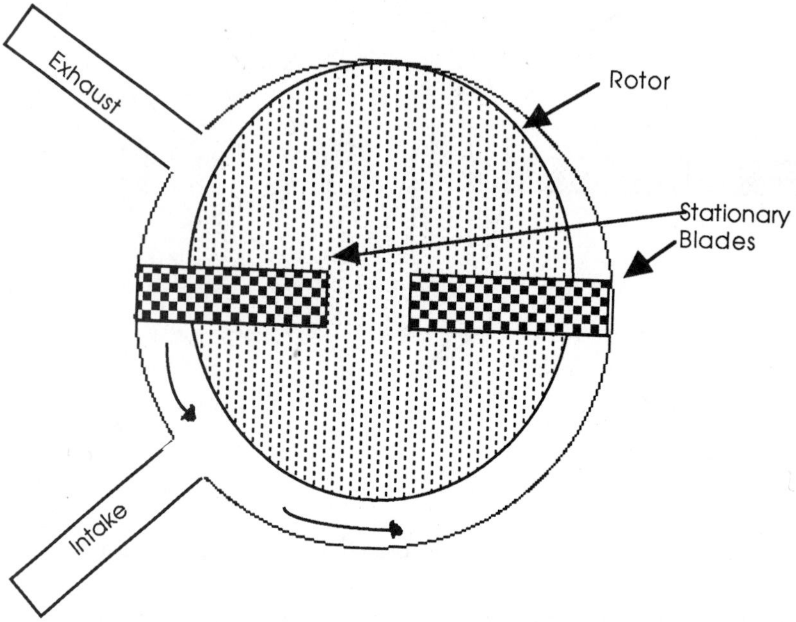

FIGURE 15-36 Rotating blade rotary compressor design.

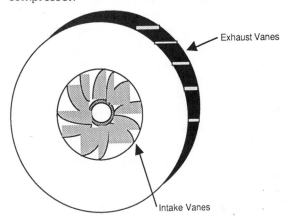

FIGURE 15-37 Impeller design of typical centrifugal compressor.

Centrifugal compressors: Large commercial refrigeration and cooling systems use centrifugal-type compressors. The design of the centrifugal refrigeration rotor is similar to that used in centrifugal water pumps, and is detailed in figure 15-37.

The rapidly spinning rotor draws vapor in through the central intake vanes. Centrifugal force throws the vapor outward. Since centrifugal force increases with speed, these units operate at high rotating speeds. Relatively few moving parts are incorporated into the centrifugal design, minimizing the service and maintenance required in these larger systems.

Many residential and most commercial, refrigeration system compressors are equipped with crankcase heaters to prevent damage during cold-start conditions. Ordinarily, the crankcase heater is energized when the compressor is not operating but power is being supplied to the system. Be sure to check the manufacturer's instructions regarding the operation of specific crankcase heating systems.

Appendix D contains an expanded troubleshooting chart that can be used to locate operating problems that occur most frequently in air-conditioning compressors. Specific service procedures will vary according to the type, age, and size of the compressor.

Refrigerant Expansion Control Devices

Expansion valves perform two basic functions in the refrigeration system: they reduce the high-pressure liquid refrigerant that comes from the condenser coils to the low-pressure liquid required for proper evaporator performance, and they keep the evaporator coils supplied with the correct amount of refrigerant for maximum operating system efficiency. There are several types of expansion devices that perform these functions.

Capillary tubes: The capillary tube was briefly discussed and illustrated in figure 15-9. This widely used device is a small-diameter tube that acts to restrict and meter the flow of refrigerant along its length. The tube contains no moving parts, and is the simplest refrigerant metering device in use. As the refrigerant flows along the tube, it begins to evaporate in the end section. By the time the refrigerant leaves the capillary tube, both its pressure and temperature are reduced to those of the evaporator. When the compressor is not running, the pressure in the system equalizes, placing minimum stress on the compressor each time it starts.

A major operating problem associated with capillary tubes is the buildup of wax or other formations in the tube that can decrease its inside diameter and thus restrict the amount of refrigerant to be metered into the evaporator coils. Capil-

lary tubes are easily replaceable. Consult the manufacturer's specifications for the proper length and diameter when tubes are being replaced.

Automatic expansion valves: Automatic expansion valves use adjustable spring pressures to open and close a valve seat that admits refrigerant into the valve and evaporator coil based on the evaporator coil pressure. The operation of an automatic expansion valve is illustrated in figure 15-38.

When the compressor is operating, a reduction in pressure in the evaporator coil causes the valve to open. The valve must be properly sized for the specific system, since the same amount of refrigerant that is being pumped by the compressor should also flow through the expansion valve assembly. The major use of automatic expansion valves is in small residential air-conditioning systems, in cold drink machines, and as replacements for defective capillary tubes.

Thermostatic expansion valves: Thermostatic expansion valves have achieved wide usage in refrigeration and air-conditioning systems. Most of these valves can be inspected in the field, and critical valve components can often be replaced without cutting into the refrigeration lines. Valve operation is based on three pressures in the system, as illustrated in figure

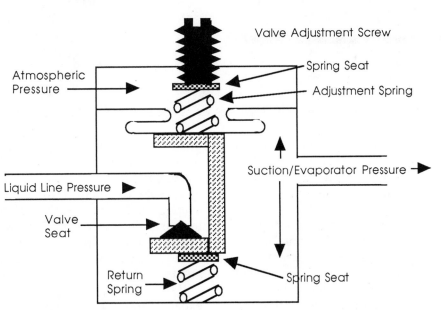

FIGURE 15-38 Cross section of typical automatic expansion valve.

Valve Adjustment Screw

Spring Seat

Adjustment Spring

Atmospheric Pressure

Suction/Evaporator Pressure

Liquid Line Pressure

Valve Seat

Spring Seat

Return Spring

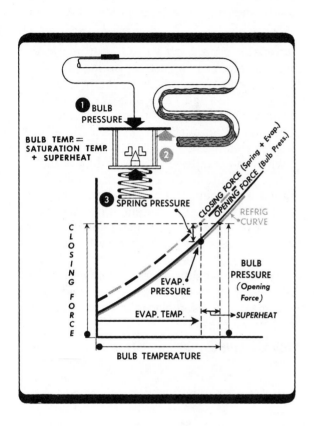

FIGURE 15-39 Thermostatic expansion valves. (Courtesy of Sporlan Valve Co.)

15-39: sensing bulb pressure, which acts to open the valve; pressure from the evaporator coils, which acts to close the valve; and spring pressure, which assists in closing the valve.

Ideally, the same refrigerant is used in both the thermostatic sensing bulb and the refrigeration system. Under these conditions, if the temperatures of the system refrigerant and sensing bulb are the same, their pressures will be identical with the valve in a neutral, closed position. As the liquid refrigerant flows through the evaporator coil, it absorbs heat, evaporates, and becomes superheated. When this superheated vapor reaches the sensing bulb, the temperature and pressure in the sensing bulb

increase. This increased pressure is transmitted to the top of the bulb side of the expansion valve, compressing the spring and forcing the valve seat to open. Refrigerant entering the evaporator causes the evaporator pressure to increase. This, in conjunction with the spring pressure in the expansion valve will cause the valve to close when the closing pressure is greater than the sensing bulb pressure.

If the supply of refrigerant through the expansion valve is insufficient, either the evaporator pressure drops or the temperature of the vapor leaving the evaporator coils increases, heating the sensing bulb, which will cause the valve to open. This action will cause the three pressures—sensing bulb, evaporator, and spring—to balance one another and close the valve. Figure 15-40 shows a typical thermostatic expansion valve together with a valve cross section. These valves are available in a variety of design configurations, with the choice depending on the size and type of cooling system to be regulated.

Electronic temperature control valves: One type of refrigerant throttling valve uses solid-state electronics to control the amount of refrigerant in the evaporator coils. Rather than using pressure-sensing devices as in automatic or thermal expansion valves, electronic valves use a solid-state thermistor-type temperature sensor that changes in electrical resistance with temperature changes. This change in electrical resistance is monitored by a solid-state control panel to open or close the control valve. An electronic control valve is shown in figure 15-41.

The greatest use for electronic control valves is for controlling temperature in commercial refrigeration and cooling applications.

Low- and high-side float controls: Float controls in refrigeration systems operate in a manner similar to the way in which the float control works in the ordinary toilet. The function of the control is to maintain a constant level of refrigerant in the evaporator coil. Float controls can be placed on either the low (suction) side or high (pressure) side of the refrigeration system.

A low-pressure float control incorporates a liquid receiver in conjuction with the float control for the storage of excess refrigerant in the system, and is detailed in figure 15-42.

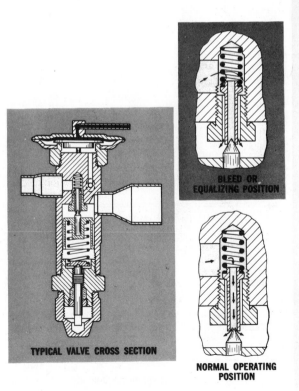

TYPICAL VALVE CROSS SECTION

BLEED OR EQUALIZING POSITION

NORMAL OPERATING POSITION

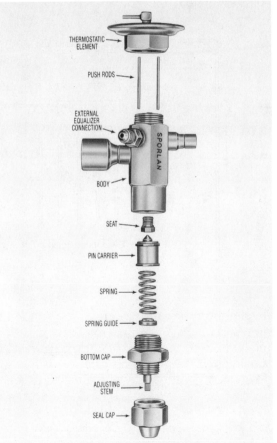

FIGURE 15-40 internal components of typical thermostatic expansion valve. (Courtesy of Sporlan Valve Co.)

FIGURE 15-41 Electronic temperature control valve. (Courtesy of Sporlan Valve Co.)

The low-side float system uses a pressure cutoff switch to operate the compressor. The flow of the refrigerant is based on the position of the float ball in the tank, which opens and closes a needle valve, depending on the level of the refrigerant.

Vaporized refrigerant from the evaporator travels through the compressor suction line to the compressor. The compressor changes the low-pressure vapor to a high-pressure vapor and discharges the vapor into the condenser coils. In the condensing coils, the refrigerant is cooled, condenses to a liquid, and flows into the receiving tank. From the receiving tank, the high-pressure liquid refrigerant flows through the float needle valve and into the evaporator. The evaporator in most of these systems is designed as a finned tray or tank. The low-pressure liquid refrigerant in the evaporator absorbs heat, vaporizes, and is drawn back to the compressor through the suction line, where

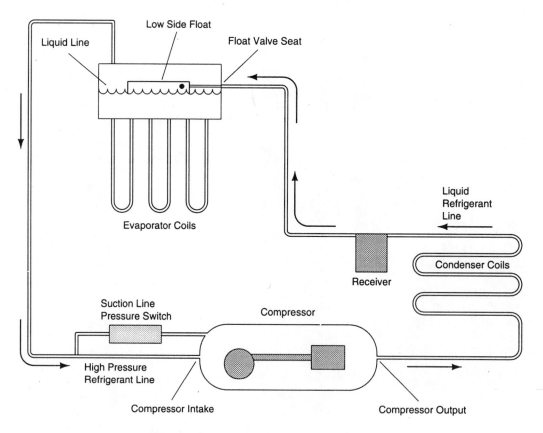

FIGURE 15-42 Low-side float control refrigeration system.

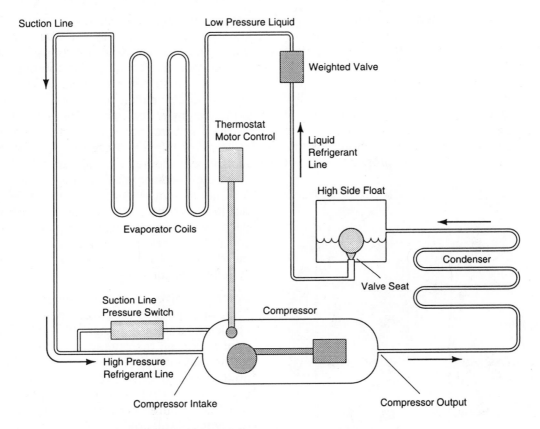

FIGURE 15-43 High-side float refrigeration system.

the cycle continues. Compressor operation is controlled by a pressure switch in the suction line. As the compressor operates, pressure in the line is reduced. When the pressure reaches a preset cut-out point, the compressor motor is shut off until pressure rises to the cut-in point on the suction line.

Low-side float systems are suitable for constant-temperature applications, such as drinking fountains and vending machines. These systems are sometimes referred to as flooded systems, since they contain a large refrigerant charge that keeps the float tank and evaporator coils almost full of refrigerant during both the on and off cooling cycles.

High-side float systems are also referred to as flooded systems. In the high-side float arrangement, the liquid receiver tank is located at the output side of the condenser coils (figure 15-43).

High-pressure vapor from the compressor enters the condenser coils, where it cools and condenses back into high-pressure liquid refrigerant, and then enters the high-pressure float tank. At a preset level, the needle valve opens, allowing the high-pressure refrigerant to flow through a capillary tube, where its pressure is reduced as it enters the low-pressure evaporator coils. Low-pressure refrigerant in the evaporator coils will quickly absorb heat, boil, and flow to the compressor through the suction lines, where its pressure is again raised to continue the cooling cycle. High-side float mechanisms have their largest application in commercial refrigeration devices.

Refrigerant Distributors

Refrigerant distributors are used in commercial systems in which several separate evaporator circuits are installed. Ordinarily, when the liquid refrigerant passes through an expansion valve, some of it will flash into a vapor, creating a vapor–iquid mixture at the valve outlet. The liquid and vapor move at different velocities, causing some evaporator circuits to receive more liquid vapor than others. Those that receive greater amounts of vapor operate with less efficiency than those sections of the evaporator that receive the proper amount of

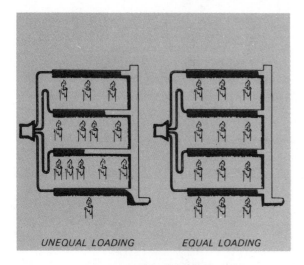

UNEQUAL LOADING   EQUAL LOADING

FIGURE 15-44 Refrigerant distributor.

liquid refrigerant. Operating with unequal amounts of refrigerant in the evaporator reduces the total effective area and efficiency of the evaporator circuit.

A refrigerant distributor is a housing that contains a nozzle and orifice assembly to in-

crease refrigerant velocity as it travels through the nozzle. The distributor also mixes the vapor and gas in equal proportions for each of the evaporator circuits. The output of the refrigerant distributor is connected to each of the evaporator circuits by a separate tube. This arrangement is depicted in figure 15-44.

The distributor nozzle is selected according to the type of refrigerant used in the system, the number of evaporator circuits to be fed by the distributor, and the manufacturer's requirements for the system. When installing or servicing distributors, consult the manufacturers of both the distributor and refrigerant systems for information concerning nozzle size, as well as any changes or modifications that might need to be made to the system or its components.

Solenoid valves: Solenoid valves control the flow of liquids and gases in a variety of applications (see Chapter 10 for additional information on the operation of these valves). In the majority of refrigeration systems, solenoid valves are used to control the flow of liquid or vapor refrigerant in either the suction or high-pressure line in the system. There are two types of solenoid valves, the normally open (NO) type and the normally closed (NC) type. The most common is the normally closed type, in which the valve passageway is closed when the valve is deenergized. When power is applied to the valve coil, the valve opens. In the normally open arrangement, the valve opens when the coil is deenergized and closes when the valve coil is energized. Solenoid valves use electromagnetic coils to raise and lower a plunger mechanism that will either open or close the valve port. Figure 15-45 illustrates four stages of operation of a normally closed pilot-operated solenoid valve. The use of a pilot port equalizes the pressure on top of the valve disk, or piston.

Solenoid valves are often installed in the liquid line to prevent the flow of refrigerant into the evaporator during the compressor off-cycle. They can also be placed in the suction line, and are suitable for high-temperature applications, such as the hot-gas defrosting system shown in figure 15-46. In this illustration, a portion of the

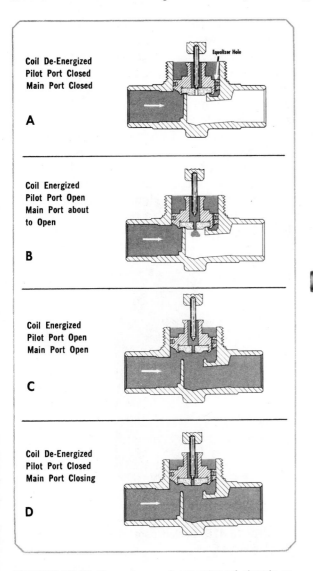

FIGURE 15-45 Four stages of operation of electric refrigerant solenoid valve. The valve seat is controlled by an electrically operated plunger moving in the solenoid core. (Courtesy of Sporlan Valve Co.)

compressor discharge gas is passed through the solenoid valve and into the evaporator, where it is cooled.

Filters, Driers, and Strainers

Why filtration devices are required: Refrigeration systems contain at least some moisture, dirt, acids, sludge, and varnish. The moisture present

can cause line freeze-ups, as well as contribute to the formation of acids, sludge, and general corrosion of the system piping and components. Dirt—consisting of everything from excess flux used in soldering piping connections to metallic particles, rust, and scale—can damage compressor cylinder walls, and also plug capillary tubes, as well as the protective screens found in expansion valves. Under severe operating conditions, acids may form as the result of chemical reactions and partial decomposition of the refrigerant gases. Sludge and varnish form as the result of high discharge temperatures acting on compressor lubricating oil.

Because some or all of these corrosive compounds may be present in any refrigeration system, the use of driers, filters, and strainers has been standard practice for many years in virtually all residential and commercial refrigeration systems. At the present time, only two standards exist for filter driers—one for water capacity, the other for flow capacity. For proper operation, the filter selected for a specific application depends on many factors: the amount of moisture expected in the system, operating temperatures in the system, allowable pressure drop through the filter assembly, bursting pressure of the filter, and the amount of acid that can be removed by the specific filter. Filters and driers are usually incorporated into one cartridge assembly and are referred to as filter/driers.

Filter/driers: Many filter/driers incorporate a filter core manufactured from a blend of dessicants. Dessicant granules adsorb moisture,

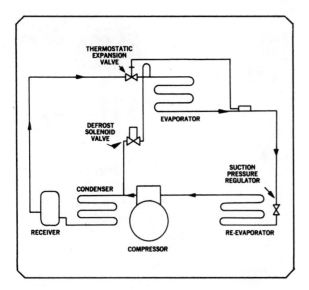

FIGURE 15-46 Use of a solenoid valve in a hot-gas discharge refrigeration system. (Courtesy of Sporlan Valve Co.)

and so should also adsorb the various acids found in the system. The construction of the filter core enables the filter to remove solid particles of solder, scale, metallic powders, carbon, sludge, and so on. The construction of the particular filtration unit also depends on

FIGURE 15-47A Replaceable core filter/drier. (Courtesy of Sporlan Valve Co.)

whether it is to be installed in the liquid or suction line and the operating pressures to which it will be subjected. Since pressure drop through the filter can be a critical factor for proper system operation, filter/driers of the correct size and capacity must be checked for suitability prior to their intallation.

Filter/driers used on smaller residential and commercial systems are usually sealed units. Those installed on large commercial air-conditioning and refrigeration systems feature replaceable filter cores. Figures 15-47 shows both a sealed filter unit and a replaceable core-type filter/drier.

FIGURE 15-47B Sealed core filter/drier. (Courtesy of Sporlan Valve Co.)

Heat pump installations merit special filtration considerations. A change in flow direction of the refrigerant could dislodge dirt and sludge particles collected by a conventionally designed filter when the refrigerant flows in the opposite direction. In the typical **heat pump filter/drier** installation in figure 15-48, the filter/drier uses check valves at either end of the filter. These direct the flow of the refrigerant in such a manner as to prevent the release of dirt collected in one operating mode of the heat pump when the refrigerant flow direction reverses.

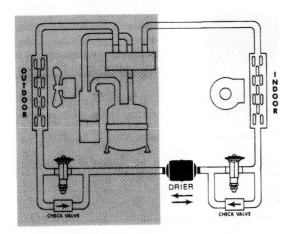

FIGURE 15-48 Use of a reversible filter/drier in a heat pump system. (Courtesy of Sporlan Valve Co.)

Filter/driers are also used to clean up a system after a hermetic motor/compressor has burned out. The filter unit can be installed directly in the suction line by removing a portion of the refrigerant line directly in front of the compressor (figure 15-49).

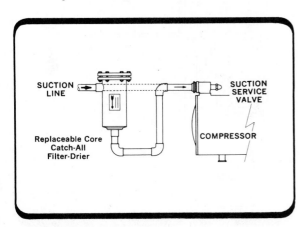

FIGURE 15-49 Installation of filter/drier in suction line for system clean-up during service procedure to install a new compressor. (Courtesy of Sporlan Valve Co.)

The type of installation shown in figure 15-49 provides a great deal of protection to a new compressor since the refrigerant–oil mixture is filtered and purified just before it returns to the compressor. Installed in this way, the filter should remove all of the contaminants remain-

ing in the system from the prior compressor burnout. If possible, an oil sample should be taken and tested for acidity. When using a replaceable-core model, new filter elements can be installed after cleanup, prior to returning the unit to full service. After two weeks of operation, the oil should be tested again to determine whether the filter elements need to be replaced. This testing procedure should be continued until it has been determined that the oil is clean and acid-free.

Strainers: Strainers are used in refrigeration systems for removing larger-scale and solid metallic particles from system circulation. The strainers are sold in screen sizes commonly ranging from 40 to 100. The size should be matched to the size of the refrigerant lines in the system. Screen sizes that are too small for the refrigerant line can cause excessive pressure drop through the strainer assembly, affecting system performance and efficiency. The strainer in figure 15-50 features a strainer access port for easy removal of the screen for cleaning and inspection without breaking the line connections.

FIGURE 15-50 Strainer valve for removing solid particulates in refrigeration system. (Courtesy of Sporlan Valve Co.)

This strainer is designed for either liquid or suction line installations, and can be used with a variety of liquids, including brine and water.

Moisture and Liquid Indicators

Commercial refrigeration and air-conditioning systems usually employ a sight glass installed on the liquid line to enable the technician to

FIGURE 15-51 Moisture and liquid-level indicator assembly. (Courtesy of Sporlan Valve Co.)

check for proper refrigerant level. Some devices incorporate liquid-level and moisture-level indication in one unit, such as the sight glass in figure 15-51 which combines the functions of indicating refrigerant level and indicating moisture in one component assembly. A shortage of system refrigerant is indicated by bubbles viewed through the built-in sight glass. The moisture level of the refrigerant, in parts per million, is indicated by comparing the color of the built-in paper indicator element with a standard color chart. The paper element is impregnated with a chemical salt that is sensitive to moisture and changes color according to the amount of moisture in the refrigerant. These elements are usually packaged in a cartridge, or fuse-type, assembly, which is replaceable if it should become damaged or contaminated. The indicator is usually installed in the liquid line, preferably after the filter/drier and ahead of the expansion device. This placement ensure accurate readings of both refrigerant levels and trapped moisture in the system.

Crankcase Pressure-Regulating Valves

Crankcase pressure-regulating valves are designed to prevent an overload of the compressor motor by limiting the pressure in the crankcase during and after a defrost cycle or after a normal shutdown of the compressor. In these instances, the pressure in the evaporator

coils may be such that the compressor motor is not able to handle the load placed on it by the high system pressures and will not be able to start. Crankcase regulating valves, sometimes also referred to as suction pressure-regulating valves, are sensitive to their outlet pressure only. In figure 15-52, note that spring force on the valve acts to keep the valve disk closed against the valve disk seat, while the outlet pressure acts as a force against the valve spring in the opening direction.

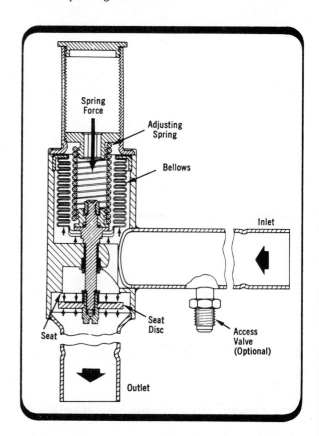

FIGURE 15-52 Compressor crankcase pressure-regulating valve. (Courtesy of Sporlan Valve Co.)

As long as the outlet pressure of the valve is greater than the valve spring setting, the valve will remain closed. As the outlet pressure is reduced, the valve will open and pass refrigerant vapor into the compressor. The valve is installed in the suction line between the evaporator and compressor as shown in figure 15-53.

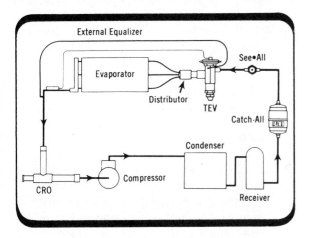

FIGURE 15-53 Installation of crankcase pressure-regulating valve in suction line. (Courtesy of Sporlan Valve Co.)

There are many selection factors that must be considered when installing a crankcase regulating valve, including the maximum allowable suction pressure, the recommended pressure drop across the valve, and the design suction pressure after compressor start-up. These valves are hermetic in construction and cannot be field serviced, and must be replaced if they become inoperative.

Oil-Level Control Systems

Oil-level controls are used in commercial refrigeration and cooling systems where multiple compressors operate in parallel. Although lubricating oil is critical for compressor lubrication, the presence of large quantities of oil mixed with the refrigerant gases greatly reduces cooling system performance. An oil-level control system is designed to remove lubricating oil from the compressor discharge gases and meter lubricating oil to all of the compressors operating in the refrigeration system, based on individual rates of compressor oil consumption. A system schematic of a typical oil-level system is detailed in figure 15-54.

Oil that is pumped by each compressor flows into a common discharge header and through an oil separator. The high-pressure refrigerant travels on to the condenser coils and the separated oil moves into an oil reservoir tank. The oil in the reservoir still contains some dissolved refrigerant gas. The refrigerant is removed from the reservoir by boiling it and passing it through a check valve and a vent line that returns it to the compressor suction header. The lubricating oil is then fed from the reservoir to individual level controls on each compressor that meter oil to the compressors based on their individual pumping rates. The oil reservoir contains a high- and low-level sight glass, which enables the technician to maintain the proper oil level in the system.

MAXIMIZING SYSTEM PERFORMANCE

Given all of the variations in air-conditioning and heat-pump system design and component configurations, most manufacturers agree on some relatively simple procedures that will maximize the operation of all air-conditioning and heat-pump systems.

Some System Efficiency Do's and Don'ts

1. Whenever possible, don't cool unused areas of a home or building. Areas in which there is a minimum of traffic, such as garages, unfinished basements, and store rooms, should have reduced supply and return airflow.

2. Avoid constant resetting of the room thermostats. Once a comfortable setting has been reached, leave the setting at that point.

3. Don't restrict air circulation. Wherever possible, air ducts should be left unobstructed in order to maximize airflow throughout the rooms in the building. Avoid placing furniture, carpets, drapes, etc., over air ducts, as this will

cause the cooling system to work harder and less efficiently in achieving the desired temperature settings.

4. When leaving a building for an extended period, set back the thermostat heat or cooling levels to minimize system operation when it is least needed. When returning after being away from the building for an extended period, the cooling system will take a little longer to reach readjusted thermostat settings.

5. Make sure that the thermostat is installed properly. Appliances that produce heat, such as lamps, or room heaters, should not be located in the vicinity of room thermostats.

6. Operate kitchen and bathroom exhaust fans when these areas are in use. Make sure that clothes driers are properly vented. All of these uses can increase indoor humidity, which will cause air conditioners and heat pumps to operate longer than necessary.

7. Prior to entertaining large groups of people, reset thermostat levels several degrees lower, since people give off considerable amounts of heat that will build up in relatively small, restricted areas of homes and buildings.

8. Keep drapes and blinds closed whenever possible to maximize the insulating effect of these devices.

9. Keep cooling temperatures as high as is practical to ensure comfort. Higher temperature levels of just a few degrees will save considerable amounts of energy and money during the average cooling season.

General Care of Air-Conditioning and Heat-Pump Systems

Periodic inspection of the air-conditioning or heat-pump system, coupled with lubrication and adjustment of some critical components, are every technician's annual system tune-up procedure. The following inspection and adjustment recommendations will help maximize system performance and minimize operating problems during the heating and/or cooling season.

1. The condensing coil must be kept free of any debris that will restrict airflow through the coil and fins. Blowing out accumulated leaves and vacuuming the coil is usually sufficient. If the coil itself becomes dirty, power to the unit should be switched off, and the coil washed out with an ordinary garden hose. The cabinet should be washed each season. A coat of ordinary car wax will reduce the likelihood of rust and corrosion of the cabinet finish.

2. Fan motors should be inspected to determine whether they are of the sealed bearing type or have oil lubrication ports. New fan motors come equipped with factory-installed lubrication that should last for at least one season. Note the lubrication schedule in figure 15-53 for relubricating fan motors. The use of a 20-weight nondetergent automobile or electric motor oil is recommended. Avoid the use of special-purpose lubricating oils, such as sewing machine, typewriter, cutting, or penetrating oils.

3. Filters should be inspected monthly during the operating season. Reusable filters should be washed with a mild detergent according to the manufacturer's specifications. Disposable filters should be replaced with new filters of the same size and type as the originals. When installing new filters, be sure that the airflow arrow points in the same direction as the airflows in the duct or filter housing.

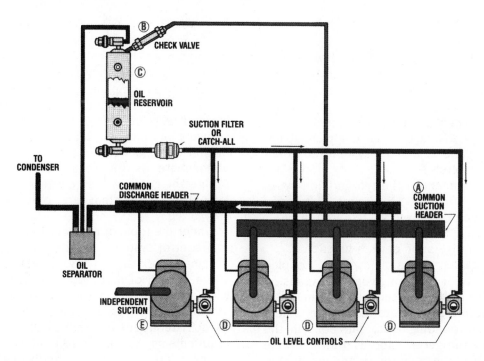

FIGURE 15-54 Oil-level control systems for multiple-compressor commercial refrigeration or cooling installation. (Courtesy of Sporlan Valve Corp.)

4. Minimum clearances to the condenser must be maintained to provide for any yard or patio improvements that may be undertaken in the vicinity of the condenser. These clearances are:

Top: 48 inches	Sides: 12 inches
Rear: 12 inches	Front: 24 inches

5. The evaporator coil drain pan or drain line must be inspected to ensure proper operation. Drain pans should be cleaned regularly to prevent odors and potential blockages. Drain lines should be inspected to make sure that they are free of obstructions and that they drain properly.

6. Belts should be inspected for proper tension. If any cracks are found, the belts should be replaced and properly retensioned.

7. All fan bearings should be lubricated where necessary, and bearings inspected for unusual wear or play, and repaired or replaced as required.

8. The system (power only) should be turned on 24 hours prior to start-up in the spring.

9. In almost all installations the use of a thermostatically controlled attic fan is recommended in order to reduce heat gain in the building during the months when the air conditioner is operating.

10. When checking a system for proper operation during the spring start-up, the superheat across the evaporator and condenser coils should be measured with an accurate thermometer. The superheat across both coils should be the same, approximately 15–18 degrees.

Following these procedures during normal maintenance or spring tune-up calls will help to minimize system malfunctions.

SUMMARY

Few appliances offer the comfort that is afforded by modern air conditioners and heat pumps. This chapter has explored the technology associated with refrigeration cycle devices, and has concentrated on the operation of typical residential and small commercial air-conditioning and heat-pump systems and their associated controls.

The development of the refrigeration cycle, as detailed in figures 15-5 through 15-8, forms the foundation of knowledge required in order to understand how this cycle and all of the associated refrigeration devices function.

Mechanical compression systems, coupled with a variety of different types of expansion devices, all function in essentially the same way. Refrigerant in an evaporator coil absorbs heat from the surrounding area, and boils in the process; the refrigerant vapor is drawn into the suction side of the compressor and discharged at high pressure and temperature on the output side. The refrigerant then travels through a set of condenser coils, where heat is given off to the surroundings and the refrigerant condenses back into a liquid. After traveling through an expansion device where its pressure is reduced and its flow is carefully metered, the refrigerant again travels through the evaporator coils to continue the cooling cycle. In this cycle, the operation of both conventional air-conditioning systems and heat pump systems is examined to highlight their similarities and differences in the same basic refrigeration cycle.

A variety of procedures and tools are available to test and evaluate refrigeration cycle performance. The manifold gauge, along with different types of recovery and recycling devices, enables complete system analysis, as well as environmentally sound practices for recovering and recycling refrigerants. Without proper recovery and recycling techniques, the environmental consequences for the ozone layer in the atmosphere as the result of freely vented refrigerant gases will surely have detrimental long-term effects that are only now being recognized.

Under the topic of refrigeration components and control systems, a variety of compressor designs and expansion devices were examined. The selection of a compressor type, together with an expansion device, must be carefully matched to the type of application, the cooling capacity required, and various other operating requirements, ranging from noise considerations to the location of individual system components. Refrigerant distributors, solenoid and pressure-regulating valves, moisture and liquid indicators, and oil-level control systems are all available to maximize system performance while minimizing system downtime for maintenance procedures. Finally, tune-up procedures that form a basic check list for all system technicians and repair personnel were reviewed.

The information presented here will form a solid foundation upon which the technician can build an understanding of the theory of system operation when dealing with system components and control functions.

PROBLEM-SOLVING ACTIVITIES

Troubleshooting Case Studies

A. After servicing an air-conditioning system, the unit is turned on, and it is observed that refrigerant is leaking from the low-pressure line where a kink in the tubing has punctured the line. Outline a procedure for repairing the leak and recharging the system to the proper specifications.

Review Questions

1. Explain the difference in function between an air-conditioner evaporator and condenser. How do the operating pressures in each of these devices relate to the properties of commercial refrigerants?

2. Using the information from figure 15-3, calculate the proper charge for a 5-ton air conditioner using R-22 refrigerant.

3. Differentiate between open and closed refrigeration systems. What compressor characteristics are needed for each type of system?

4. Explain the function of the expansion device in a refrigeration system. Highlight the differences in operating characteristics among capillary tubes and automatic and thermostatic expansion valves.

5. What are two major differences between a residential air-conditioning system and a heat pump system?

    a. Diagram the refrigerant flow in a heat pump during the cooling cycle.

    b. Diagram the refrigerant flow in a heat pump during the heating cycle.

    c. How does a heat pump provide space heating when outdoor temperatures are too low for extracting heat from the outdoor air economically?

6. Cite one major factor to bear in mind when considering the conversion of a conventional hot-air furnace to provide both heat and air conditioning.

7. Diagram the installation of a manifold gauge assembly when:

    a. Evacuating a refrigeration system.

    b. Charging a refrigeration system.

    c. Testing suction pressure in a refrigeration system.

8. Explain the function of superheat in determining whether a refrigeration system is properly charged with refrigerant. Outline two procedures for using superheat measurements to determine proper refrigerant charge.

9. Explain the difference between hermetic and semihermetic compressors.

    a. What is one major advantage of a full hermetic compressor over a semihermetic unit?

b. What is one major advantage of a semihermetic compressor over a full hermetic unit?

10. Although float controls provide an expansion function in the refrigeration system, they do this in a manner different from that of conventional expansion valves and capillary tubes. Detail the operation of a low-side float control system and show how this system differs from an automatic expansion valve arrangement.

11. Differentiate between strainers and filter/driers. Identify the use of each type of device in the refrigeration system.

APPENDIXES

APPENDIX **A**

International Units of Measure
SI System

Name	Quantity Measured	Symbol
Amp	Electrical current (I)	A
Candela	Intensity of illumination	dc
Coulomb	Unit of electrical charge	Q
Farad	Electrical capacitance	f
Henry	Electrical inductance	h
Joule	Energy	J
Kelvin	Absolute temperature scale	K
Kilogram	Mass (weight)	kg
Liter/second	Rate of flow (liquids & gases)	l/s
Liter	Volume (liquids & gases)	l
Meter	Length (distance)	m
Meter/second	Velocity (speed)	m/s
Mole	Molecular weight	mol
Newton/meter	Force x length & torque	N/m
Newton	Force	N
Ohm	Electrical resistance (R)	
Pascal	Pressure	Pa
Volt	Electrical force (E)	V
Watt	Electrical power & Joule/second	W

APPENDIX **B**

Conversions from English to Metric and Metric to English

CONVERSION TABLES

Length
Cubic volume
Liquid volume
Weight, force, and mass
Force × distance and torque
Pressure
Energy
Power
Velocity
Rate of flow

DIRECTIONS FOR CONVERSION TABLE USE

All of the tables in this appendix are used in the same way. Find the unit of the measure that is to be converted *from* in the row to the left, then follow that row across to the column containing the unit to be converted *to*. Multiply the measure by the conversion factor in the box. The conversion factors that include an "E" are in scientific notation. The number following

the "E" indicates decimal point adjustment (−minus to the left and + plus to the right). For example 6.2137E-6 becomes .0000062137 and 1E+9 becomes 1,000,000,000.

EXAMPLE FOR CONVERSION TABLE USE

Problem: Convert 50 yards of cable to meters. Find yards on the left of the length conversion table and follow across under meters to find the conversion factor .9144. Multiply 50 yards times .9144 and solve for 45.72 Meters of cable. Note the solution in bold.

IMPORTANT SAFETY NOTE

Critical conversion applications should be double checked with other conversion tables or charts to affirm accuracy. Greater accuracy and numerical rounding in a more conservative direction are often found in manufacturer supplied charts and tables.

LENGTH

	Inch	Foot	Yard	Mile	Millimeter	Centimeter	*Meter*	Kilometer
Inch	1	.0833	.02778	.000016	25.4	2.54	.0254	.000025
Foot	12	1	.333	.000189	304.8	30.48	.3048	.000305
Yard	36	3	1	.000568	914.4	91.44	*.9144*	.000914
Mile	63360	5280	1760	1	1609344	160934.4	1609.344	1.609344
Millimeter	.03937	.003281	.001094	6.2137E-7	1	.1	.001	1E-6
Centimeter	.3937	.03281	.010936	6.2137E-6	10	1	.01	.00001
Meter	39.37	3.28	1.094	.000621	1000	100	1	.001
Kilometer	39370	3281	1094	.62137	1000000	100000	1000	1

CUBIC VOLUME

Cubic	Inch	Foot	Yard	Mile	Millimeter	Centimeter	Meter	Kilometer
Inch	1	.00058	.000021	3.931E-15	16387	16.387	.000016	1.6387E-14
Foot	1728	1	.03704	6.79E-12	2.83E+7	28317	.028317	2.8317E-11
Yard	46656	27	1	1.834E-10	7.65E+8	764555	.764555	7.65E-10
Mile	2.54E+14	1.472E+11	5.45E+9	1	4.17E+18	4.17E+15	4.17E+9	4.17
Millimeter	.000061	3.53E-8	1.31E-9	2.399E-19	1	.001	1E-9	1E-18
Centimeter	.061	.000035	1.31E-6	2.399E-16	1000	1	1E-6	1E-15
Meter	61024	35.315	1.308	2.399E-10	1E+9	1000000	1	1E-9
Kilometer	6.1E+13	3.53E+10	1.3E+9	.2399	1E+18	1E+15	1E+9	1

LIQUID VOLUME

	Ounce	Pint	Quart	Gallon	Milliliter	Liter
Ounce	1	.0625	.03125	.0078125	29.57	.02957
Pint	16	1	.5	.125	473	.473
Quart	32	2	1	.25	946.4	.9464
Gallon	128	8	4	1	3785.4	3.7854
Milliliter	.033814	.0021134	.001057	.0002642	1	.001
Liter	33.814	2.1134	1.0567	.26417	1000	1

WEIGHT, FORCE, AND MASS

	Ounce	Pound	Gram	Kilogram	Newton
Ounce	1	.0625	28.35	.02835	.278
Pound	16	1	453.6	.4536	4.448
Gram	.035274	.002205	1	.001	.0098066
Kilogram	35.274	2.205	1000	1	9.8066
Newton	3.597	.2248	101.972	.101972	1

FORCE x DISTANCE AND TORQUE

	Inch-ounce	Foot-pound	Meter-kilogram	Newton-Meter
Inch-ounce	1	.0052	.00072	.00706
Foot-pound	192	1	.1383	1.3558
Meter-kilogram	1388.74	7.233	1	9.807
Newton-meter	141.612	.73756	.1020	1

PRESSURE

	Pounds/ sq. Inch	Pounds/ sq. foot	Pascal	Kilo/pascal	Bar	Kilogram/sq. centimeter
Pounds/ sq.inch	1	144	6894.8	6.8948	.068948	7030695
Pounds/ sq.foot	.006945	1	47.882	.0479	.000479	48826.49
Pascal	.000145	.02088	1	.001	.00001	1019.72
Kilo/pascal	.145	20.884	1000	1	.01	1.0197E+6
Bar	14.5	2088.4	100000	100	1	1.0197E+8
Kilogram/ sq.centimeter	1.422E-7	.00002	.000981	9.807E-7	9.807E-9	1

ENERGY

	Joule (N/m)	Watt-hour	Kilowatt-hour	Btu	Calorie
Joule (N/m)	1	.000278	2.7778E-7	.000947	.23866
Watt-hour	3599.3	1	.001	3.4095	859.18
Kilowatt-hour	3599345	1000	1	3409.5	859184
Btu	1055.87	.293297	.000293	1	251.996
Calorie	4.19002	.001164	1.1639E-6	.00397	1

POWER

	Watt (joule sec.)	Kilowatt	Btu/hour	Horsepower
Watt (joule sec.)	1	.001	3.4121	.001341
Kilowatt	1000	1	3412.1	1.341
Btu/hour	.293	.000293	1	.000393
Horsepower	746	.7457	2544.4	1

VELOCITY

	Foot/second	Mile/hour	Meter/second	Kilometer/hour
Foot/second	1	.68182	.3048	1.0973
Mile/hour	1.4667	1	.44704	1.6093
Meter/second	3.2808	2.237	1	3.6
Kilometer/hour	.91134	.6214	.2778	1

RATE OF FLOW

	Cubic foot/ minute	Gallons/ minute	Cubic meters/ second	Liters/ minute
Cubic foot/ minute	1	24.54	.00155	9.2903
Gallons/ minute	.040746	1	.000063	.37854
Cubic meters/ second	645.8	15850.4	1	6000
Liters/minute	.10764	2.6417	.000167	1

APPENDIX C

Miscellaneous Conversions

TEMPERATURE

Unlike other unit conversions, zero degrees is not the same on all temperature scales. As a result simple multiplier conversion factors will not work. The following procedures will result in correct conversions.

From Fahrenheit to Celsius:

1. 5/9 or .5555 times degrees Fahrenheit.
2. Subtract 32 from the result of step 1 for answer.

From Fahrenheit to Rankine:

Add 459.67 to degrees Fahrenheit to find answer.

From Celsius to Fahrenheit:

1. 9/5 or 1.8 times degrees Celsius.
2. Add 32 to the result of step 1 for answer.

From Celsius to Kelvin:

Add 273.15 to degrees Celsius to find answer.

CONVERTING FRACTIONS TO DECIMAL

Use a calculator and divide the numerator of the fraction by the denominator. For example to convert 13/64 to decimal enter 13, then divide by 64. The correct result is .203125.

APPENDIX **D**

Air-Conditioning Systems
(Source: The Williamson Corporation)

TROUBLE ANALYZER TABLE

Relevant Service Procedure is Referenced in Parentheses ( ).

COMPRESSOR WILL NOT OPERATE

PILOT CIRCUIT

Symptom	Cause	Remedy
1. Contactor open	Check out items 2 to 10, this table	
2. Transformer dead	a. Defective	a. Replace
	b. Overloaded	b. Install proper size (SP-25)
	c. Short-circuited	c. Check continuity
	d. Improperly wired	d. Check wiring diagram (see section F)
	e. Improper voltage	e. Check power supply ± 10% rating on transformer
3. Thermostat circuit open	a. Improperly wired	a. Check wiring diagram (see section F)
	b. Loose or broken wiring	b. Repair or replace
	c. Wrong or defective subbase	c. Refer to manufacturer's installation instructions or repair
	d. Defective thermostat	d. Replace with new
	e. Improper cooling anticipator	e. Replace subbase
	f. Contacts dirty or pitted	f. Clean or replace
	g. Wrong size wire	g. Minimum 18 AWG (see section F)

Symptom	Cause	Remedy
4. Relay malfunction	a. Improperly wired	a. Check wiring diagram (see section F)
	b. Defective	b. Replace
5. Wired improperly	Error	Check wiring diagram (see section F)
6. Thermal overloads open—control center or compressor	a. Low-voltage power circuit	a. If wiring is O.K., check transformer capacity and load (SP-25)
	b. Defective overload	b. Replace (SP-46)
	c. Low suction pressure	c. Refer to item 79, this table
	d. Compressor locked	d. Replace compressor (SP-57, 58)
	e. Short in winding	e. Replace compressor (SP-57, 58)
6A. Compressor internal thermostat open permanently	a. Defective	a. Replace compressor (SP-57, 58)
7. Low-pressure switch open	a. Low suction pressure	a. Refer to item 79, this table
	b. Defective bellows	b. Replace control—follow manufacturer's instructions
	c. Defective switch	c. Replace (SP-17)
	d. Dirty or pitted contacts	d. Replace control
8. Contactor faulty	Defective coil	Replace coil or contactor
9. Loose connections	Vibration or error	Check all connections and tighten
10. High-pressure switch open	a. High discharge pressure	a. Refer to item 39, this table
	b. Refrigerant overcharged	b. Release excess refrigerant (SP-17)
	c. Defective control	c. Replace (SP-17)

POWER CIRCUIT

11. Open contactor—no power	a. Blown fuses	a. Check for proper size and type (see section F)
	b. Switch open	b. Close disconnect switch and check voltage

Symptom	Cause	Remedy
12. Fuses blown	a. Wrong size or type	a. Replace with proper size and type (see section F)
	b. High amperage	b. Check with amprobe (SP-39)
	c. Short-circuited	c. Continuity check (SP-36)
	d. Unit shorted to ground	d. Continuity check (SP-36)
	e. Improperly wired	e. Check wiring diagram (see section F)
	f. Defective potential relay	f. Refer to item 52, this table (SP-49)
	g. Defective starting capacitor	g. Refer to item 22, this table (SP-50)
	h. Defective running capacitor	h. Refer to item 54, this table (SP-51)
	i. Defective contactor (see item 24)	i. Refer to item 23, this table (SP-28)
	j. Defective fan motor	j. Replace
	k. Defective fan capacitor	k. Replace (SP-51)
	l. Defective compressor	l. Replace (SP-57, 58)
	m. Disconnect switch malfunction	m. See item 13
	n. Loose terminal wires	n. See item 14
	o. Loose fuses	o. Replace
13. Disconnect switch malfunction	Defective	Repair or replace
14. Loose terminal wires	Vibration or error	Check and tighten
15. Wired improperly	Error	Check wiring diagram (see section F)
16. Contactor closed	Check out items 17~19, this table	
17. Loose connections at compressor	Vibration or error	Tighten
18. Compressor motor windings open	Defective motor	Replace (SP-57, 58)
19. Loose connections at overload switch	Vibration or error	Check and tighten

Symptom	Cause	Remedy
20. Cycling overload switches	a. Low voltage	a. Check wire sizes with power company
	b. Loose connections	b. Inspect and tighten
	c. Defective capacitor	c. Replace relay (SP-49)
	d. Defective overload	d. Replace capacitor (SP-50, 51)
	e. Defective overload	e. Replace switch (SP-55)
	f. Locked compressor	f. Replace compressor (SP-57, 58)
21. Locked compressor	a. Defective	a. Replace compressor (SP-57, 58)
	b. Low or out of oil	b. Replace compressor (SP-57, 58)
	c. Foreign material	c. Replace compressor (SP-57, 58)
22. Starting capacitor malfunction	a. Moisture	a. Relocate to dry location
	b. Wrong capacitor	b. Replace (SP-54)
	c. Defective potential relay	c. Replace (SP-53)
	d. Defective	d. Replace (SP-54)
	e. Low voltage	e. Check power company and wire size
23. Contactor malfunction	a. Improperly wired	a. Check wiring diagram (see section F)
	b. Low voltage	b. Check with power company and wire size
	c. Short-circuited or grounded	c. Continuity check and correct
	d. Defective coil	d. Replace contactor
	e. Burned contacts	e. Replace contacts
	f. Defective control	f. Replace controls
24. Condenser fan relay malfunction	a. Improperly wired	a. Check wiring diagram (see section F)
	b. Low voltage	b. Check with power company and wire size
	c. Short-circuited or grounded	c. Continuity check and correct
	d. Defective coil	d. Replace coil
	e. Burned contacts	e. Replace contacts
	f. Defective control	f. Replace control

Symptom	Cause	Remedy

COMPRESSOR WILL NOT RESTART AFTER STOPPING

Symptom	Cause	Remedy
25. Contactor open—no power	a. Blown fuses	a. Check for proper size and type (see section F)
	b. Switch open	b. Close disconnect switch and check for voltage
26. Fuses blown	a. Wrong size or type	a. Replace with proper size and type (see section F)
	b. High amperage	b. Check with amprobe (SP-22)
	c. Short-circuited	c. Continuity check (SP-23)
	d. Unit shorted to ground	d. Continuity check (SP-24)
	e. Improperly wired	e. Check wiring diagram (see section F)
	f. Defective potential relay	f. Replace
	g. Defective starting capacitor	g. Replace (SP-54)
	h. Defective running capacitor	h. Replace (SP-54)
	i. Defective contactor	i. Replace
	j. Defective fan motor	j. Replace
	k. Defective fan capacitor	k. Replace (SP-54)
	l. Defective compressor	l. Replace (SP-57, 58)
	m. Loose fuses	m. Tighten fuse holder
	n. Loose wire	n. Tighten
27. Contactor open—power	Check out pilot circuit, items 1 to 10, this table	
28. Condenser fan relay malfunction	a. Low voltage	a. Check wire size and with power company
	b. Short-circuited or grounded	b. Continuity check and correct
	c. Defective coil	c. Replace coil
	d. Burned contacts	d. Replace contacts
	e. Defective control	e. Replace control
29. Contactor coil open	Shorted, grounded, or open coil	Repair or replace
30. Open overload switch	a. Low voltage	a. Check wire size and with power company
	b. Loose connections	b. Inspect and tighten
	c. Defective potential relay	c. Replace relay (SP-53)
	d. Defective capacitor	d. Replace capacitor (SP-54)

Symptom	Cause	Remedy
	e. Defective overload	e. Replace (SP-55)
	f. Locked compressor	f. Replace compressor (SP-57, 58)
31. High-pressure cut-out open	Refer to item 41, this table	Refer to item 39, this table
32. Low-pressure cut-out open	a. Low suction pressure	a. Refer to item 79, this table
	b. Defective bellows	b. Replace control
	c. Defective switch	c. Replace control
	d. Dirty or pitted contacts	d. Replace control
33. Relay malfunction	a. Improperly wired	a. Check wiring diagram (see section F)
	b. Defective	b. Replace
34. Potential relay faulty	a. Defective contacts	a. Replace control (SP-53)
	b. Open coil	b. Replace
	c. Improper wiring	c. Check wiring diagram (see section F)
	d. Bad starting capacitor	d. Replace (SP-54)
	e. Improper mounting	e. Correct--refer to compressor manufacturer's instructions
	f. Wrong relay	f. Refer to compressor manufacturer's instructions
35. Locked compressor	a. Defective	a. Replace compressor (SP-57, 58)
	b. Low or out of oil	b. Replace compressor (SP-57, 58)
	c. Foreign material	c. Replace compressor (SP-57, 58)
36. Starting capacitor malfunction	a. Moisture	a. Relocate to dry location
	b. Wrong capacitor	b. Replace
	c. Defective potential relay	c. Replace (SP-53)
	d. Defective	d. Replace (SP-54)
	e. Low voltage	e. Check power company and wire size
37. Running capacitor malfunction	a. Moisture	a. Relocate to dry location
	b. Wrong capacitor	b. Replace
	c. Defective	c. Replace (SP-54)
	d. Low voltage	d. Check power company and wire size

Symptom	Cause	Remedy
38. Pilot circuit malfunction	Check out pilot circuit, items 1 to 10, this table	
38A. PTCR malfunction	a. Cycling too rapid	a. Allow at least 5 minutes between cycles
	b. Low voltage	b. Check power company and wire size
	c. Defective	c. Replace (SP-60

COMPRESSOR SHORT CYCLES

Symptom	Cause	Remedy
39. High discharge pressure with recycling-type control	a. Dirty condenser	a. Clean
	b. Condensing fan belt broken or slipping	b. Repair or replace (SP-59)
	c. High ambient at condensing unit	c. Check air recirculation
	d. Insufficient air	d. Correct RPM, or remove restriction of air
	e. Partially closed discharge valve	e. Back seat
	f. Restriction in liquid line	f. Check liquid filter drier--kinks, etc. (SP-15)
	g. Overcharge	g. Release excess refrigerant (SP-10)
	h. Noncondensibles	h. Eliminate (SP-8)
	i. Damaged condenser coil	i. Repair or replace
	j. Defective condensing fan motor	j. Repair or replace
40. High-pressure switch open—condenser fan running	Check out items 43 to 47	
41. condenser airflow restricted	a. Obstructed location	a. Remove obstruction or relocate (See Installation Procedures--Sites, section C)
	b. Dirty or damaged coil	b. Clean, repair, or replace
	c. Fan or motor improper RPM	c. Adjust
	d. Motor running wrong way	d. Reverse
	e. Belt slipping	e. Tighten or replace (SP-59)
42. Recirculating discharge air	Location of unit or obstruction of discharge air	Relocate, baffle, or remove obstruction (refer to Installation Instructions--Sites, section C)

Symptom	Cause	Remedy
43. Noncondensibles in system	Air	(SP-8)
44. Overcharge of refrigerant	a. Not using measuring method b. Charging at low ambient	a. Release excess refrigerant (SP-10) b. Release excess refrigerant (SP-10)
45. Condensing fan malfunction	a. Variable pulley too loose b. Set screw loose	a. Tighten or replace b. Tighten or replace
46. High-pressure switch closed—condenser fan stopped	Check out items 47 to 50	
47. Loose electrical connections	Vibration or error	Tighten
48. Frozen condenser fan bearings	Insufficient lubrication	Replace fan bearings
49. Condenser fan motor cycles on overload	a. Low voltage b. Loose connections c. Defective capacitor d. Defective overload	a. Check with power company b. Inspect and tighten c. Replace capacitor (SP-54) d. Replace switch (SP-55)
50. Burned-out condenser fan motor	a. Insufficient lubrication b. Wrong size motor c. Improper belt adjustment or alignment d. Wrong wire size	 b. Replace with proper size (SP-59)
51. Normal pressures	Check out items 52 to 61	
52. Potential relay malfunction	a. Defective contacts b. Open coil c. Improper wiring d. Bad starting capacitor e. Improper mounting f. Wrong way	a. Replace control (SP-53) b. Replace control c. Check wiring diagram (see section F) d. Replace (SP-54) e. Correct--refer to compressor manufacturer's instructions f. Replace--refer to compressor manufacturer's instructions

Symptom	Cause	Remedy
53. Starting capacitor malfunction	a. Moisture b. Wrong capacitor c. Defective potential relay d. Defective e. Low voltage	a. Relocate to dry location b. Replace (see section F) c. Replace (SP-53) d. Replace (SP-53) e. Check wire size and with power company
54. Running capacitor malfunction	a. Moisture b. Wrong capacitor c. Defective d. Low voltage	a. Relocate to dry location b. Replace (see section F) c. Replace (SP-54) d. Check wire size and with power company
55. Low voltage	a. Inadequate wire sizes b. Overloaded circuit c. Insufficient power supply	a. Check wiring diagram (see section F) b. Remove overload or rewire c. Check with power company
56. Defective time delay switch	a. Maladjustment b. Improper voltage	a. Adjust or replace b. Correct
57. Loose connections	Vibration or error	Check all connections and tighten
58. Defective overload	Defective	Replace (SP-55)
59. Defective contact points	Dirty or pitted contacts	Replace
60. Dual-pressure-control malfunction	Defective	Replace (SP-17)
61. Low suction pressure	Refer to item 79, this table	
62. Low-pressure switch opens and closes—evaporator fan running	Check out items 63 to 72, this table	
63. Restricted filter drier	a. Moisture b. Foreign material	a. Replace (SP-15) b. Replace (SP-15)
64. Expansion valve malfunction	a. Foreign material in valve b. Power head defective	a. Remove and clean or replace (SP-19) b. Replace (SP-19

Symptom	Cause	Remedy
65. Undercharge of refrigerant	a. Leaks b. Insufficiently charged	a. Test for leaks, repair (SP-11) b. Recharge (SP-10)
66. Partly closed suction or receiver valve	a. Valve not backseated b. Defective valve	a. Back-seat valve b. Replace
67. Evaporator airflow restricted	a. Dirty air filters b. Fan running too slowly c. Dirty blower, squirrel cage d. Dirty evaporator coil e. Fan running backwards f. Restriction or obstructions in return air g. Insufficient return air h. Damaged coil	a. Clean or replace b. Adjust (see CFM chart, section G) c. Clean d. Clean e. Reverse f. Locate and remove g. Refer to NESCA manuals h. Repair or replace
68. Frosted evaporator coil	Suction pressure low	Correct—see item 79
69. Dirty air filters	Improper maintenance	Clean or replace frequently
70. Pipe or duct dampers closed	Loose	Open and secure dampers
71. Restriction of duct	a. Obstruction b. Improper duct and pipe application	a. Remove or replace b. Refer to NESCA manuals
72. Improper air circulation	a. Improper installation b. Bypass damper on humidifier open	a. Refer to NESCA manuals b. Close
73. Low-pressure switch opens and closes—evaporator fan stopped	Check out items 74 to 78	
74. Loose connections in evaporator fan motor	Visual inspection	Tighten
75. Evaporator fan motor overload open	a. Improper alignment of belt or pulleys b. Bearings seized c. Undersized motor d. Improperly wired	a. Align (SP-59) b. Replace c. Replace with correct size d. Check wiring diagram (see section F)

Symptom	Cause	Remedy
	e. Defective motor	e. Replace
	f. Short-circuited	f. Check continuity (SP-23)
	g. Defective capacitor	g. Replace
76. Evaporator fan motor burned out	a. Insufficient lubrication	
	b. Wrong size motor	b. Replace with proper size (SP-59)

COMPRESSOR RUNS—BUT INSUFFICIENT AIR CONDITIONING

Symptom	Cause	Remedy
77. Evaporator fan belt broken	a. Too tight	a. Adjust (SP-59)
	b. Improperly aligned	b. Align (SP-59)
	c. Deteriorated or worn	c. Replace (SP-59)
78. Evaporator fan relay malfunction	a. Improperly wired	a. Check wiring diagram (see section F)
	b. Defective control	b. Replace or repair (SP-28)
	c. Wrong fan relay	c. Check with manufacturer
	d. Overloaded	d. Check amps and replace (SP-22)
	e. Short-circuited	e. Check continuity (SP-23)
	f. Defective subbase or thermostat	f. Replace
79. Suction pressure low	1. Lack of refrigerant	1. Add sufficient charge (SP-10)
	a. Leaks	a. Fix leaks and add sufficient charge (SP-10)
	b. Defective TX valve	
	c. Restricted cap tube	b. Replace TX valve (SP-19)
	d. Restricted liquid drier	c. Replace cap tube (SP-13)
	e. Liquid or suction line precharged coupling not completely open	d. Replace drier
		e. Open for inspection. Remove restriction
		Note: Precharged couplings can be opened for inspection and retightened.
	2. Insufficient air	2. Refer to section G, Operating Pressure Charts
	a. Dirty air filters	a. Replace filters
	b. Dirty evaporator coil	b. Clean

Symptom	Cause	Remedy
	c. Humidifier bypass damper open	c. Close damper for cooling season
	d. Dirty squirrel cage	d. Clean
	e. Broken or slipping evaporator blower belt or pulleys	e. Replace or correct (SP-59)
	f. Fan running backward	f. Reverse
	g. Wrong size motor or improper RPM	g. Replace or adjust
	h. Thermostat set too low	h. Set to recommended comfort range
	i. Covered or restricted return air openings	i. Remove obstruction
	3. Lower than normal return air temperature	3. Check air distribution
80. Low-pressure switch malfunction	a. Low suction pressure	a. Refer to item 81, this table
	b. Defective bellows	b. Replace bellows
	c. Defective switch	c. Replace (SP-17)
	d. Dirty or pitted contacts	d. Replace or clean contacts (SP-26)
	e. Improper setting	e. See item 81, this table
81. Improper low-pressure switch setting	Settings too low	Adjust to recommended settings:
		a. Cut out at 40 PSI on low side—R-22
		b. Cut in at 70 on low side—R-12
		c. Cut out at 20 PSI (low side)—R-12
		d. Cut in at 35 (low side)—R-12
82. High suction pressure	a. Overcharge of refrigerant	a. Purge (SP-10)
	b. Incorrect superheat adjustment	b. Adjust to correct setting (SP-5)
	c. Defective compressor valves	c. Check and replace (SP-34)
	d. Noncondensibles in system	d. Eliminate (SP-8)

Symptom	Cause	Remedy
83. Low head pressure	a. Low refrigerant charge	a. Recharge (SP-10)
	b. Leak in system	b. Repair leak and recharge unit (SP-11, SP-10)
	c. Defective compressor valves	c. Check and replace (SP-34)
	d. Low ambient temperature at condenser	d. Install low ambient kit (SP-16)
	e. Defective expansion valve	e. Replace valve (SP-19)
	f. Leaking valve in oil separator	f. Repair or replace valve
84. Suction pressure slightly low	a. Foreign material	a. Remove (SP-7)
	b. Improper superheat setting	b. Adjust or replace (SP-5)
85. Dirty air filters	Improper maintenance	Clean or replace frequently
86. Restriction in refrigeration system	a. Restricted liquid line filter-drier	a. Replace (SP-15)
	b. Thermoexpansion valve obstructed or stuck	b. Inspect and clean valve (SP-19)
	c. Restricted capillary tubes	c. (See SP-13)
87. Expansion valve malfunction	a. Foreign material in valve	a. Remove and clean or replace (SP-19)
	b. Power head defective	b. Replace (SP-19)
88. Evaporator coil partly frosted	a. Lack of refrigerant	a. Add sufficient charge (SP-10)
	b. Restricted capillary tubes	b. Replace (SP-13)
	c. Dirty or defective expansion valve	c. Repair or replace (SP-19)
	d. Dirty evaporator	d. Clean
	e. Damaged coil	e. Repair or replace
89. Refrigerant undercharge	a. Leaks	a. Test for leaks, repair (SP-11)
	b. Insufficiently charged	b. Recharge (SP-10)
90. Inadequate air distribution	a. Obstruction	a. Remove or replace
	b. Improper duct and pipe application	b. Refer to NESCA manuals
	c. Bypass damper on humidifier open	c. Close

Symptom	Cause	Remedy
91. Dampers partly closed	Error or loose	Balance system and secure dampers
92. Restricted condenser airflow	a. Obstructed location	a. Remove obstructions or relocate (See Procedures—sites, section C)
	b. Dirty or damaged coil	b. Clean, repair or replace
	c. Fan or motor improper RPM	c. Adjust (see CFM charts, section G)
	d. Motor running wrong way	d. Reverse
	e. Belt slipping	e. Tighten or replace (SP-59)
	f. Broken pulley	f. Replace
	g. Location of unit or obstruction	g. Relocate, baffle, or remove obstruction (see Installation Procedures—sites, section C)
93. Undersized unit	Improperly sized	Refer to NECA manuals

COMPRESSOR RUNS—BUT EXCESSIVE AIR CONDITIONING

Symptom	Cause	Remedy
94. Thermostat out of calibration	Improperly calibrated	Recalibrate or replace
95. Control circuit short-circuited	Defective wiring	Inspect and repair
96. Contactor stuck closed	Burned contacts	Replace
97. Pressure control stuck	Burned contacts	Replace

MISCELLANEOUS

Symptom	Cause	Remedy
98. Noisy condensing unit	a. Compressor not riding free on external springs	a. Remove shipping hold-down devices
	b. Tubing rattles	b. Locate and correct
	c. Foreign matter on fan blades	c. Remove
	d. Condenser not level	d. Correct
	e. Condenser fan motor noisy	e. Replace
	f. Condenser fan belt too tight	f. Adjust (SP-59)
	g. Condenser fan mounting loose	g. Tighten

Symptom	Cause	Remedy
99. Compressor loses oil	h. Pulleys improperly aligned or broken	h. Align or replace (SP-59)
	i. Casing rattles	i. Locate and correct
	j. Bearings or shafts worn	j. Align or replace
	k. Propeller fan out of balance or loose hub	k. Replace
	l. Improper base	l. See Installation Procedures—Bases, section C
	m. Components loose	m. Tighten
	n. Electrical components hum or chatter	n. Correct or replace
	o. Internal compressor springs broken or detached	o. Replace compressor (SP-58)
	p. Unit location—acoustical	p. Relocate or baffle
	q. Condenser fan running too fast	q. Adjust
	r. Refrigerant floodback—expansion valve open	r. Expansion valve malfunction—suction line improperly trapped (see Installation Procedures—traps, section C)
	s. Compressor low on oil	s. Units with oil level indicators, check, and if necessary, add oil
	a. Leaks in system	a. Check for leaks and repair (SP-11)
	b. Improper trapping in suction line	b. Refer to Installation Procedures—Traps, section C
	c. Suction line too large	c. See pipe size chart, section C
	d. Condensing unit out of level	d. Relevel
	e. Liquid refrigerant in crankcase	e. Install crankcase heater and properly trap (see manufacturer's instructions)
	f. Cap tube system overcharge	f. See (SP-10)
	g. Expansion bulb improperly located	g. See (SP-19)

Symptom	Cause	Remedy

100. Frosted or excessively cold suction line	1. With capillary tube
	    a. Refrigerant floodback or overcharge
	    b. See Suction Pressure Low, item 79 Insufficient air
	2. With TX valve
	    a. Bulb improperly located
	    b. Defective TX valve flooding back liquid
		a. Remove refrigerant (SP-10)
		
		a. Relocate bulb (SP-19)
		
		b. Replace valve

101. Frosted liquid line	Restriction in liquid line	Remove restriction (SP-15)
102. Hot liquid line	High discharge pressure	Refer to item 39, this table

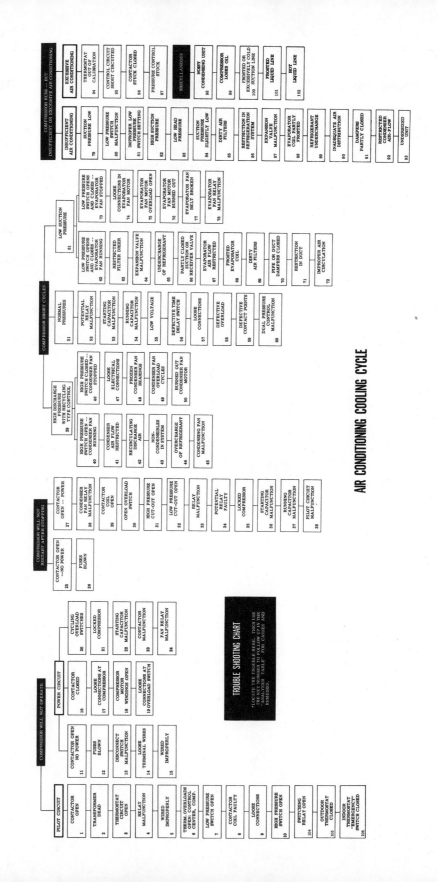

AIR CONDITIONING COOLING CYCLE

APPENDIX E

Electrical and Electronic Symbols

Symbols that represent components in circuit diagrams are often confusing. Though standard symbols are available, the meanings of many are not obvious. As a result, manufacturers may represent components in a more picture-like fashion. Some schematic diagrams are a mixture of standard symbols and pictorial representations. Well done schematics resolve this problem by including a legend that defines the symbols used. Many new appliances include both a schematic and a perspective drawing of the electrical wiring in an envelope fastened inside the case. A particularly confusing schematic can be clarified through a telephone call to the manufacturer or supplier. As you progress in the HVAC profession, build a library of reference material. These "paper tools" are important tools of the trade. The appendix that follows provides the most common symbols with a few brief clarifying notes.

symbols a: circuit, electric sources, switches and circuit breakers
symbols b: sensor switches, resistive devices, inductors and coils and relays
symbols c: motors and transformers
symbols d: capacitors, diodes, power diodes, transistors and integrated circuits

Circuit

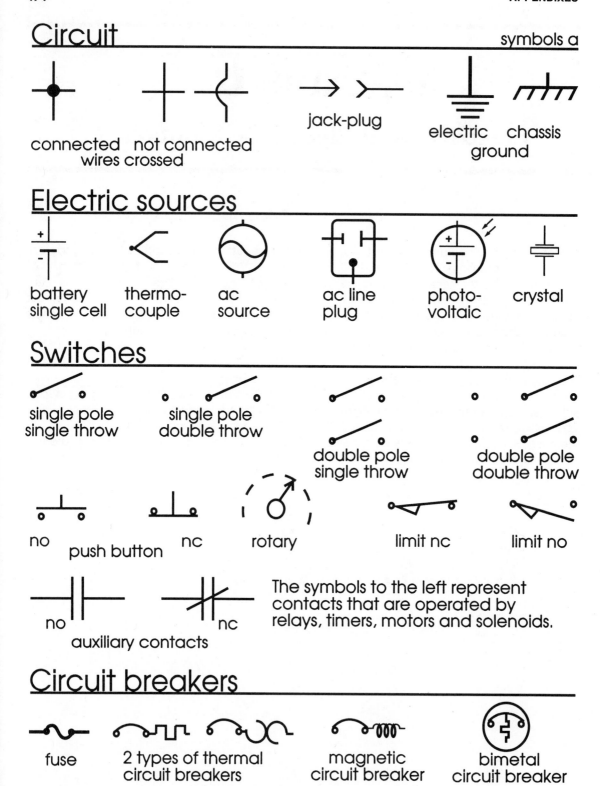

connected　not connected
wires crossed

jack-plug

electric　chassis
ground

Electric sources

battery
single cell

thermo-
couple

ac
source

ac line
plug

photo-
voltaic

crystal

Switches

single pole
single throw

single pole
double throw

double pole
single throw

double pole
double throw

no

push button

nc

rotary

limit nc

limit no

no

nc

auxiliary contacts

The symbols to the left represent
contacts that are operated by
relays, timers, motors and solenoids.

Circuit breakers

fuse

2 types of thermal
circuit breakers

magnetic
circuit breaker

bimetal
circuit breaker

Sensor switches

symbols b

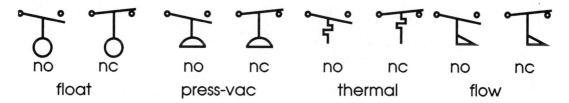

no   nc   no   nc   no   nc   no   nc

float   press-vac   thermal   flow

Resistive devices

resistor

variable
resistor

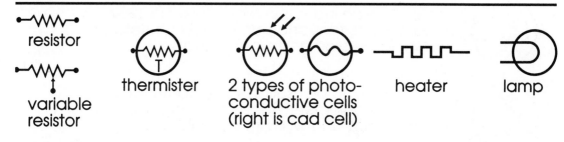

thermister

2 types of photo-
conductive cells
(right is cad cell)

heater   lamp

Inductors and coils

inductor
air core

inductor
iron core

motor
timer
solenoid
relay

general coil
device symbol

Relays

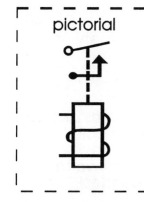

pictorial

A relay may be represented
as a pictorial if contacts
and coil are close in the
schematic. A relay may be
shown as a symbol, especially
if the contacts and relay coil
are not close on the
schematic.

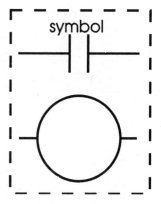

symbol

Motors

 brush type ac-dc motor: used in portable tools and appliances. The high power, direction control and speed control are assets while the brushes and commutator limit reliability.

 ac induction motor, often shaded pole: used for small fans and pumps. The coil is frequently excluded.

 ac split phase induction motor: used in 120v compressors, fans and pumps. Circuit details will show the start and run method.

 ac 3 phase motor: used in large commercial HVAC units and equipment. Many multi-coil variations of both delta and wye are available.

Transformers

 voltage step-down: used in low voltage control circuits

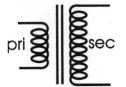 voltage step-up: used in fuel ignition circuits

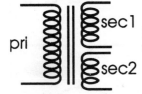 multiple secondary: used to provide a variety of voltages

 auto transformer: used in lab power supplies and a step-up version is the automobile ignition coil. Danger - the auto transformer does not provide isolation from the hot ac source. Even low voltage settings can result in lethal shock if grounded.

Capacitors

symbols d

 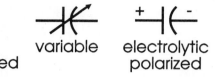 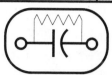

plastic
non-polarized

variable

electrolytic
polarized

oil or electrolytic start
and run ac capacitor
The resistor shows normal
leakage of the insulation.

Diodes

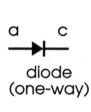 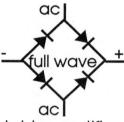

diode
(one-way)

bridge rectifier

zener
diode

photo
diode

light
emitting
diode

Power diodes

diac

scr

photo
scr

triac

Transistors

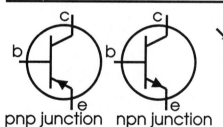   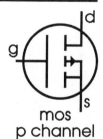

pnp junction   npn junction   npn photo   mos
n channel

mos
p channel

Integrated circuits

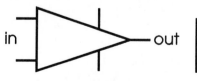 

in   out

general analog
amp: includes op amps
and sensor amps

16 pin ic chip
digital or
analog

key to letters
a anode
b base
c cathode (diodes)
c collector (transistors)
e emitter
g gate
s source

INDEX

Page numbers followed by *f* denote a figure or table. Page numbers followed by *a* denote appendixes.